Contents

Get the most from this book		iv
Acknowledgements		vi
1	Energy transfer	1
2	Nutrient cycles	29
3	Response	48
4	Nervous coordination	69
5	Muscles and movement	86
6	Internal control	102
7	Genes, alleles and inheritance	122
8	Gene pools, selection and speciation	149
9	Populations in ecosystems	166
10	The control of gene expression	188
11	Gene cloning and gene transfer	206
12	Using gene technology	225
13	Developing mathematical skills	240
14	Developing practical skills	258
15	Exam preparation	271
Index		277
Free online resources		281

Get the most from this book

Welcome to the **AQA A-level Biology Year 2 Student's Book**. This book covers Year 2 of the AQA A-level Biology specification.

The following features have been included to help you get the most from this book.

Prior knowledge

This is a short list of topics that you should be familiar with before starting a chapter. The questions will help to test your understanding.

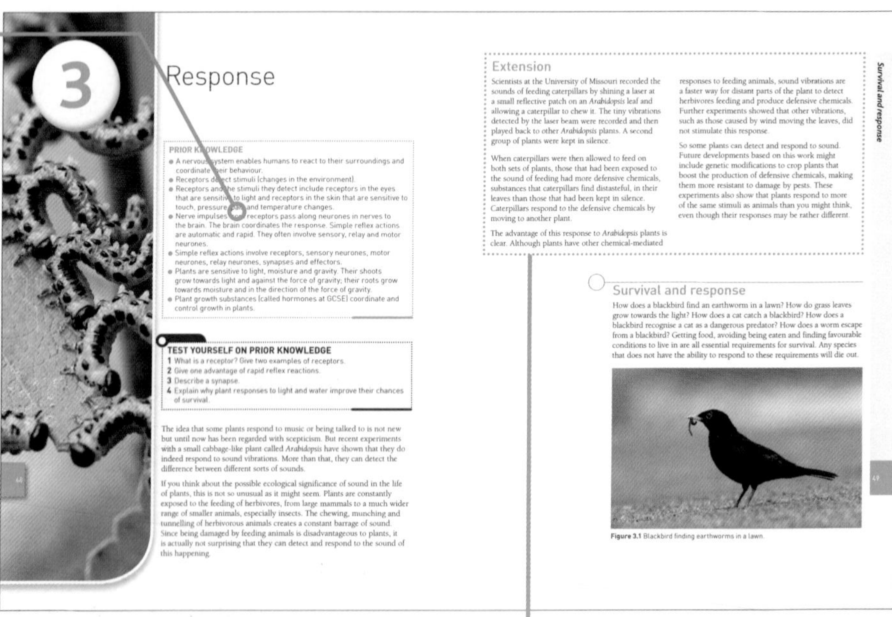

Extension

Throughout the book you will also find Extension boxes, which contain extra material to deepen your understanding of a topic.

Tips

These highlight important facts, common misconceptions and signpost you towards other relevant topics.

Key terms and formulae

These are highlighted in the text and definitions are given in the margin to help you pick out and learn these important concepts.

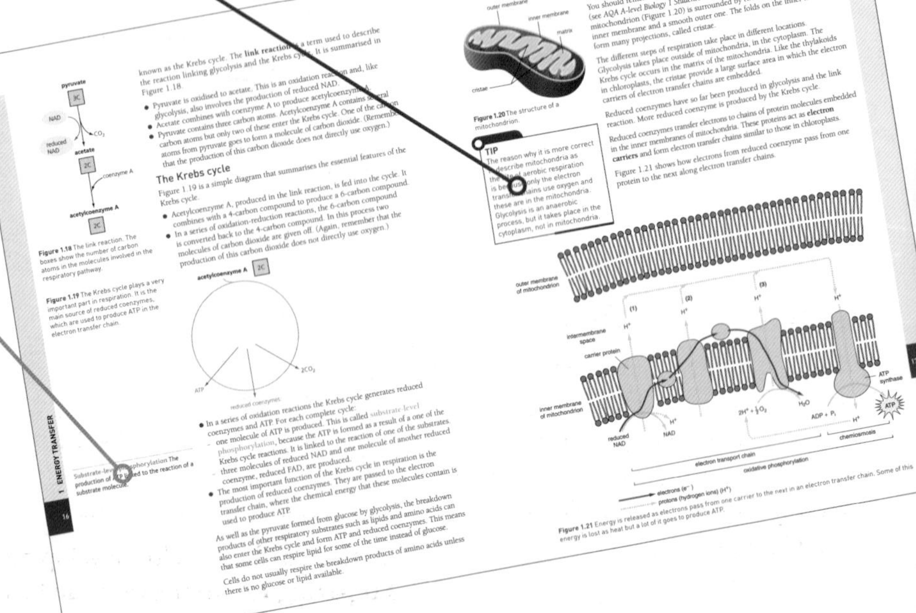

Examples

Examples of questions and calculations feature full workings and sample answers.

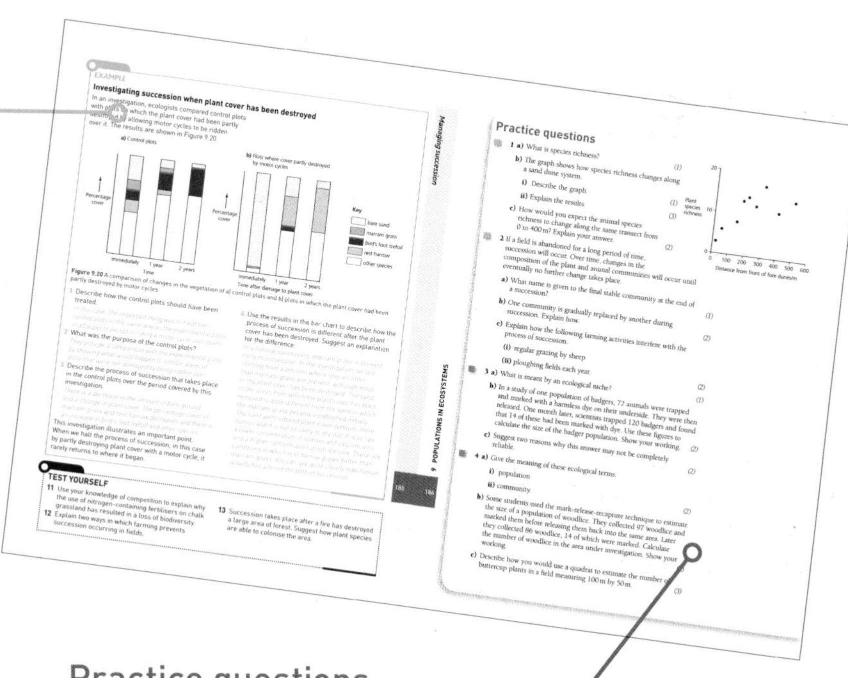

Practice questions

You will find Practice questions at the end of every chapter. These follow the style of the different types of questions you might see in your examination, including multiple-choice questions, and are colour coded to highlight the level of difficulty. Test your understanding even further, with Maths questions and Stretch and challenge questions.

- Green – Basic questions that everyone should be able to answer without difficulty.
- Orange – Questions that are a regular feature of exams and that all competent candidates should be able to handle.
- Purple – More demanding questions which the best candidates should be able to do.
- Stretch and challenge – Questions for the most able candidates to test their full understanding and sometimes their ability to use ideas in a novel situation.

Test yourself questions

These short questions, found throughout each chapter, are useful for checking your understanding as you progress through a topic.

Activities and Required practicals

These practical-based activities will help consolidate your learning and test your practical skills. AQA's required practicals are clearly highlighted.

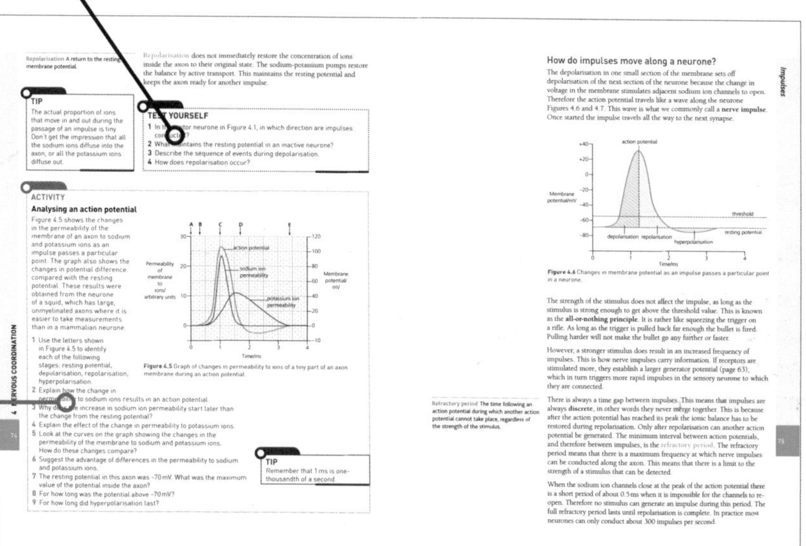

Dedicated chapters for developing your **Maths** and **Practical skills** and **Preparing for your exam** can be found at the back of this book.

Acknowledgements....

The Publisher would like to thank the following for permission to reproduce copyright material:

p.1 © Yi Liu – Fotolia.com; **p.4** © Dr.Jeremy Burgess/Science Photo Library; **p.5** © Mark Smith; **p.13** © tsach – Fotolia.com; **p.18** © Bill Indge; **p.20** © Bill Indge; **p.24** © birdiegal – Fotolia.com; **p.29** © Sylvie Bouchard – Fotolia.com; **p.30** © Gary Braasch/Getty Images; **p.34** © Bill Indge; **p.37** © Dr Jeremy Burgess/Science Photo Library; **p.41** © SOLLUB – Fotolia.com; **p.43** © blickwinkel/Alamy; **p.48** © tilzit – Fotolia.com; **p.49** © Erni – Fotolia.com; **p.52** © Piotr Naskrecki/Getty Images; **p.55** © Graham Jordan/Science Photo Library; **p.61** © Image Source/Alamy; **p.63** © CNRI/Science Photo Library; **p.69** © Ed Reschke/Getty Images; **p.70** © Science Pictures Limited/Science Photo Library; **p.78** © Dr. Don Fawcett/Getty Images; **p.80** © Manfred Kage/Science Photo Library; **p.86** © jirapong – Fotolia.com; **p.87** *l* © Luke Horsten/Getty Images, *r* © Woodfall/Photoshot; **p.89** *l* © Lehtikuva OY/REX, *r* © Mark Dadswell/Getty Images Sport; **p.92** *l* © Steve Gschmeissner/Science Photo Library, *r* © Biology Media/Science Photo Library; **p.93** © Don Fawcett/Science Photo Library; **p.100** © Visuals Unlimited/Corbis; **p.102** © Ed Reschke – Oxford Scientific/Getty Images; **p.105** © Steve Gschmeissner - Science Photo Library/Getty Images; **p.122** © Carolina Biological/Visuals Unlimited/Corbis; **p.125** © Biosphoto/Biosphoto; **p.130** © University of Wisconsin-Madison; **p.149** © Chris Mattison/FLPA; **p.150** © giedriius – Fotolia.com; **p.151** *l* © Piotr Filbrandt – Fotolia.com, *r* © daibui – Fotolia.com; **p.157** © Eye Of Science/Science Photo Library; **p.159** *tl* © kjekol – Fotolia.com, *tr* © JohanSwanepoel – Fotolia.com, *bl* © zuzule – Fotolia.com, *br* © Reddogs – Fotolia.com; **p.160** © Steve Hopkin/Ardea; **p.162** © Gail Johnson – Fotolia.com; **p.166** © Soru Epotok – Fotolia.com; **p.167** © Mark Smith; p. 172 © arolina66 – Fotolia.com; **p.178** *l* © pisotckii – Fotolia.com, *r* © Andrea Wilhelm – Fotolia.com; **p.179** © Nazzu – Fotollia.com; **p.181** *l* © Bill Indge, *r* © Bill Indge; **p.182** © alekswolff – Fotolia.com; **p.184** © videnko – Fotolia.com; **p.188** © DenisNata – Fotolia.com; **p.190** © Najlah Feanny/CORBIS SABA; **p.197** © romaneau – Fotolia.com; **p.198** © Mark Thomas/Science Photo Library; **p.206** © MBI/Alamy; **p.218** © Maximilian Stock Ltd/Science Photo Library; **p.219** *t* © Naturepix/Alamy, *b* © Nick Cobbing/Rex Features; **p.225** © Tek Image – Science Photo Library/Getty Images; **p.226** © David Parker/Science Photo Library; **p.240** © Martin Shields/Alamy; **p.245** *t* © Biology Media/Science Photo Library, *b* © Dr. Gladden Willis – Visuals Unlimited/Getty Images; **p.251** © josephotographie.de – Fotolia.com; **p.258** © Mark Smith; **p.259** *t* © Mark Smith, *b* © Wikipedia; **p.260** © Martyn F. Chillmaid/Science Photo Library; **p.262** *tl* © Alan John Lander Phillips – E+/Getty Images, *tr* © Steve Gschmeissner/Science Photo Library, *b* © Serpil_Borlu – iStock via Thinkstock; **p.263** © Dr. Don Fawcett/Getty Images; **p.264** © Mark Smith; **p.265** *t* © Courtesy of Dominikmatus via Wikipedia, *b* © Mark Smith; **p.266** © Mark Smith; **p.267** Mark Smith; **p.268** © Dr Brian Beal; **p.269** *t* © Mark Smith, *b* © Mark Smith; **p.271** © milanmarkovic78 – Fotolia.com.

t = top, *b* = bottom, *l* = left, *r* = right, *m* = middle

Every effort has been made to trace all copyright holders, but if any have been inadvertently overlooked, the Publisher will be pleased to make the necessary arrangements at the first opportunity.

1 Energy transfer

PRIOR KNOWLEDGE

- Radiation from the Sun is the source of energy for most communities of living organisms.
- Green plants and algae absorb a small amount of the light that reaches them. The transfer from light energy to chemical energy occurs during photosynthesis. This energy is stored in the substances that make up the cells of the plants and algae.
- Respiration supplies the energy needs for living processes, including movement. Much of this energy is eventually transferred to the surroundings.
- In all cells, carbohydrate or lipid molecules are broken down during respiration to release the energy required to resynthesise ATP. In some cells, ATP can also be made using light energy during photosynthesis.
- The amounts of material and energy contained in the biomass of organisms are reduced at each successive stage in a food chain because some materials and energy are always lost in organisms' waste materials.

TEST YOURSELF ON PRIOR KNOWLEDGE

1. Name two substances that make up the cells of a plant.
2. A mouse is completely at rest in a comfortable environmental temperature. Name **two** processes during which it uses ATP.
3. Name the enzyme involved in resynthesising ATP.
4. In what form is energy released by respiration eventually transferred to the surroundings?

Introduction

Sugar cane is grown as a crop in more than 70 countries and provides around 80% of the world's sugar. It is one of the most efficient crop plants in cultivation. Originally from south Asia, sugar cane is a tropical grass. Along with some other tropical grasses such as elephant grass, it has evolved a way of photosynthesising that allows it to make the most of the tropical sun.

Most plants use a form of photosynthesis called C_3 photosynthesis. In these plants, the first stable product of photosynthesis is a three-carbon molecule. The reaction is catalysed by an enzyme called ribulose bisphosphate carboxylase (rubisco for short). But rubisco has a surprising flaw. Although it catalyses the reaction with carbon dioxide, it will also catalyse a reaction

TIP
You are not required to know about C4 photosynthesis, but it is an example of the kind of material you might be expected to analyse or interpret.

with oxygen instead, resulting in a different outcome and inhibiting the enzyme's contribution to photosynthesis.

When plants close their stomata in hot, dry conditions, the carbon dioxide concentration in their leaves falls and oxygen competes more successfully with carbon dioxide for rubisco. The efficiency of photosynthesis is then reduced. Plants in tropical conditions, especially grasses, face hot, dry conditions more often than those in other areas. Some, like sugar cane, evolved a form of photosynthesis, called C_4 photosynthesis, that reduces this problem.

Sugar cane does not grow as effectively in the UK. But you may have seen fields of elephant grass being grown in the UK as a biofuel for electricity generation. Elephant grass is also a C_4 plant and its rapid growth and high annual biomass yield makes it useful as a renewable energy source. Unlike sugar cane and elephant grass, most plant species in the UK carry out the form of photosynthesis you will learn about in this chapter.

Life on Earth depends on the continuous transfer of energy through photosynthesis, feeding and respiration. Plants and other chlorophyll-containing organisms photosynthesise, absorbing light. In this process, some of the energy of light is conserved in the production of ATP and ultimately in carbohydrates and other biological molecules.

These biological molecules can be used directly by the plants, mostly in respiration and to make the other biological molecules they require, or they can be transferred to other organisms by the animals that feed on them and by saprobionts that decompose them.

During respiration, various respiratory substrates are oxidised and some of their chemical energy is conserved in the production of ATP. ATP can then be used in various forms of biological work, such as movement, synthesis of large biological molecules and active transport.

Photoautotrophic organisms Organisms that synthesise their own biological molecules using light energy.

These energy transfers are fundamental to life. Respiration is common to all organisms and photosynthesis is common to all **photoautotrophic organisms**. The fact that these processes are so widespread suggests that these organisms all evolved from common ancestors. This is indirect evidence for evolution.

Energy and efficiency

Respiratory substrates Biological molecules used as fuel in respiration.

Biomass The mass of carbon in biological molecules or dry mass of tissue per given area per given time.

Neither photosynthesis nor respiration is totally efficient. During respiration, for example, not all the chemical energy from a molecule of **respiratory substrate** is transferred into molecules of ATP. Some of the energy will inevitably be lost as heat. This is obviously important when we come to look at the transfer of **biomass** and its chemical energy from one organism to another along a food chain or through a food web.

Figure 1.1 summarises the three ways in which energy is transferred within and between different organisms.

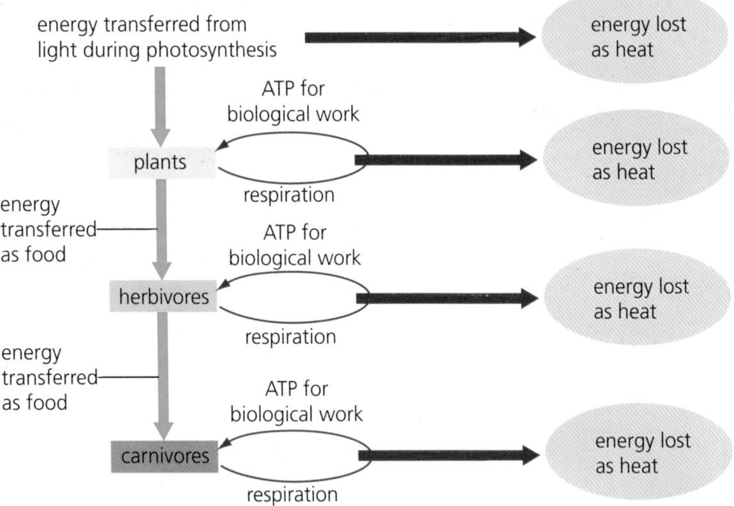

Figure 1.1 A summary of the ways in which energy is transferred within and between organisms. Note that, each time a transfer occurs, some energy is lost as heat.

Photosynthesis

Organisms must have a continuous supply of biological molecules for respiration and with which to build new cells and tissues. For animals, these biological molecules come from food: this is sometimes from other animals but ultimately from plants. It is only by photosynthesis that light energy can be transferred to chemical energy.

Photosynthesis is a complex process involving a number of separate reactions. It is useful to get an idea of the overall process before we look at the detail. Figure 1.2 shows the two basic steps.

- Light-dependent reaction; light energy is absorbed by chlorophyll and some of this energy is transferred to chemical energy in ATP. A second substance is also produced. This is reduced NADP (see page 15 for a definition of reduction and oxidation). In order to produce these substances, a molecule of water is split and oxygen is given off as a waste product.
- Light-independent reaction: ATP and reduced NADP are involved in the use of carbon dioxide to make carbohydrate.

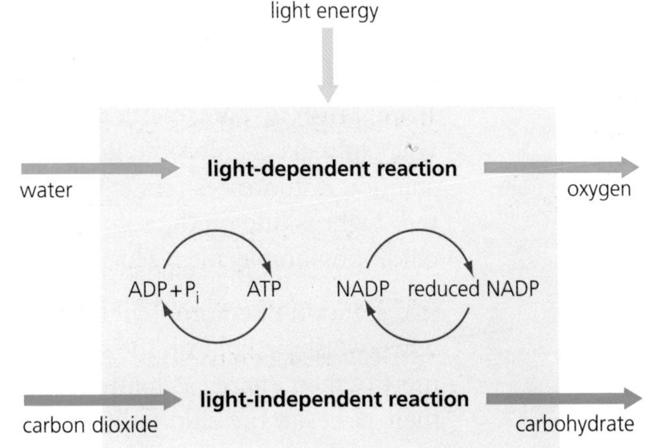

$$\text{carbon dioxide} + \text{water} \xrightarrow{\text{light energy}} \text{carbohydrate} + \text{oxygen}$$

Figure 1.2 The main steps in photosynthesis. The substances entering and leaving the leaf can be arranged to give the basic equation for photosynthesis.

Structure of chloroplasts

Thylakoids Disc-shaped, membrane-bound structures found inside chloroplasts.

Grana Groups of thylakoids arranged in stacks.

You should remember from the first year of the course that chloroplasts are the site of photosynthesis (see *AQA A-level Biology 1 Student's Book*, Chapter 3 page 42). Each chloroplast (Figure 1.3 below) is surrounded by two membranes. Both the outer and inner membranes of a chloroplast are smooth. Inside the chloroplast there are a series of disc-shaped, membrane-bound structures called **thylakoids**. In some places the thylakoids are arranged in stacks called **grana**. The membranes that form the grana provide a very large surface area for chlorophyll molecules and other light-absorbing pigments.

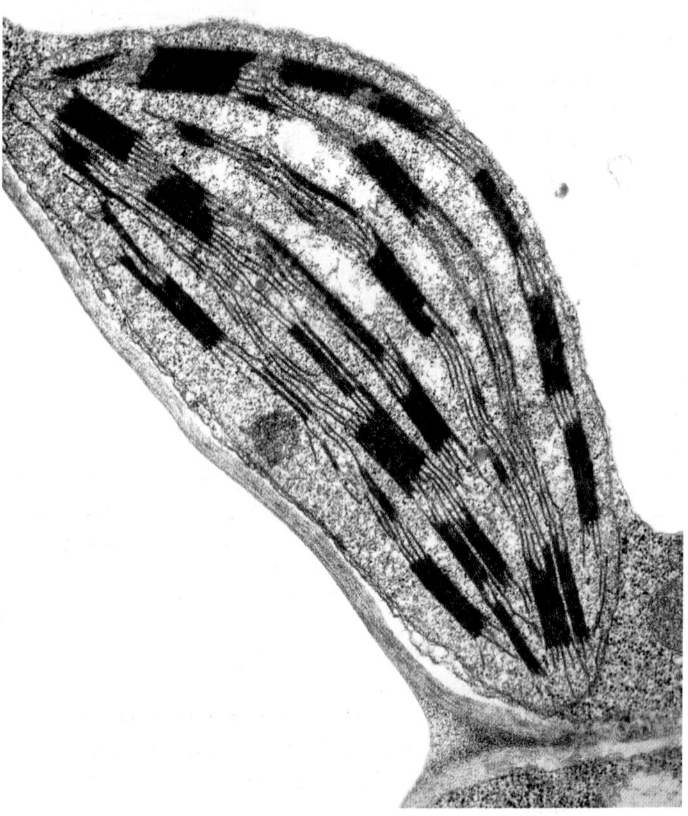

Figure 1.3 A transmission electron micrograph of a chloroplast. This chloroplast is approximately 5 μm in diameter.

The light-dependent reaction

Photoionisation The process by which a chlorophyll molecule becomes positively charged as a result of losing two electrons when it absorbs light.

If you crush up some nettle leaves in an organic **solvent** such as ethanol you can make a chlorophyll solution. If you shine a bright light on this solution it fluoresces (emits light), but, instead of appearing green, it looks red. Light falling on the solution causes electrons to leave some of the chlorophyll molecules. This is because of **photoionisation**.

In a solution of chlorophyll the electrons have nowhere to go. This is why, when we shine light on the solution, it fluoresces red. The electrons lose most of their energy as light of a different wavelength as they fall back into their places in the chlorophyll molecules.

In a chloroplast, however, these electrons do not return to the chlorophyll molecule from which they came. They pass down a series of electron carriers, losing energy as they go. In chloroplasts, this energy is conserved in the production of ATP and reduced NADP.

REQUIRED PRACTICAL 7

Use of chromatography to investigate the pigments isolated from leaves of different plants

This is just one example of how you might tackle this required practical.

A chlorophyll solution contains a mixture of chlorophyll and other pigments. The different pigments can be separated using chromatography.

Figure 1.4 shows the result of using thin layer chromatography (TLC) to separate the pigments extracted from nettle leaves.

1 What is a solvent?
2 Suggest why the origin line was placed a little way above the base of the plate.
3 How could the origin spot have been made sufficiently concentrated?

The different pigments in the mixture can be identified by finding their R_f values. The R_f is an example of a ratio and is calculated as:

$$R_f = \frac{\text{distance moved by the component spot}}{\text{distance moved by the solvent from the origin}}$$

The standard R_f values for these pigments using this solvent are:

chlorophyll *a*	0.31	chlorophyll *b*	0.15
carotene	0.96	xanthophyll	0.63

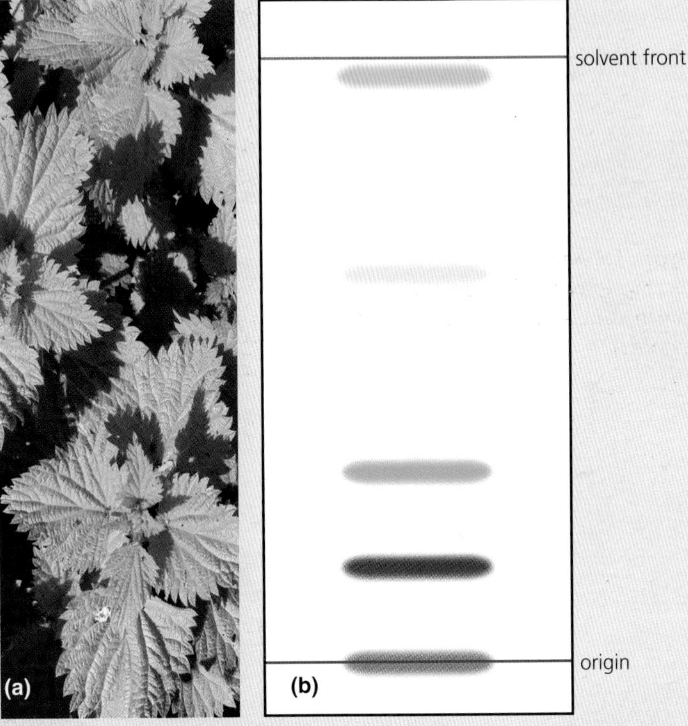

Figure 1.4 (a) Nettles. (b) Thin layer chromatography plate following separation of nettle leaf extract.

TIP
You do not need to know about R_f values, but using them is a good way to practise your maths skills.

TIP
Look at Chapter 15 in *AQA A-level Biology 1 Student's Book* for information on how to carry out thin layer/paper chromatography.

4 Identify the different pigments on the thin layer chromatography plate in Figure 1.4b by measuring the distances with a ruler and calculating their R_f values.
5 Give one possible source of error in finding the R_f values.

The light-dependent reaction is described on the next page and is summarised in Figure 1.5 below.

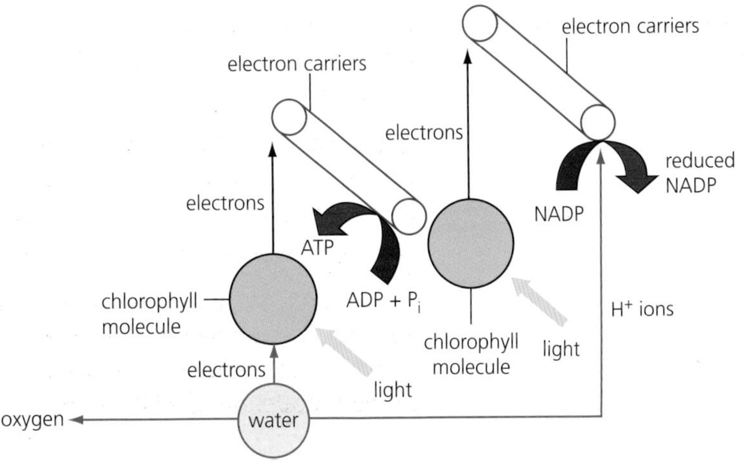

Figure 1.5 A summary of the light-dependent reaction of photosynthesis.

Electron transfer chain A series of electron carriers arranged within a membrane that can accept and then pass on electrons.

- Light strikes a chlorophyll molecule, causing photoionisation. Two electrons leave the chlorophyll molecule and pass to an electron carrier.
- The electrons are transferred along a series of electron carriers, which forms an **electron transfer chain** within the thylakoid membrane. The electrons lose energy as they are passed from one electron carrier to the next. This energy is used to produce ATP.
- **Photolysis** takes place. This is the breakdown of a water molecule to release protons, electrons and oxygen.

$$\underset{\text{water}}{H_2O} \rightarrow \underset{\text{protons}}{2H^+} + \underset{\text{electrons}}{2e^-} + \underset{\text{oxygen}}{\tfrac{1}{2}O_2}$$

- The two electrons from photolysis replace those lost from the chlorophyll molecule during photoionisation. Oxygen is released as a waste product.
- Light strikes a second chlorophyll molecule, causing photoionisation. Two electrons leave the second chlorophyll molecule and pass to an electron carrier. They pass along another electron transfer chain. They are used, together with the protons from the photolysis of water, to produce reduced NADP (see page 15 for a definition of reduction and oxidation).

$$NADP + 2H^+ + 2e^- \rightarrow \text{reduced NADP}$$

TIP

Notice that the oxygen produced in photosynthesis comes from photolysis of water, rather than from the carbon dioxide that plants take up from their environment.

Electron transfer chains and ATP production

As each pair of electrons passes along an electron transfer chain, a small amount of energy is released. This energy enables carrier proteins (see *AQA A-level Biology 1 Student's Book*, Chapter 3 page 46) within the thylakoid membranes to actively transport protons from the stroma across the thylakoid membrane and into the spaces between the thylakoids. This develops a higher concentration of protons inside the thylakoid spaces than in the stroma. As a result, protons diffuse down their concentration gradient from the thylakoid spaces to the stroma. They diffuse through molecules of ATP synthase embedded in the thylakoid membranes, and the resulting change in environment results in a change in the protein, causing the ATP synthase molecules to spin. This spinning provides energy for the synthesis of ATP from ADP and inorganic phosphate.

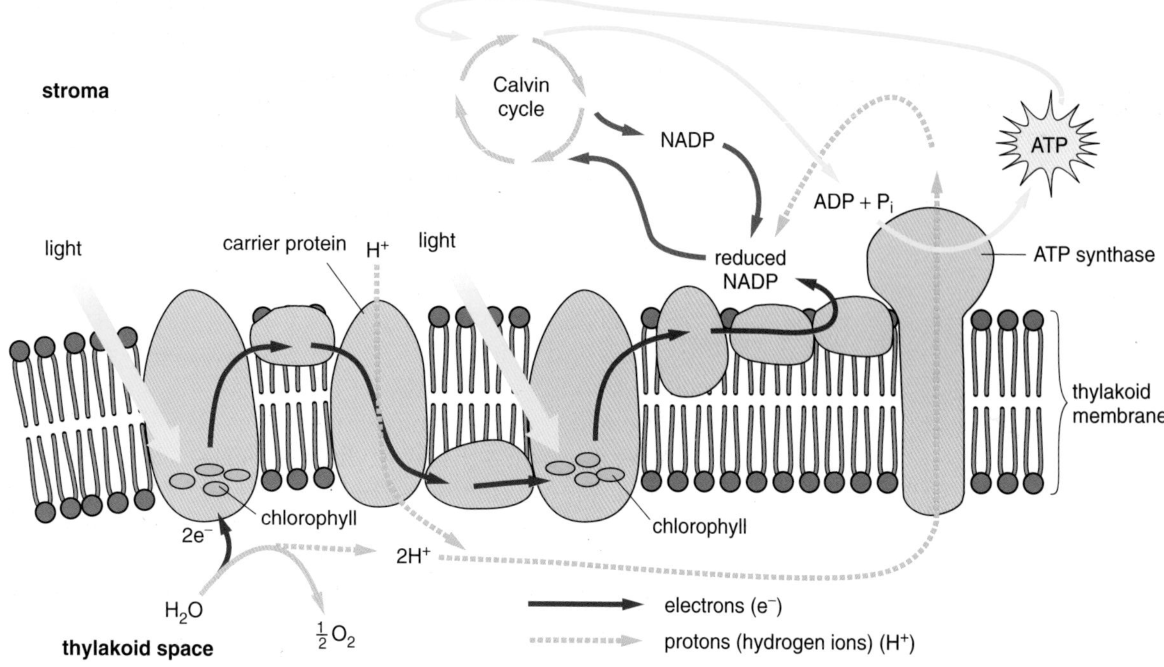

Figure 1.6 How the electron transfer system and ATP synthase molecules are embedded in the thylakoid membrane.

Chloroplasts isolated from plant leaves can be used to investigate the light-dependent reaction. If isolated chloroplasts are placed into a dilute buffer, they absorb water by osmosis and burst, releasing their thylakoids into the solution. If a blue dye called dichlorophenolindophenol (DCPIP) is added, electrons from active electron transfer chains are transferred to DCPIP molecules, reducing them rather than NADP molecules. When DCPIP is reduced, it changes colour from blue to colourless.

REQUIRED PRACTICAL 8

Investigation into the effect of a named factor on the rate of dehydrogenase activity in extracts of chloroplasts

This is just one example of how you might tackle this required practical.

Figure 1.7 shows the results for isolated chloroplasts mixed with dilute buffer and DCPIP and placed into different conditions.

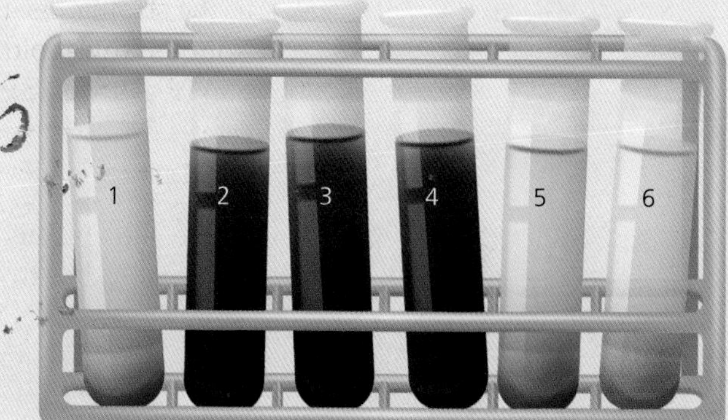

Figure 1.7 Results of an investigation using spinach chloroplast extract with DCPIP.

The tubes were set up as follows:

Tube 1: chloroplast extract, buffer, DCPIP
Tube 2: chloroplast extract, buffer, DCPIP, completely wrapped in foil
Tube 3: boiled chloroplast extract, buffer, DCPIP
Tube 4: boiled chloroplast extract, buffer, DCPIP, completely wrapped in foil
Tube 5: chloroplast extract, buffer
Tube 6: chloroplast extract, buffer, completely wrapped in foil

All were then placed under a bright light for several hours.

Using what you now know about the light-dependent reaction, explain the results.

Rubisco, or ribulose bisphosphate carboxylase. The enzyme that catalyses the reaction between carbon dioxide and ribulose bisphosphate in the Calvin cycle.

The light-independent reaction

The light-independent reaction of photosynthesis comprises a cycle of reactions called the **Calvin cycle** (see Figure 1.8). The main steps in the cycle are as follows:

- Carbon dioxide reacts with **ribulose bisphosphate** (**RuBP**) to form two molecules of **glycerate 3-phosphate (GP)**. This reaction is catalysed by an enzyme called ribulose bisphosphate carboxylase (**rubisco**).
- GP is then reduced to **triose phosphate**. This is a reduction reaction and requires the two substances formed during the light-dependent reaction: reduced NADP and ATP. The reduced NADP is used to reduce GP. ATP provides additional energy for the reaction.
- Triose phosphate is a simple sugar. Some triose phosphate is converted to useful organic substances, such as sucrose for transport or cellulose for storage, or into amino acids and triglycerides.
- Most of the triose phosphate is converted into glucose and used by the plant as a respiratory substrate.
- Some triose phosphate is used to regenerate more RuBP.

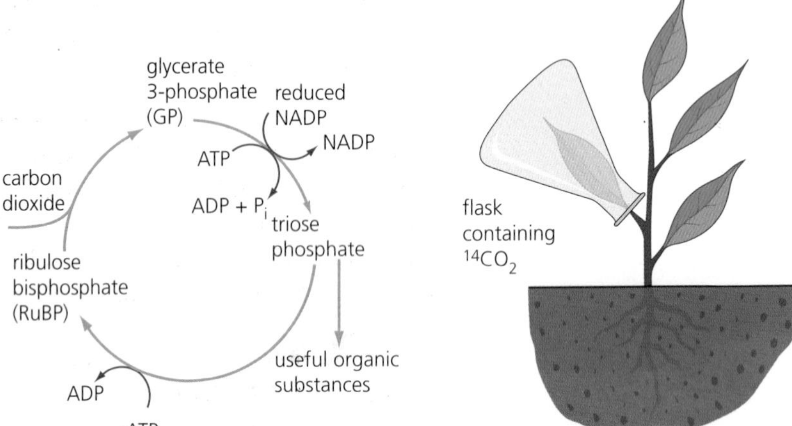

Figure 1.8 The main steps in the Calvin cycle.

Figure 1.9 Supplying radioactive carbon dioxide to a plant leaf.

The Calvin cycle can be investigated by supplying a plant with radioactively labelled carbon dioxide. A plant leaf can be enclosed in a flask containing radioactive carbon dioxide (Figure 1.9). A series of leaves can be left in radioactive carbon dioxide for different amounts of time. The leaves can then be removed from the plant and analysed for radioactive substances.

The results of such an investigation are shown as a graph in Figure 1.10. The first radioactive substance detected in the leaf is GP. RuBP is detected more slowly. This indicates that the radioactive carbon dioxide is used to form GP and that the GP is then used to form RuBP.

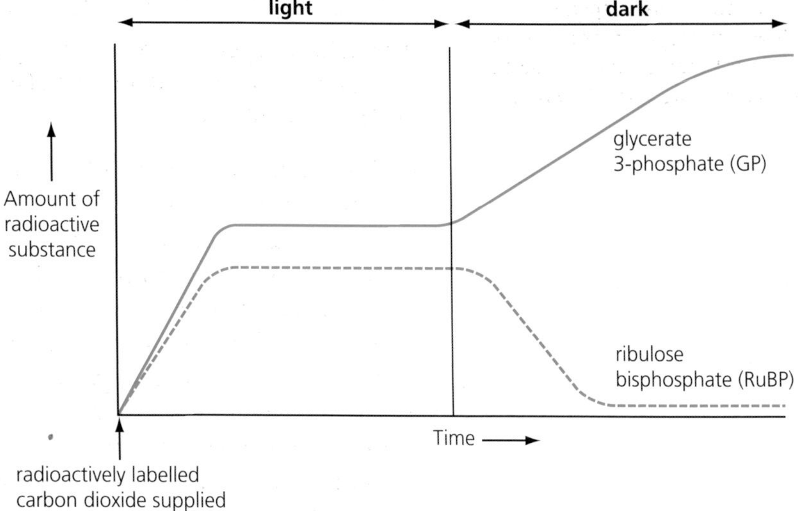

Figure 1.10 Graph showing changes in the amount of some radioactive substances in a leaf in light and dark conditions, after being given radioactively labelled carbon dioxide.

After a certain time in light, the amount of radioactive GP and RuBP both become constant. This is because GP and RuBP are being formed in a cycle. As fast as GP is being formed from carbon dioxide and RuBP, it is being used to regenerate RuBP.

If the light is switched off, the amount of GP increases while the amount of RuBP decreases. This is because the light-dependent reaction can no longer produce ATP and reduced NADP. If you look at Figure 1.8, you will see that ATP and reduced NADP are required to convert GP into triose phosphate but not to convert RuBP into GP. In the dark, RuBP is converted to GP, so the amount of RuBP decreases while the amount of GP increases.

TEST YOURSELF

1 What is photoionisation?
2 In the light-dependent reaction of photosynthesis, what happens to the electrons that come from:
   a) a water molecule
   b) the first chlorophyll molecule struck by light
   c) the second chlorophyll molecule struck by light?
3 Name the three-carbon sugar produced in the Calvin cycle.
4 Look at Figure 1.10. Use the information to explain the changes in the amounts of radioactive substances in the dark.

Limiting factors and photosynthesis

Yield The biomass of the part of the crop that is harvested.

Plants rely on photosynthesis to produce their respiratory substrates and other biological molecules they need to form their biomass. The greater a plant's rate of photosynthesis, the greater its rate of growth and, if we are considering crop plants, the higher the yield. Among the environmental factors that may affect the rate of photosynthesis are:

- light intensity
- carbon dioxide concentration
- temperature
- availability of water in the soil.

The graph in Figure 1.11 shows the effects of light intensity, carbon dioxide concentration and temperature on the rate of photosynthesis.

> **TIP**
> The shape of curve A in Figure 1.11 is very similar to the one for the rate of an enzyme-catalysed reaction against substrate concentration (see *AQA A-level Biology 1 Student's Book*, Chapter 2 page 32) because both involve limiting factors.

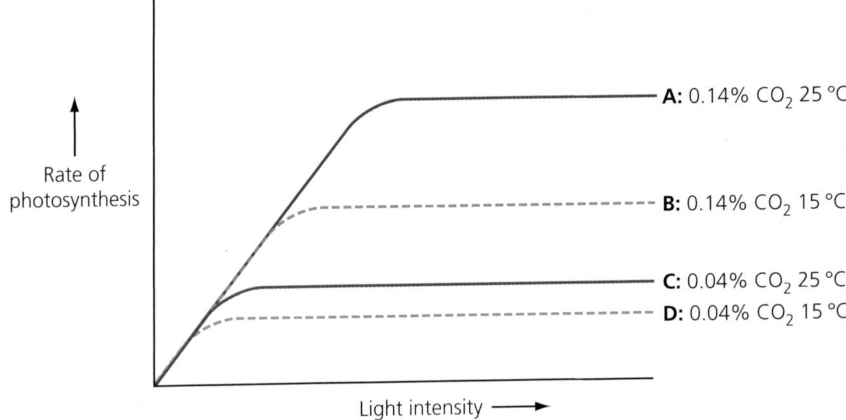

Figure 1.11 Limiting factors: the effects of light intensity, carbon dioxide concentration and temperature on the rate of photosynthesis.

Over the first part of the curve A, the rate of photosynthesis is directly proportional to light intensity. Because of this, we can describe light as limiting the rate of photosynthesis over this part of the curve (it is a **limiting factor**). As light intensity continues to increase, the curve starts to flatten out: light intensity is no longer limiting the rate of photosynthesis.

However, the graph also shows two other environmental factors that may interact with light intensity. For example, increasing carbon dioxide concentration or temperature will not increase the rate of photosynthesis when light intensity is low, such as on a winter's day in the UK.

Carbon dioxide limits the rate of photosynthesis in bright conditions. Increasing the concentration of carbon dioxide in the atmosphere from its normal level of approximately 0.04% to 0.14% and keeping the temperature constant has a much greater effect on the rate of photosynthesis than increasing the temperature from 15 °C to 25 °C and keeping the carbon dioxide concentration constant.

> **TIP**
> The amount of water available to a plant can limit the rate of photosynthesis if lack of water causes the stomata to close. The rate of photosynthesis is actually then limited by reduced carbon dioxide availability, not by a lack of water for photosynthesis itself.

You can also see from the graph that increasing the temperature from 15 °C to 25 °C increases the rate of photosynthesis. The effect is greatest when neither light intensity nor carbon dioxide concentration are limiting.

By understanding how environmental factors can limit photosynthesis, farmers are able to take steps to overcome their effects and improve the yield of their crops. This is easier in closed environments such as glasshouses, but some simple practices, such as limiting the shade from high hedges or irrigating crops, can be worthwhile for field crops too.

The chemical energy stored in the dry biomass can then be estimated by calorimetry. In this case, GPP would be measured in $kJ\,m^{-2}\,day^{-1}$ (or $kJ\,m^{-3}\,day^{-1}$ if it were aquatic algae).

Net primary production The chemical energy stored in plant biomass after losses to respiration have been taken into account.

Some of the substances formed during photosynthesis are not, however, used to form new cells and tissues. They are used in respiration. The difference between GPP and respiration is **net primary production** (NPP). We can calculate net primary production from the equation:

$$\text{NPP} = \text{GPP} - R$$

where R represents the energy loss through respiration.

Net primary production is important because it represents the amount of energy available to the primary consumers (herbivores) and decomposers at other trophic levels in a food web (see page 22).

Efficiency of energy transfer by heather

Figure 1.14 Heather.

Figure 1.14 shows heather plants. Scientists have found heather particularly useful in studying the efficiency of photosynthesis for the following reasons.

- In moorland areas, heather grows in large clumps called stands. Each stand is almost pure heather, so we do not have to consider what fraction of the total visible-light energy is used by other species.
- In many moorland areas, heather is managed. This means that it is burned at regular intervals. Old woody plants are replaced by young plants, which provide a better food supply for game birds, such as grouse. Estate managers keep records so we usually know the age of a particular stand of heather.
- Many investigations have been carried out using heather. By sharing the findings of their research, scientists have been able to replicate and further test their work. This increases the reliability of the conclusions that they draw.

We can calculate the efficiency of energy transfer in heather from the formula:

$$\text{Efficiency of energy transfer} = \frac{\text{chemical energy increase in an area of heather plants in a year}}{\text{light energy falling on this area of heather plants in a year}}$$

To calculate the figure on the top line of the equation we need to multiply the chemical energy in 1 g of heather by the biomass of heather produced in $g\,m^{-2}\,year^{-1}$.

The chemical energy in 1 g of heather can be found by using a calorimeter (see overleaf).

TIP

This is just an example of calculating the efficiency of energy transfer. You may be asked to calculate the efficiency of transfer in other situations, such as in farm crops or animals.

EXAMPLE

Finding the chemical energy in 1 g of heather

We can find this by burning a sample of heather in a **calorimeter** (Figure 1.15) and measuring the energy released as heat.

Look at Figure 1.15 and answer the questions.

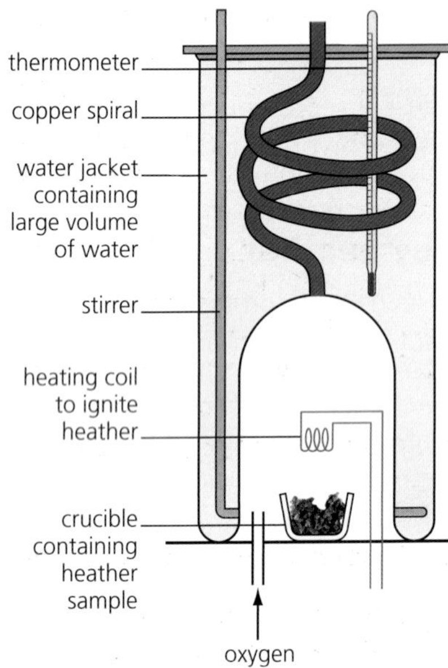

Figure 1.15 A calorimeter used to measure the energy released when a sample of heather is burned.

1 An oxygen supply is connected to the apparatus. What is the advantage of burning the heather in oxygen and not in air?
Oxygen ensures that the sample of heather is burned completely.

2 A copper spiral is attached to the top of the combustion chamber. What is the function of this copper spiral?
The copper spiral provides a large surface area for heat exchange with the water.

3 The water jacket contains a large volume of water. The total rise in temperature of the water in this jacket when the heather is burned is only small. Explain the advantage of a small rise in temperature.
A small temperature rise means less heat will be lost to the surroundings.

Table 1.2 Shows some typical results obtained from the apparatus in Figure 1.15.

Table 1.2 Results obtained from burning a sample of heather in a calorimeter.

Volume of water in water jacket/cm^3	650
Temperature of water before the heather sample was burned/°C	18
Temperature of water after the heather sample was burned/°C	23
Mass of heather/g	0.5

4 The amount of energy needed to raise the temperature of 1 cm^3 of water by 1 °C is 4.2 joules. Calculate the amount of energy released by burning 1 g of heather.
In this investigation, the energy released has raised the temperature of 650 cm^3 of water by 5 °C. So, the total amount of energy released by the heather sample is 4.2 × 650 × 5 J.
This is the amount of energy released by 0.5 g of heather. We need to divide the total amount of energy released by 0.5 to get the amount of energy released per gram:

$$4.2 \times 650 \times 5/0.5\,J\,g^{-1}$$

This gives us a figure of 27300 $J\,g^{-1}$ or 27.3 $kJ\,g^{-1}$.

5 Do you think that the figure that we have calculated is an overestimate or an underestimate? Explain your answer.
It is probably an underestimate. There are several reasons for this. Not all the heather will have been burned. The ash that is left in the crucible may contain some chemical energy that has not been released.
In addition, it is unlikely that all the heat from the combustion will have been transferred to the water in the water jacket.

TEST YOURSELF

5 Give three reasons for the relatively low efficiency of photosynthesis.
6 Describe how the dry mass of a sample of plant material would be found.
7 What is the equation for finding net primary production?
8 Suggest suitable units for the net primary production of algae in a lake.
9 Explain how the net primary production of trees is made available to decomposers in a wood.

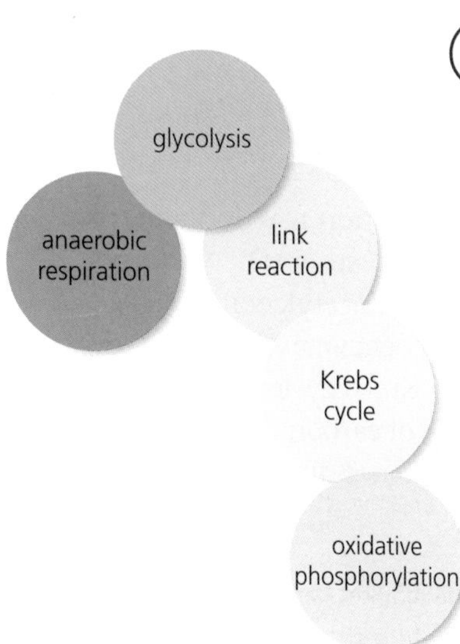

Figure 1.16 The breakdown of glucose in respiration. Aerobic respiration takes place in the presence of oxygen. Respiration can continue when there is no oxygen. This is anaerobic respiration.

Phosphorylation The addition of a phosphate group to a molecule.

Respiration

Respiration takes place in all living cells, providing indirect evidence for evolution. It is a biochemical process in which biological molecules called respiratory substrates are used as fuel. They are broken down in a series of stages and the chemical energy they contain is transferred to ATP. We can summarise the process with the equation:

$$C_6H_{12}O_6 + 6O_2 \rightarrow 6CO_2 + 6H_2O + \text{energy}$$

Unfortunately this equation is misleading in a number of ways. It shows the fuel as glucose. In many cells the main fuel is glucose, but fatty acids, glycerol and amino acids are also respiratory substrates and can be used for respiration. The equation also shows that oxygen is required: this is correct when **aerobic respiration** is occurring but respiration can also take place anaerobically. **Anaerobic respiration** means respiration without oxygen. It is also important to understand that the oxygen is not directly used to make the carbon dioxide shown in the equation. Finally, the equation shows respiration as a single reaction. It isn't a single reaction. It involves a number of reactions in which the respiratory substrate is broken down in a series of steps, releasing a small amount of energy each time. The steps involved in the respiration of glucose are summarised in Figure 1.16.

Glycolysis

Glycolysis (Figure 1.17) is the first step in the biochemical pathway of respiration.

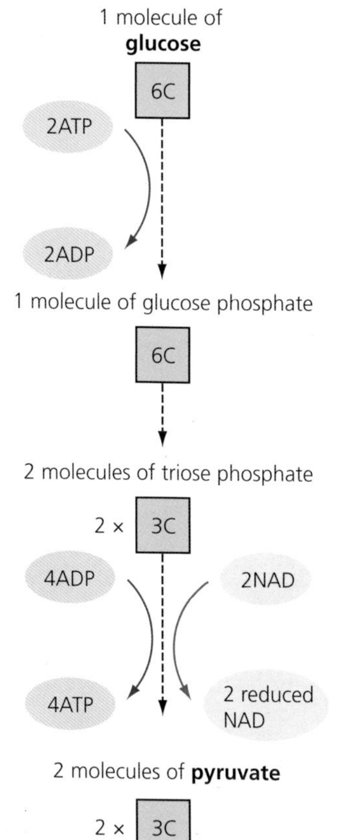

Figure 1.17 A summary of glycolysis, the first step in respiration. The boxes show the number of carbon atoms in the different molecules. There is a net gain of two molecules of ATP from each molecule of glucose.

- Glucose contains a lot of chemical energy. In order to release this energy, some additional energy from ATP is required to achieve the activation energy for the reaction (see *AQA A-level Biology 1 Student's Book*, Chapter 2 page 20). In the first stage of glycolysis, a molecule of glucose is converted into glucose phosphate. This requires two molecules of ATP and is called **phosphorylation**.
- Each molecule of glucose phosphate is then oxidised to two molecules of triose phosphate.
- Each molecule of triose phosphate is then converted to pyruvate. This reaction produces ATP. A total of four molecules of ATP are produced, two for each triose phosphate molecule. During glycolysis, then, there is a net gain of two molecules of ATP for each molecule of glucose.
- The conversion of triose phosphate to pyruvate is an oxidation reaction and involves the removal of hydrogen to reduce a coenzyme called NAD. NAD is converted to reduced NAD as a result.

NAD, oxidation and reduction

Oxidation is sometimes represented as the addition of oxygen to a substance but, more accurately, it is any reaction in which electrons are *removed*. **Reduction**, on the other hand, involves the gain of electrons. Whenever one substance is oxidised another must be reduced. In simple terms, if one substance loses electrons, another must gain them. We often use the term **oxidation–reduction reaction** for a reaction in which one substance is oxidised and another is reduced. In glycolysis, the conversion of triose phosphate to pyruvate is an oxidation reaction in which pyruvate loses electrons and NAD gains them, becoming reduced NAD.

The link reaction

Pyruvate still contains a lot of chemical energy. When oxygen is available, this energy can be made available in a series of reactions known as the Krebs cycle. The **link reaction** is a term used to describe the reaction linking glycolysis and the Krebs cycle. It is summarised in Figure 1.18.

- Pyruvate is oxidised to acetate. This is an oxidation reaction and, like glycolysis, also involves the production of reduced NAD.
- Acetate combines with coenzyme A to produce acetylcoenzyme A.
- Pyruvate contains three carbon atoms. Acetylcoenzyme A contains several carbon atoms but only two of these enter the Krebs cycle. One of the carbon atoms from pyruvate goes to form a molecule of carbon dioxide. (Remember that the production of this carbon dioxide does not directly use oxygen.)

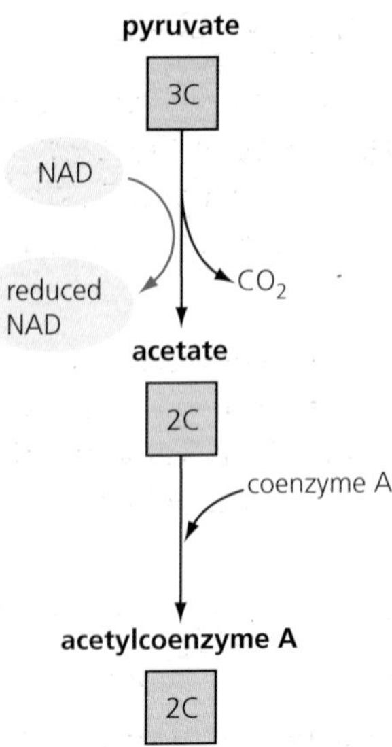

Figure 1.18 The link reaction. The boxes show the number of carbon atoms in the molecules involved in the respiratory pathway.

The Krebs cycle

Figure 1.19 is a simple diagram that summarises the essential features of the Krebs cycle.

- Acetylcoenzyme A, produced in the link reaction, is fed into the cycle. It combines with a 4-carbon compound to produce a 6-carbon compound.
- In a series of oxidation-reduction reactions, the 6-carbon compound is converted back to the 4-carbon compound. In this process two molecules of carbon dioxide are given off. (Again, remember that the production of this carbon dioxide does not directly use oxygen.)

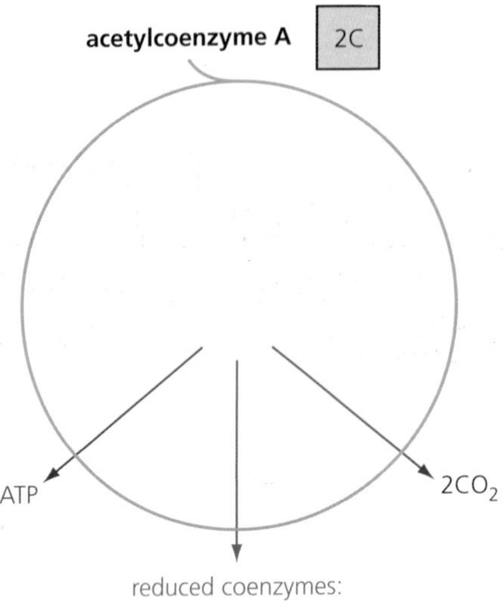

Figure 1.19 The Krebs cycle plays a very important part in respiration. It is the main source of reduced coenzymes, which are used to produce ATP in the electron transfer chain.

- In a series of oxidation reactions the Krebs cycle generates reduced coenzymes and ATP. For each complete cycle:
  - one molecule of ATP is produced. This is called **substrate-level phosphorylation**, because the ATP is formed as a result of a one of the Krebs cycle reactions. It is linked to the reaction of one of the substrates.
  - three molecules of reduced NAD and one molecule of another reduced coenzyme, reduced FAD, are produced.
- The most important function of the Krebs cycle in respiration is the production of reduced coenzymes. They are passed to the electron transfer chain, where the chemical energy that these molecules contain is used to produce ATP.

Substrate-level phosphorylation The production of ATP linked to the reaction of a substrate molecule.

As well as the pyruvate formed from glucose by glycolysis, the breakdown products of other respiratory substrates such as lipids and amino acids can also enter the Krebs cycle and form ATP and reduced coenzymes. This means that some cells can respire lipid for some of the time instead of glucose.

Cells do not usually respire the breakdown products of amino acids unless there is no glucose or lipid available.

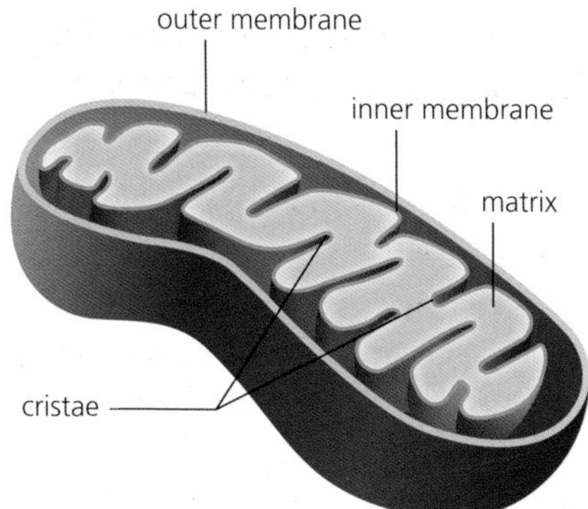

Figure 1.20 The structure of a mitochondrion.

TIP

The reason why it is more correct to describe mitochondria as the site of aerobic respiration is because only the electron transfer chains use oxygen and these are in the mitochondria. Glycolysis is an anaerobic process, but it takes place in the cytoplasm, not in mitochondria.

Mitochondria and electron transfer chains

You should remember that **mitochondria** are the site of aerobic respiration (see *AQA A-level Biology 1 Student's Book*, Chapter 3, page 41). Each mitochondrion (Figure 1.20) is surrounded by two membranes, a folded inner membrane and a smooth outer one. The folds on the inner membrane form many projections, called cristae.

The different steps of respiration take place in different locations. Glycolysis takes place outside of mitochondria, in the cytoplasm. The Krebs cycle occurs in the matrix of the mitochondria. Like the thylakoids in chloroplasts, the cristae provide a large surface area in which the electron carriers of electron transfer chains are embedded.

Reduced coenzymes have so far been produced in glycolysis and the link reaction. More reduced coenzyme is produced by the Krebs cycle.

Reduced coenzymes transfer electrons to chains of protein molecules embedded in the inner membranes of mitochondria. These proteins act as **electron carriers** and form electron transfer chains similar to those in chloroplasts.

Figure 1.21 shows how electrons from reduced coenzyme pass from one protein to the next along electron transfer chains.

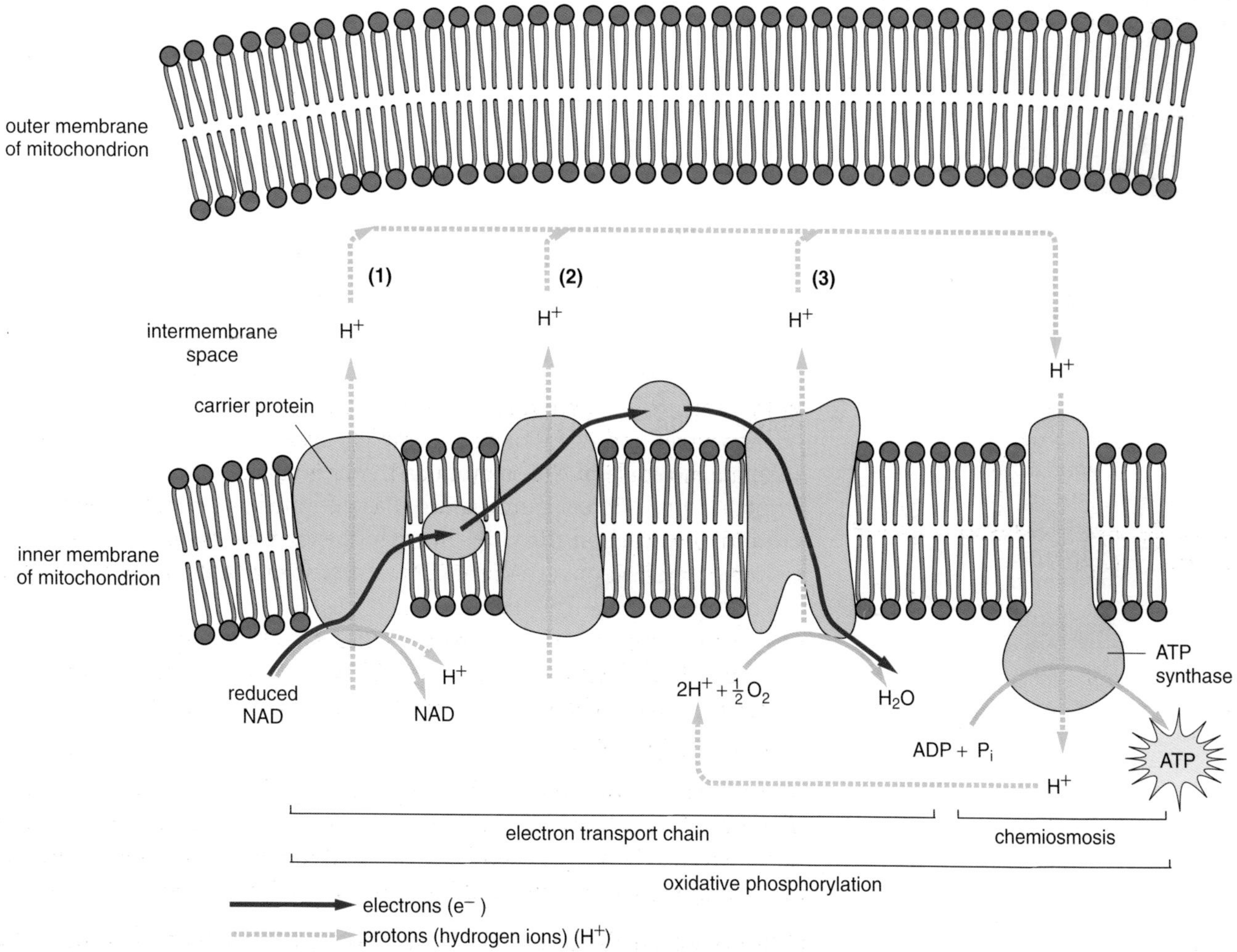

Figure 1.21 Energy is released as electrons pass from one carrier to the next in an electron transfer chain. Some of this energy is lost as heat but a lot of it goes to produce ATP.

TIP
Remember that oxygen is not used to make carbon dioxide; it is used as the terminal electron acceptor in electron transfer chains.

Oxidative phosphorylation The production of ATP by an electron transfer chain using oxygen as the final electron acceptor.

When these electrons are transferred, energy is released. Some of this energy is lost as heat but some is used by carrier proteins in the active transport of protons across the inner mitochondrial membrane into the space between the inner membrane and the outer membrane. This develops a higher concentration of protons in the space between the membranes than there is in the matrix of the mitochondrion.

As a result, protons diffuse down their concentration gradient from the space between the membranes to the matrix (this is an example of facilitated diffusion; see *AQA A-level Biology 1 Student's Book*, Chapter 3 page 46). They diffuse through molecules of ATP synthase embedded in the inner mitochondrial membrane, causing the ATP synthase molecules to spin. This spinning provides energy for the synthesis of ATP from ADP and inorganic phosphate. The last molecule in the electron transfer chain is oxygen. Oxygen combines with protons and electrons to produce water. ATP production by the electron transfer chain is called **oxidative phosphorylation**.

The heat released during respiration can be used to good effect by organisms. Endothermic animals use it to raise their body temperature above that of their environment. Although plants have much slower rates of respiration and produce less heat, they sometimes do this too. An interesting example is shown in Figure 1.22.

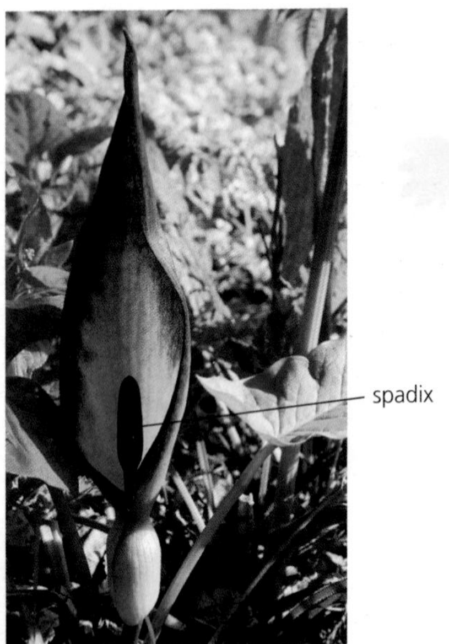

Figure 1.22 When a cuckoo pint flowers, the temperature of the spadix increases until it is as much as 15 °C above the temperature of the environment. This increase in temperature causes molecules of a substance similar to substances found in faeces and decaying bodies to be released. This attracts the flies that pollinate cuckoo pint flowers.

Anaerobic respiration

Sometimes there is not enough oxygen for an organism to respire only aerobically using the pathways just described, and also some organisms are adapted to live without oxygen. Under these conditions ATP is produced by anaerobic respiration. The only stage in the anaerobic pathway that produces ATP is glycolysis, so although it is fast the process is not as efficient as aerobic respiration, because there is an incomplete breakdown of glucose.

Look back at Figure 1.17. You will see that during glycolysis the coenzyme NAD is reduced. Reduced NAD is normally converted back to oxidised NAD when its electrons are passed to the electron carriers in the electron transfer chain. This can only happen when oxygen is present. Obviously, if all the oxidised NAD in a cell was converted to reduced NAD, the process of respiration, would stop.

In anaerobic respiration in animals pyruvate is converted to **lactate**. In plants and microorganisms, such as yeast, it is converted to **ethanol** and **carbon dioxide**. In both of these pathways (Figure 1.23), reduced NAD is converted back to oxidised NAD. This allows glycolysis to continue.

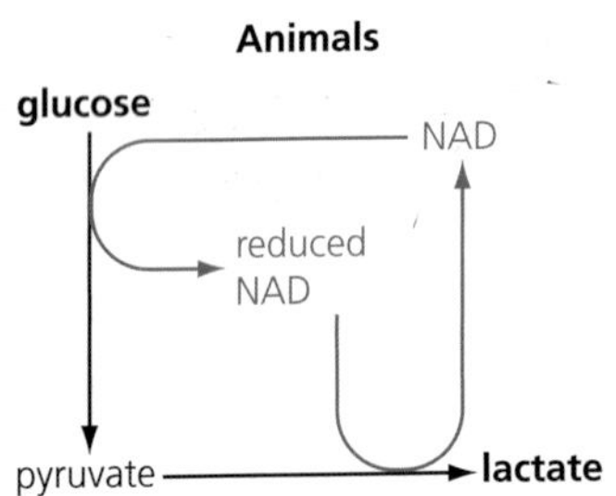

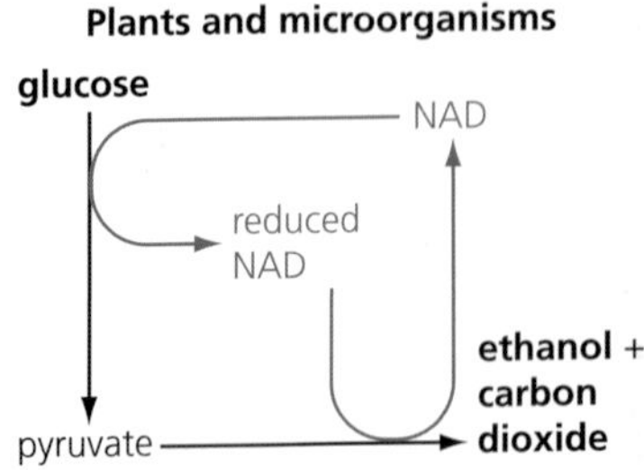

Figure 1.23 Anaerobic respiration allows organisms to produce ATP in the absence of oxygen.

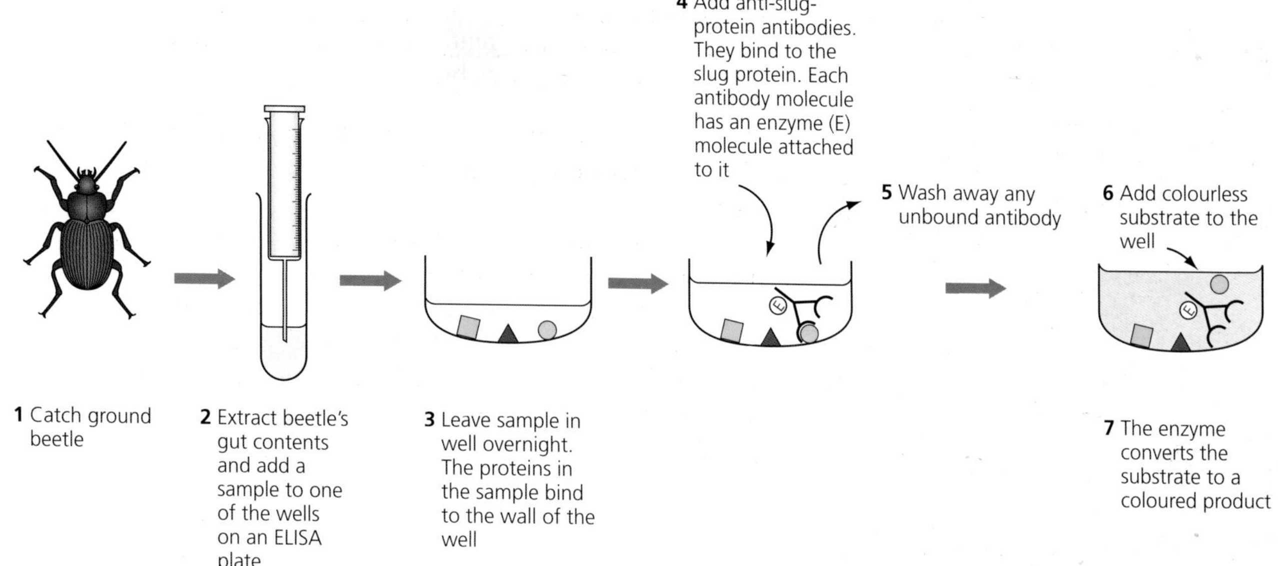
4 Add anti-slug-protein antibodies. They bind to the slug protein. Each antibody molecule has an enzyme (E) molecule attached to it
5 Wash away any unbound antibody
6 Add colourless substrate to the well
E
1 Catch ground beetle
2 Extract beetle's gut contents and add a sample to one of the wells on an ELISA plate
3 Leave sample in well overnight. The proteins in the sample bind to the wall of the well
7 The enzyme converts the substrate to a coloured product

Figure 1.25 Using an ELISA test to confirm that ground beetles eat slugs.

Before we look at the diagram in detail, there are two important principles about which we must remind ourselves. We first met these principles in *AQA A-level Biology 1 Student's Book*.

- Different species contain different proteins (see *AQA A-level Biology 1 Student's Book*, Chapter 12 page 221). Suppose our ground beetle had been feeding on slugs and, say, earthworms. Some of the proteins found in the slugs would probably be very similar to proteins found in the earthworms so we probably couldn't tell which animal the protein concerned came from. Some, however, would be different. These specific slug proteins would have specific sequences of amino acids and their molecules would have specific tertiary structures.
- Antibodies are also proteins. They have specific binding sites. These binding sites mean that they will only bind to molecules that have a complementary shape (see *AQA A-level Biology 1 Student's Book*, Chapter 6 page 91). An anti-slug-protein antibody will only bind to one particular slug protein. It won't bind to proteins from any other species unless they are identical to the slug protein.

Steps 1 and 2 in Figure 1.25 should be easy enough to understand but we may need to explain some of the other steps. We will start with step 3 where the sample of gut contents has been left overnight in one of the wells on the ELISA plate.

1 Three protein molecules are shown attached to the wall of the well. Why do these protein molecules have different shapes?

Each shape represents a different protein, with a different amino acid sequence, so there are three different proteins shown here. All three could be slug proteins, but it is possible that one or two of them might have come from other animals that the ground beetle had eaten.

2 Step 4 shows anti-slug-protein antibody binding only to a slug protein. Why does this antibody bind only to slug protein?

This is the point made earlier. Antibodies are specific and an anti-slug-protein antibody will only bind to the protein shown as a blue circle. This is a slug protein.

3 Why is the unbound antibody washed away (step 5)?

If we don't wash the unbound antibody away, it will remain in the well. The enzyme on the unbound antibody will result in the coloured product being formed even if no slug protein is present.

4 Not all ground beetles eat slugs. Explain how you would be able to tell if the ground beetle from which you had obtained the gut sample had not been eating slugs.

There would be no slug proteins attached to the wall of the well to bind to the anti-slug-protein antibodies. Therefore there would be no enzyme to catalyse the reaction in which the colourless substrate was converted to a coloured product.

5 How could you use an ELISA plate to find out whether slugs were important items of food for ground beetles?

You could add samples from different ground beetle guts to different wells on the ELISA plate. By counting the number of wells where there was a colour change, you could find the percentage of ground beetles that ate slugs.

Trophic levels

Trophic levels The feeding positions organisms occupy in a food web.

In any ecosystem different organisms gain their food in different ways (Figure 1.26). They feed at different trophic levels. Green plants are primary producers. They produce biological molecules from carbon dioxide, water and mineral ions. They rely on photosynthesis to transfer light energy to chemical energy in biological molecules. The other organisms that make up the community rely either directly or indirectly on the biological molecules produced by the producers. Primary consumers (**herbivores**) feed on producers. Secondary consumers feed on primary consumers and tertiary consumers feed on secondary consumers. Organisms that are not eaten eventually die. Another group of organisms, the **saprobiotic decomposers**, digest dead tissues and use the biological molecules that make up these tissues as a source of chemical energy.

TIP
Some organisms, such as the dark green bush cricket, feed at different trophic levels and some organisms feed on different sources of food when they are larvae and when they are adults. The caterpillar of the peacock butterfly, for example, eats nettle leaves. The adult butterfly feeds on nectar produced by flowers.

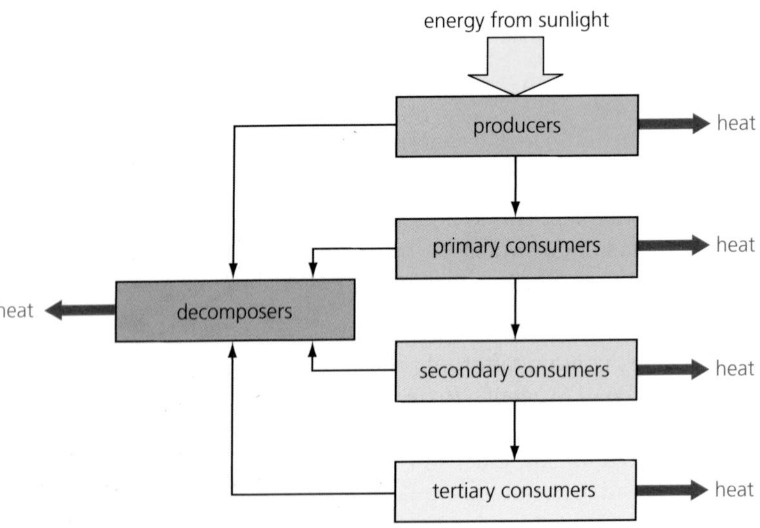

Figure 1.26 The transfer of energy in an ecosystem. The boxes represent trophic levels. The arrows show the direction in which energy is transferred.

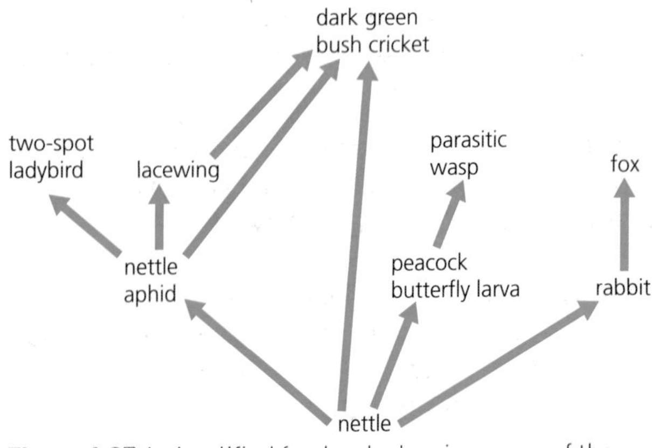

Figure 1.27 A simplified food web showing some of the organisms that feed on nettles.

We often talk about **food chains**, suggesting perhaps that we frequently encounter situations where animal B feeds only on plant A. In turn, animal C only eats animal B, and animal D only eats animal C. This hardly ever happens under natural conditions. Food chains are linked to each other to form complex **food webs**. Figure 1.27 shows a possible food web associated with a nettle patch.

Many farming practices are based on an understanding of the energy losses between trophic levels and attempt to reduce them. For example, pests that eat crops divert energy away from the human food chain. Reducing pest populations on crops by the use of chemical **pesticides** (see *AQA A-level Biology 1 Student's Book*, Chapter 13 pages 235–236) minimises the energy losses. This increases the yield.

TIP
Refresh your memory about the impacts of pesticides in Chapter 13 on biodiversity in *AQA A-level Biology 1 Student's Book*.

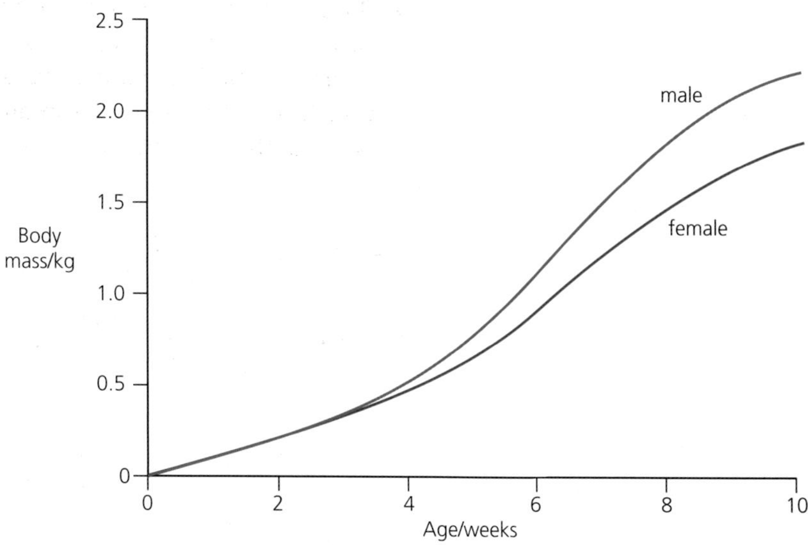

Figure 1.30 Growth curves for male and female broilers.

TIP

Growth and food consumption in broiler chickens is simply an example to illustrate the concept of net production in consumers. You do not need to memorise this example.

You can see from this graph that intensively reared broilers, fed with high-energy, high-protein food, grow very rapidly. A modern broiler may be ready for marketing 7–8 weeks after hatching. At this age it is still growing and may not have even reached its maximum growth rate. The reason for slaughtering birds at this age can be seen if you look at the data in Table 1.4. Look at the figures for male birds. You can see from the last column that the mean mass of food eaten per kilogram gain in mass rises steadily. In other words, efficiency of food conversion falls. This is mainly because the bird produces less protein-rich muscle and more body fat as it gets older.

Table 1.4 also shows that, by the time they have reached 10 weeks, female broilers have a smaller mean body mass and their efficiency of food conversion is lower than that of males. The difference in efficiency of food conversion may not seem very much – 2.5 kg of food per kilogram gain in body mass compared with 2.4 kg in males – but differences like this have a considerable influence on overall profit.

Table 1.4 Growth and food consumption in broiler chickens.

Sex	Age/weeks	Mean body mass/kg	Mean cumulative mass of food eaten/kg	Mean mass of food eaten per kilogram gain in mass/kg
Male	2	0.2	0.3	1.5
	6	1.0	2.0	2.0
	10	2.2	5.4	2.4
Female	2	0.2	0.3	1.5
	6	0.9	1.8	2.0
	10	1.8	4.5	2.5

TEST YOURSELF

15 State the equation for the net production of consumers.

16 Explain how energy is lost in faeces.

17 It is rare for there to be more than five trophic levels in a food web. Explain why.

18 Trout grown in fish farms are fed special pelleted food. Give two reasons why pelleted food enables farmed fish to convert food into new tissue more efficiently than free-range chickens, which find their own food.

19 Intensively reared broilers are kept under temperature-controlled conditions. Suggest how controlling temperature may increase net production.

Practice questions

1 The diagram shows some of the steps in respiration.

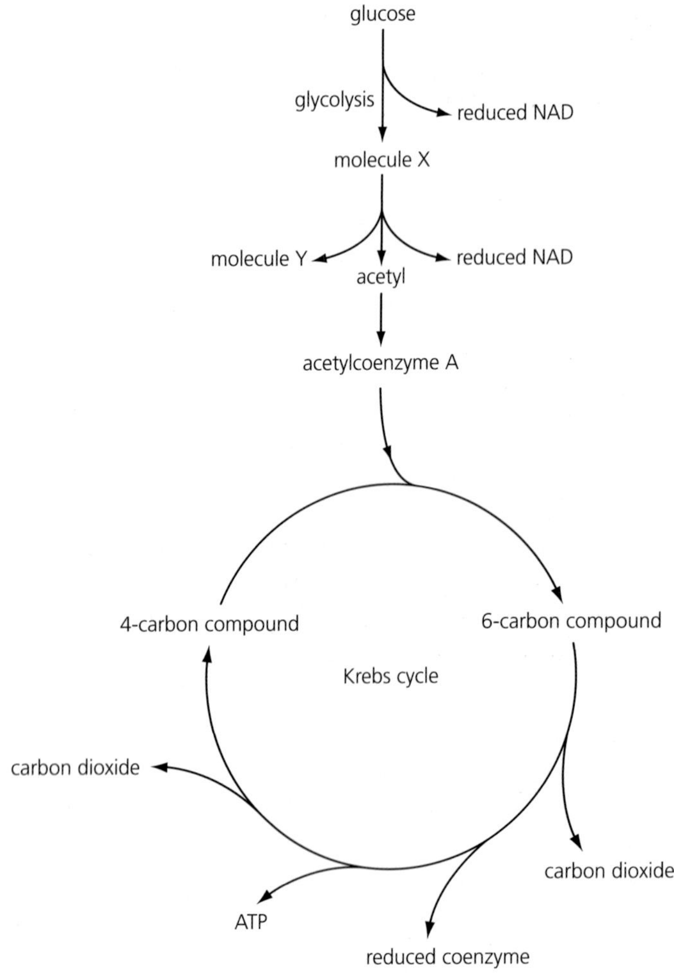

a) Where in a cell does glycolysis occur? *(1)*

b) i) Name molecule X. *(1)*

ii) Name molecule Y. *(1)*

c) Name the step in respiration that produces acetylcoenzyme A. *(1)*

d) What type of phosphorylation produces the ATP in the Krebs cycle? *(1)*

e) Describe how the reduced coenzyme produced by the Krebs cycle is used. *(2)*

2 The diagram summarises the steps involved in photosynthesis.

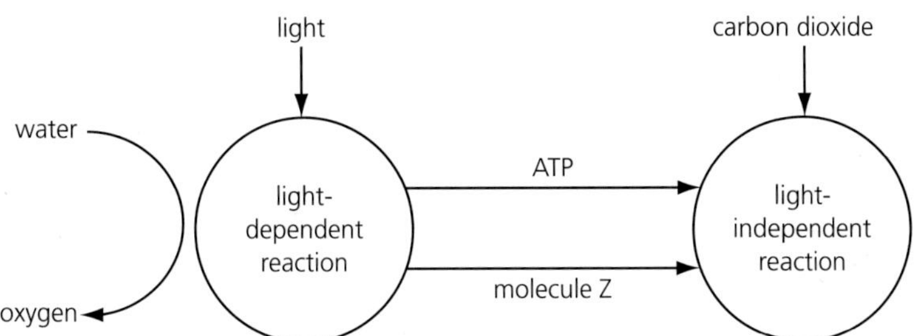

a) Name molecule Z. *(1)*

b) Name the three-carbon sugar produced by the light-independent reaction. *(1)*

c) i) Which enzyme catalyses the reaction between carbon dioxide and ribulose bisphosphate in the Calvin cycle? *(1)*

ii) What is the product of this reaction? *(1)*

d) Describe and explain what happens to the ribulose bisphosphate concentration if a plant is placed in the dark. *(2)*

3 The trophic levels in a food web can be numbered, starting with 1.0 for primary producers, 2.0 for primary consumers and so on. The table shows the mean trophic levels of marine fish caught for human food in the period 1950 to 2000.

Year	Mean trophic level
1950	3.37
1960	3.36
1970	3.39
1980	3.29
1990	3.26
2000	3.21

a) i) Phytoplankton consists of single-celled photosynthetic organisms that float in the surface water. A species of fish feeds only on phytoplankton. What would be the trophic level of this species of fish? *(1)*

ii) Another species of fish has a trophic level of 3.0. Is this fish a primary consumer, a secondary consumer or a tertiary consumer? *(1)*

b) i) Describe how the mean trophic level of the marine fish catch has changed over the period 1950 to 2000. *(2)*

ii) Suggest an explanation for the change you described in your answer to question 3b(i). *(2)*

c) Over the same period of time, more cattle and sheep have been fed on protein concentrates as well as grass. These protein concentrates are often made from animal material. Suggest how the mean trophic levels of farm animals have changed over the period shown in the table. *(2)*

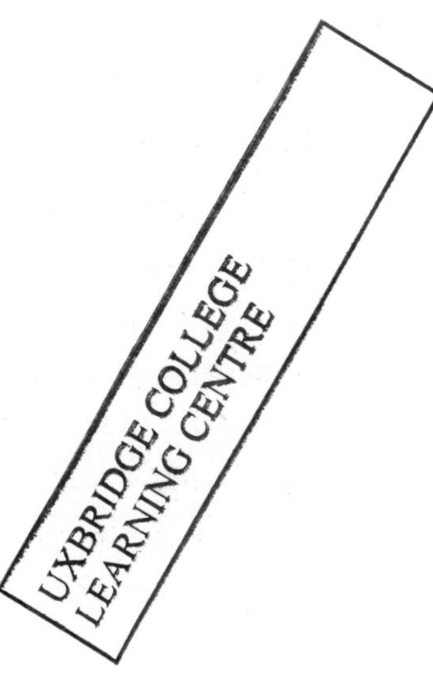

4 The graph shows the mean biomass of heather from plants of different ages growing on an area of moorland in Yorkshire.

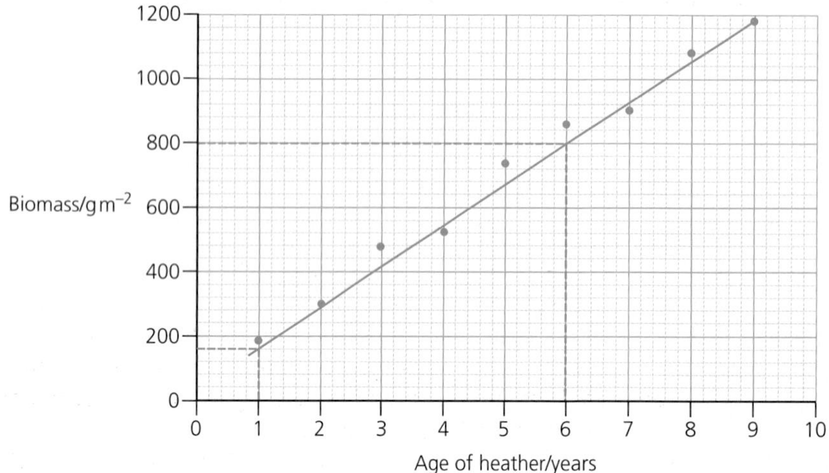

a) Give two ways that biomass can be measured. *(1)*

b) The biomass of the heather is given as dry mass.

i) Describe how you would measure the dry mass of a heather sample. *(2)*

ii) How would you make sure that your value for dry mass was valid? *(1)*

iii) What is the advantage of measuring biomass as dry mass? *(1)*

c) Describe how you would use the graph to calculate the mean annual increase in dry biomass of the heather plants. *(2)*

d) What is the evidence from the graph that the heather did not increase in biomass by the same amount each year? *(1)*

e) Suggest why the heather did not increase in biomass by the same amount each year. *(1)*

Scientists estimated the total amount of light energy falling on 1 m^2 of the heather moorland in Yorkshire where the samples were collected to be 1 415 000 kJ. They based this estimate on the amount of light falling on the heather in the growing season, so it does not include the winter months when the temperatures are too low for growth.

f) Use all the figures in this section to calculate the efficiency of photosynthesis in these heather plants. *(3)*

Stretch and challenge

5 Examine the structure of marine ecosystems around deep-sea volcanic vents. Discuss how chemical potential energy in large molecules is transferred from one organism to another through a food web, even though there are no photoautotrophs present.

6 Some plants have alternative photosynthesis pathways, e.g. crassulacean acid metabolism (CAM) and C4 photosynthesis. Contrast these with the C3 photosynthesis that most plants use. To what extent are these alternative pathways an adaptation to the environment?

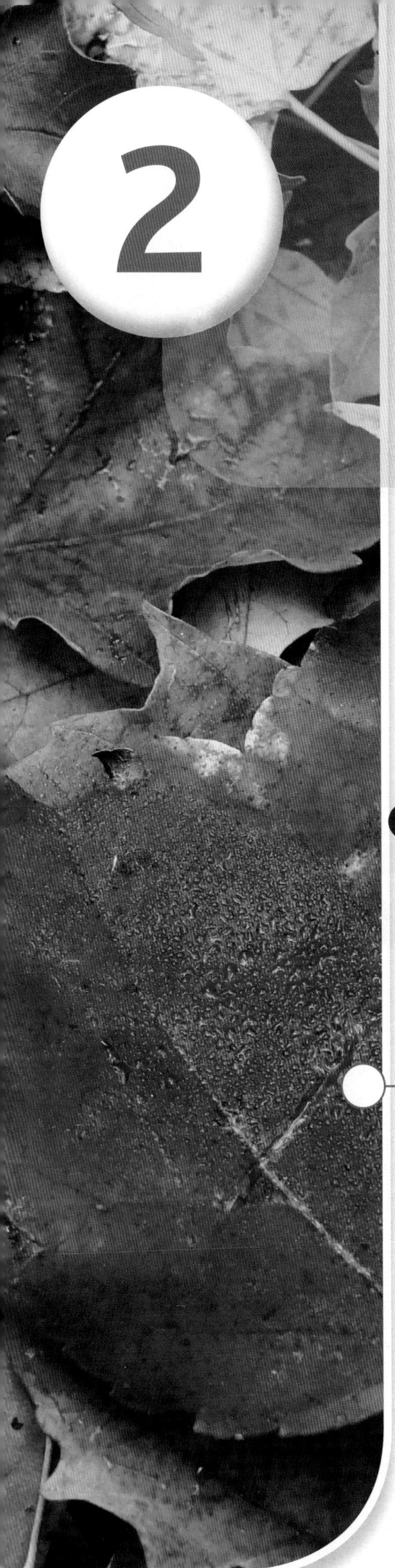

Nutrient cycles

PRIOR KNOWLEDGE

- Living things remove materials from the environment for growth and other processes. These materials are returned to the environment either in waste materials or when living things die and decay.
- Materials decay because they are broken down (digested) by micro-organisms. Microorganisms are more active and digest materials faster in warm, moist, aerobic conditions.
- The decay process releases substances that plants use in growth.
- In a stable community, the processes that remove materials are balanced by processes that return materials. The materials are constantly cycled.
- To produce proteins, plants also use nitrate ions that are absorbed from the soil.
- Phosphate ions are components of both DNA and ATP.

TEST YOURSELF ON PRIOR KNOWLEDGE

1 What conditions accelerate decay?
2 Give two types of microorganism that are involved in decay.
3 Other than proteins, name one group of biological molecules that contain nitrogen.
4 Describe the way in which phosphate groups form part of the structure of ATP.

Introduction

One of the problems that almost all human communities face is how to get rid of the enormous quantities of waste they produce. Some of this waste can be composted. Composting uses microorganisms to digest organic material. The compost that is produced can be added to the soil as a natural fertiliser, providing useful nutrients to plants. Composting has been carried out for hundreds of years – the ancient Romans left written records of composting – but scientific research has led to discoveries that mean that we can now make compost more efficiently and on a much larger scale (Figure 2.1).

Figure 2.1 Composting on a commercial scale.

Most commercial composting involves a similar process. Figure 2.2 shows how compost is made.

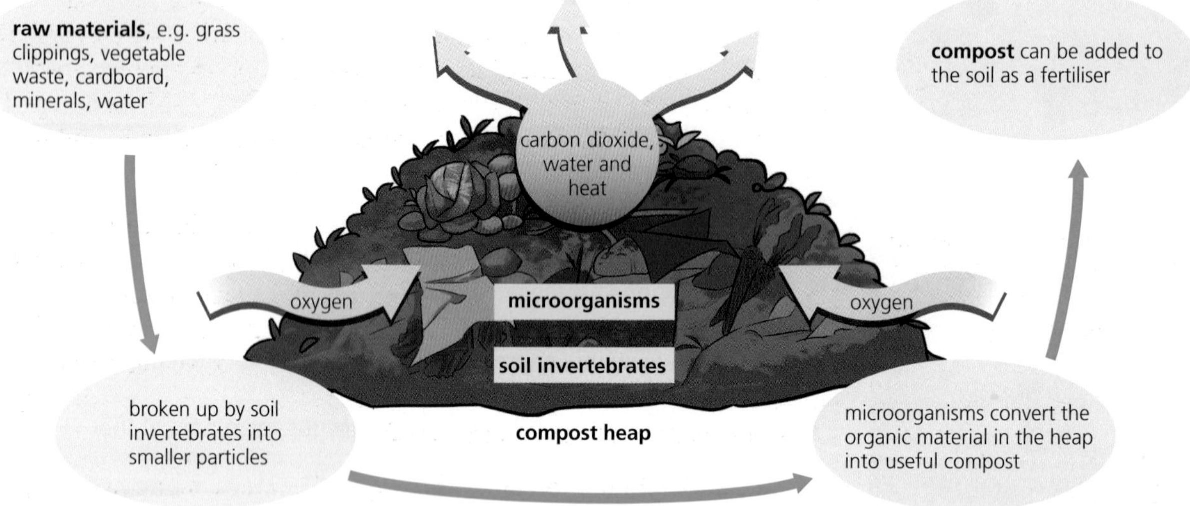

Figure 2.2 How compost is made.

- Suitable organic material, for example grass clippings, vegetable waste and even coffee grounds and cardboard, is collected into a heap. In small-scale garden compost heaps, soil invertebrates such as worms, slugs and millipedes break this material up into smaller particles. In large-scale composting, it is broken up by mechanical grinders and choppers.
- Soil microorganisms colonise the heap. They are called mesophils because they live at temperatures of between 10 °C and 45 °C. The compost heap heats up as they multiply and respire.
- The mesophils are gradually replaced by thermophils. Thermophils are microorganisms that live in conditions where the temperature is high; in this case, between approximately 50 °C and 65 °C. This is the active phase of composting and these thermophilic microorganisms are mainly responsible for converting organic material in the heap into useful compost.
- The last stage is the curing phase. The temperature falls and mesophils again colonise the heap. During the curing phase the compost matures until it finally becomes suitable for adding to the soil.

TIP

You do not need to be able to recall the steps involved in making compost.

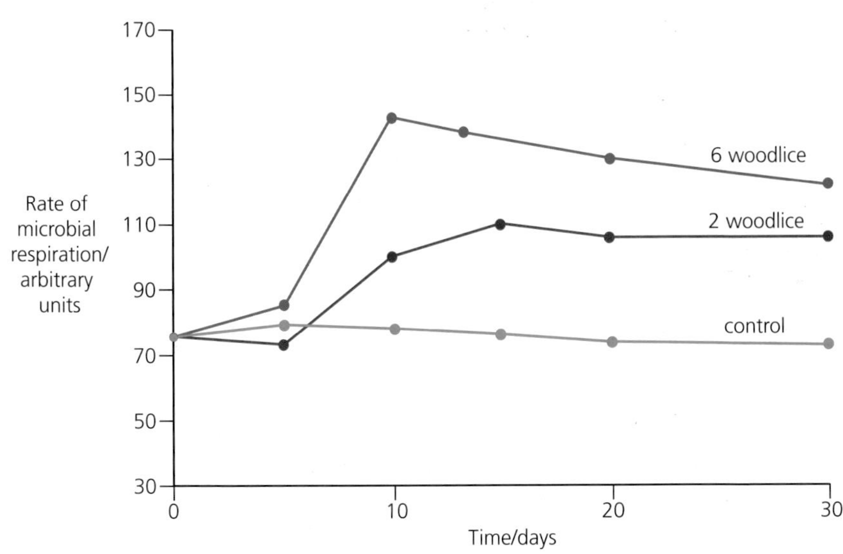

Figure 2.4 The effect of woodlouse activity on the microbial breakdown of leaves.

4 Describe how the control container should have been set up in this investigation.

It should be set up exactly the same as the other containers and exposed to the same environmental conditions. The only difference should have been that it did not contain any woodlice.

5 What does Figure 2.4 show about the effect of woodlouse activity on the microbial breakdown of leaves?

To answer this question you need to compare the containers in which woodlice were added with the control container. You can see that microbial respiration in the control chamber remains more or less constant. After 5 days, the rate of microbial respiration in both of the containers with woodlice increases and is higher than in the control. In addition, there is a greater rate of microbial respiration when six woodlice have been added than when there are only two woodlice in the container.

6 Suggest how the presence of woodlice could have caused the increase in the microbial respiration rate that occurred between 5 and 10 days.

The woodlice break up the leaves into smaller fragments by chewing them so there is more surface area to be colonised by microorganisms. This means that they can feed faster and gain respiratory substrates more quickly.

TEST YOURSELF

1 Why is nutrient recycling in natural ecosystems vital for plant growth?
2 Explain what is meant by decomposition.
3 What is a saprobiont?
4 Describe how the method of digestion used by saprobionts results in some nutrients being made available to plant roots.

Nutrient cycles

Living organisms, such as animals and plants, require many different chemical elements. Plants take up many of these elements as ions from the soil. Inside the plant they are involved in various chemical reactions and are eventually incorporated into biological molecules that form plant cells and tissues. Consumers obtain most of their supplies of these elements from plants or from other animals that feed on plants. A snail, for example, digests the biological molecules in its plant food and absorbs the products through its gut wall (see *AQA A-level Biology 1 Student's Book* Chapter 8). In this way, elements are passed from organism to organism along the food chains that make up a food web (Figure 2.5).

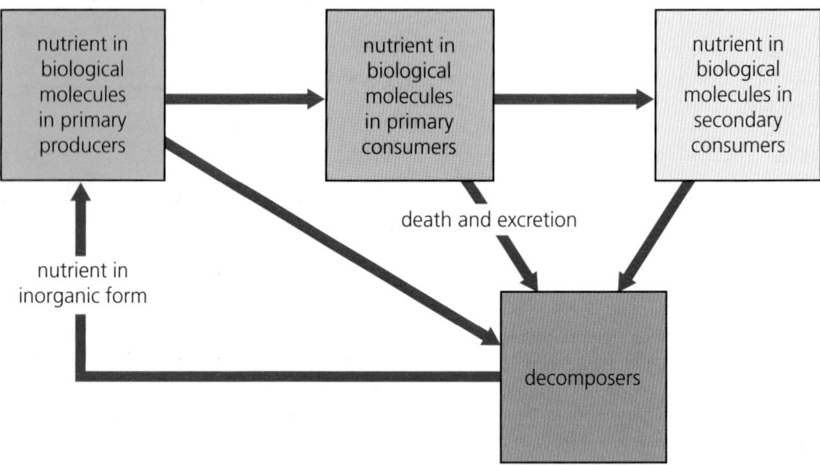

Figure 2.5 A simple nutrient cycle. All elements including nitrogen, phosphorus and iron are cycled in this basic way. The detail of the process is different in each nutrient cycle.

Those organisms, or parts of organisms, that are not eaten as food eventually die and are decomposed. Saprobionts, such as fungi and bacteria, break down the biological molecules that form the dead material (Figure 2.6). They absorb some of the products but the rest are released, some as mineral ions that can be taken up by plants.

Nutrient cycle How a chemical element moves from the abiotic environment into living organisms and then back into the abiotic environment.

This is the basic **nutrient cycle**. Elements are taken up by plants as ions and some are incorporated into biological molecules and pass from organism to organism in the various trophic levels. Death and decomposition result in microorganisms making these elements available to plants again.

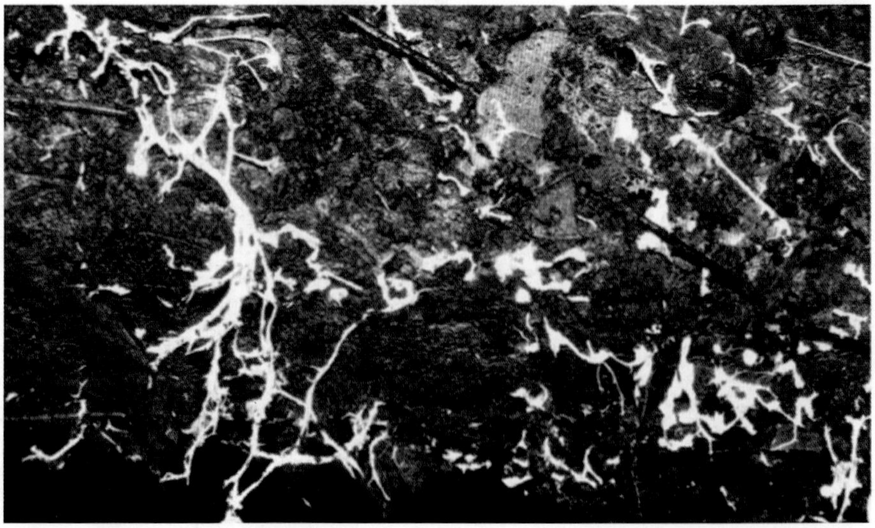

Figure 2.6 These white thread-like structures are fungal hyphae covering the surface of a piece of rotting wood. The fungi are saprobionts. Their hyphae secrete enzymes that hydrolyse biological molecules such as lignin in wood.

The phosphorus cycle

The general features of the **phosphorus cycle** differ very little from the basic nutrient cycle shown in Figure 2.5. Phosphate ions are released from rocks on land by chemical weathering and washed into soils by rain. They are absorbed by plant roots by **active transport** and are used to produce ATP and nucleic acids in plant cells.

Primary consumers feed on plants and the biological molecules in their food are digested to smaller phosphate-containing molecules such as nucleotides. Phosphate ions may also be present in their food. Nucleotides and phosphate ions are absorbed in the small intestine and are used to produce nucleic acids and ATP in animal cells. The same thing happens when a secondary consumer eats a primary consumer.

When organisms produce faeces or die, the phosphorus-containing substances in their tissues or faeces are digested by **saprobiotic** bacteria releasing phosphate ions, which can then be taken up again by plants. Animal urine also contains excreted phosphate ions. Phosphorus can cycle like this within terrestrial communities for centuries. You can trace this terrestrial phosphate loop in Figure 2.7.

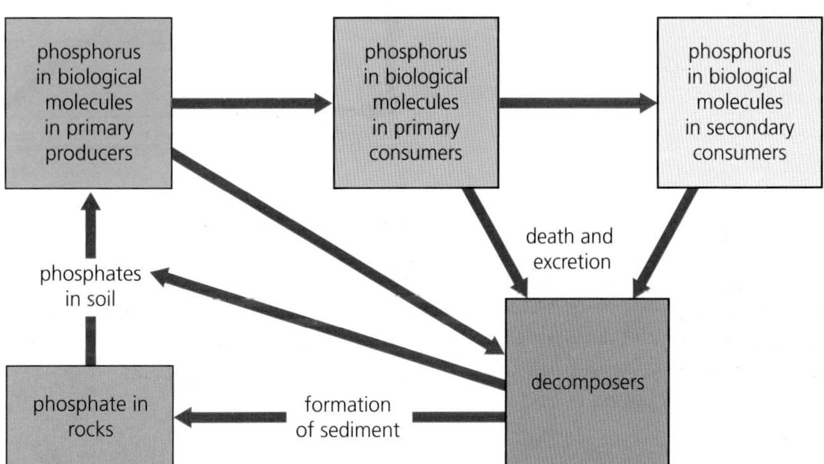

Figure 2.7 The phosphorus cycle.

Some phosphate ions are washed from soils into streams, and eventually reach lakes or the sea. Here they may be taken up by algae near the surface and incorporated into their biological molecules. The algae may be eaten by fish, passing the phosphorus from one trophic level to another. But when aquatic organisms die they tend to sink to the bottom of lakes or the sea, where they decompose and phosphate ions become trapped in aquatic sediments.

Over very long periods of geological time, aquatic sediments form sedimentary rocks and may eventually become exposed again on land, where chemical weathering releases phosphate back into the soil. This part of the cycle is extremely slow and takes millions of years.

The nitrogen cycle

The nitrogen cycle is more complex than the phosphorus cycle. Plants take up nitrate ions from the soil. The nitrates are absorbed into the roots by **active transport** and are used to produce amino acids and then

Ammonification The decomposition of amino acids in proteins, releasing ammonia as a product.

Nitrification The two-step oxidation of ammonium ions firstly to nitrite ions and then to nitrate ions.

TIP
Although the product of ammonification is ammonia (NH_3), ammonia ionises in soil water and in aquatic environments to give ammonium ions (NH_4^+).

proteins and other nitrogen-containing substances in plant cells, including the nitrogenous bases in nucleotides. Primary consumers feed on plants and the proteins in their food are digested to release amino acids. These amino acids are absorbed from the gut and built up into the proteins that form the tissues of the primary consumers. The same thing happens when a secondary consumer eats a primary consumer. In this way, nitrogen is passed from one trophic level to the next through the food web.

When organisms die, the nitrogen-containing substances they contain are digested by saprobiotic bacteria. Nitrogen from consumers is also made available to saprobiotic bacteria through nitrogen-containing excretory products, such as urea in urine. These saprobiotic bacteria release ammonia, so the process is called ammonification. Another group of bacteria, the **nitrifying** bacteria, then convert ammonia to nitrites and nitrates in a process called nitrification. The complete nitrogen cycle is summarised in Figure 2.8.

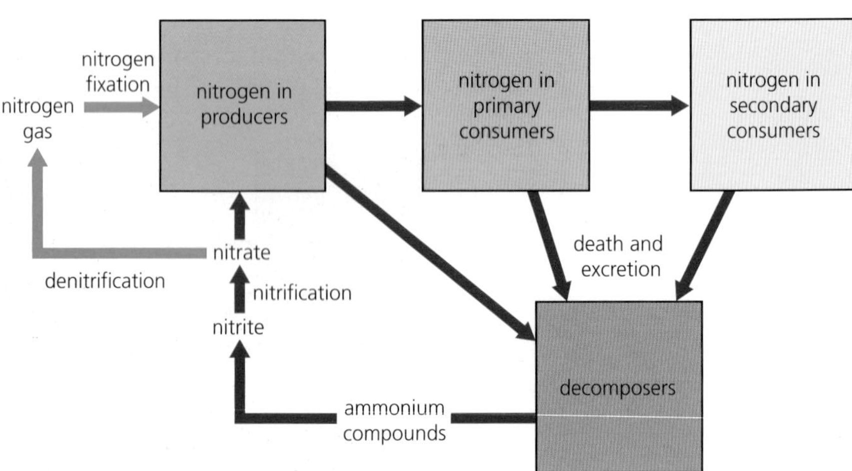

Figure 2.8 The nitrogen cycle.

Denitrification The reduction of nitrate ions to nitrogen gas.

Under anaerobic conditions, such as those that are found when soil becomes water-logged, **denitrifying** bacteria are found in large numbers. These bacteria are able to use nitrate in place of oxygen as an electron acceptor in the respiratory pathway. This reaction, called denitrification, involves reduction of nitrate to nitrogen gas. This nitrogen escapes into the atmosphere, so it is no longer available to plants.

Nitrogen fixation The reduction of nitrogen gas to ammonia.

Nitrogen gas may be made available to plants again by nitrogen fixation.

Some species of microorganism are able to fix nitrogen to form ammonia. Some live free in the soil. Others are associated with the roots of leguminous plants, such as peas and beans, clover and lupins. The biochemical reactions associated with nitrogen fixation are complex, but they can be summarised by the following equation.

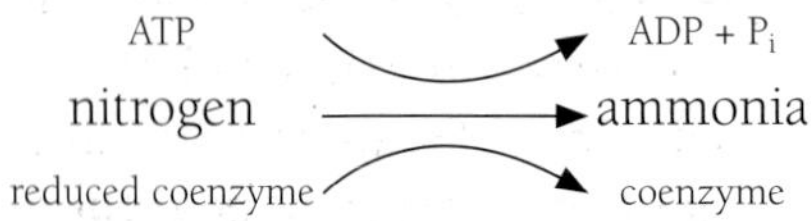

Nitrogen is reduced to ammonia. The reaction is catalysed by the enzyme nitrogenase. Nitrogenase, however, does not function in the presence of oxygen and many nitrogen-fixing organisms have adaptations that ensure that anaerobic conditions exist in the parts of cells involved in nitrogen fixation.

Mycorrhizae and ion uptake

So far in this chapter we have seen how microorganisms have key roles in decomposition and in the different processes in nutrient cycles. There is a third way in which microorganisms are crucial in recycling chemical elements, by assisting plants with their uptake of inorganic ions and water.

Mycorrhizae Associations between the plant roots and beneficial fungi found in nearly all plants on Earth.

The root systems of most plant species have fungi, called **mycorrhizae** (see Figure 2.9), growing in and around them. The fungi are often highly specific to the plant species and have evolved with them. This is an example of mutualism, a relationship between two species where both gain a nutritional advantage.

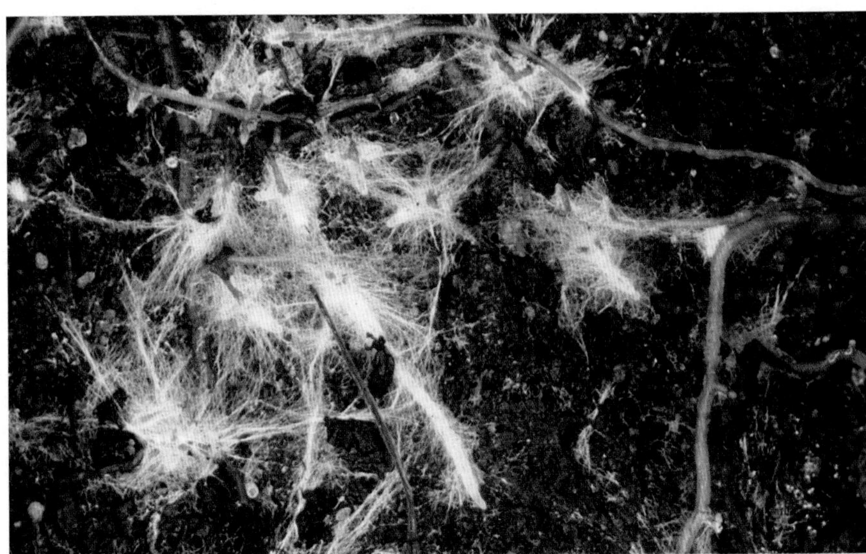

Figure 2.9 Mycorrhizae around the roots of a young pine tree.

Each fungus consists of microscopic threads called hyphae. The fungus colonises the roots from the soil by hyphae growing on, and often into, the root tissues. Once established, many hyphae extend out from the root surface into the surrounding soil. Because there are so many hyphae, they vastly increase the surface area of the plant for uptake of water and ions, including phosphates and nitrates. This is shown in Figure 2.10.

The hyphae absorb ions and water from the soil and transport them into the plant roots. In return, the part of the fungus growing inside the root is able to obtain carbohydrates translocated from the plant leaves. This supplements their food supply.

Because fungi are saprobiotic, their hyphae still secrete enzymes and hydrolyse biological molecules in leaf litter and other organic detritus in the soil. This releases ions, which can then be absorbed and transported back along the hyphae to the plant roots. Scientists have shown that in some circumstances plant roots would be incapable of accessing phosphate ions at all without the help of mycorrhizae.

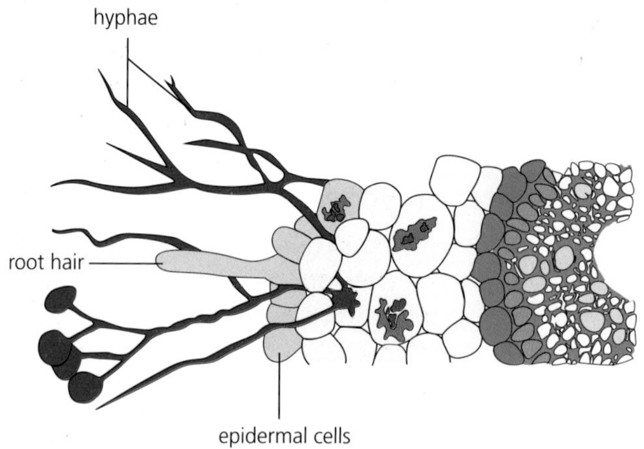

Figure 2.10 Fungal hyphae extend into the soil and increase the root surface area.

TEST YOURSELF

5 Describe the ways that the nitrogen cycle shown in Figure 2.8 is similar to the carbon cycle that you already know about.
6 Describe the ways that the phosphorus cycle shown in Figure 2.7 is different from the nitrogen cycle.
7 What is nitrification and why is it important for plants?
8 What are mycorrhizae and how do they benefit many plant species?

Fertilisers and net primary production

When organisms living in natural ecosystems die, they decompose. Soil microorganisms convert, for example, the nitrogen in organic substances such as proteins and nucleic acids to nitrates, and these are taken up by the producers. There is a continual recycling of nutrients. In agricultural ecosystems, however, a large part of the biomass that is produced is harvested and removed. This is true whether we are considering crop plants, such as wheat or potatoes, or animals, such as dairy cattle and sheep. Unless the mineral ions in this biomass are replaced, their concentration in the soil decreases and crop yield or milk yield falls.

Fertilisers Materials added to soil to provide nutrients for plants.

Fertilisers can be used to add mineral ions such as those of nitrogen, phosphorus and potassium to the soil. We can either use **artificial fertilisers** or **natural fertilisers**, such as farmyard manure.

Artificial fertilisers:

- have a guaranteed composition, making it easier to determine rates of application and predict the effect on crop yield
- are concentrated sources of nutrients and can therefore be applied in smaller amounts; this saves on transport costs and on the damage done by heavy machinery compacting the soil and crushing the crop
- are clean and convenient to handle and apply evenly.

Natural fertilisers, on the other hand:

- are mixtures of substances and may contain trace elements, substances that are important to plants in small amounts
- add organic matter to the crop; this may improve soil structure, reducing erosion and improving water-holding properties
- release the nutrients they contain over a longer period of time.

It is clear from various investigations that regular dressings with farmyard manure can benefit a crop and produce increases in yield similar to those obtained with artificial fertiliser. Table 2.1 shows the results obtained from a number of different investigations.

The first column in the table gives the name of the scientist or scientists responsible for carrying out the investigation and publishing the results. Comparing the work of different scientists allows us to see that similar findings have been reported by others. This helps to make sure that the conclusions that we draw are robust, or reliable.

Table 2.1 Yields of crops with long-term applications of farmyard manure or artificial fertiliser.

		Yield in tonnes per hectare		
Investigation	**Crop**	**Control**	**Farmyard manure**	**Artificial fertiliser**
Dyke (1964)	Wheat	2.08	3.50	3.11
Trist and Boyd (1966)	Wheat	1.28	2.38	2.43
	Barley	1.03	2.03	2.26
Johnston and Poulton (1977)	Barley	1.59	3.03	2.87
Warren and Johnston (1962)	Sugar beet	3.80	15.60	15.60
	Mangolds	3.80	22.30	30.90

Look at the data on yield in the other columns. You can see that in all cases the addition of either farmyard manure or artificial fertiliser led to an increase when compared to the control.

How much fertiliser should be applied?

Farmers can improve the net primary production (see page 13) of a crop by adding fertiliser, but they must apply the right amount at the right time. There are recommendations available for how much fertiliser to apply to a particular crop, but these are only general estimates. Factors such as the previous crop that was grown and the type of soil mean that these recommendations have to be modified if they are going to be applied to a particular crop growing in a particular field.

Figure 2.11 shows the effect of adding different amounts of two types of nitrogen-containing fertiliser on the yield of maize in a number of trial plots.

Curve B shows the effect of adding a fertiliser containing nitrogen but no potassium. The shape of this curve is typical of a yield response to added fertiliser. The yield increases with increasing application of fertiliser. It reaches a peak and then falls. This pattern is sometimes referred to as the law of diminishing returns, because beyond a certain point the addition of more fertiliser results in very little extra gain, or even reduction.

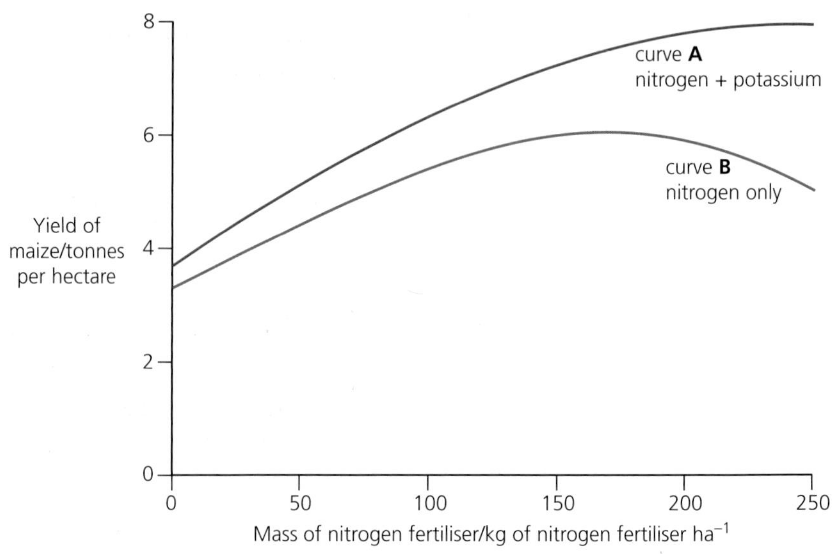

Figure 2.11 The effect of adding nitrogen-containing fertiliser on the yield of maize.

Now look at the curve between applications of 0 and 180 kg of nitrogen per hectare. You have seen curves with this shape frequently during your A-level course. What it is showing you here is that, as the curve rises, the amount of nitrogen added in the fertiliser is limiting the yield. At approximately 160 kg of nitrogen per hectare, the curve flattens out. Something else is limiting the yield. Curve A suggests that this may be potassium.

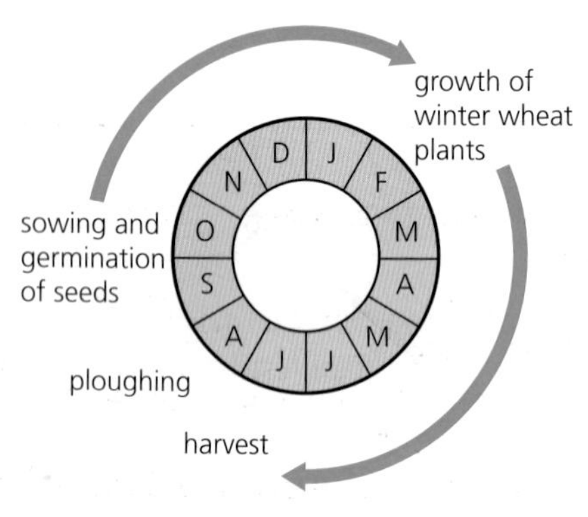

Figure 2.12 Growth, development and harvest of a winter wheat crop. The letters in the circle correspond to the months of the year.

When should fertiliser be applied to a crop?

In the UK, winter wheat is an important crop. Figure 2.12 shows when a winter wheat crop is sown, grown and harvested.

We will look at some of the issues relating to the application of nitrogen-containing fertiliser to this crop. The timing of any application must take the following points into consideration.

- Maximum uptake of soil nitrates occurs early in the growth of the plants.
- Time is required for nutrients in fertilisers to dissolve and reach the roots.
- Wastage through loss from the soil must be kept to a minimum.
- Weather conditions need to be taken into account, and fertilisers should not be applied while it is raining.

EXAMPLE

Timing the application of artificial fertiliser

Most of the nitrogen in artificial fertiliser is supplied either as ammonium ions or as nitrate ions.

1 Use your knowledge of the nitrogen cycle to explain why there is a rapid decrease in the concentration of the ammonium ions added in the fertiliser.

Nitrifying bacteria oxidise ammonium ions to nitrates and then to nitrates.

2 Nitrate added to the soil at the time of ploughing may be lost rapidly by denitrification. At this time of the year, heavy rain is likely. Explain why the rate of loss through denitrification is rapid:

a) after a period of heavy rain

Denitrifying bacteria are more active in anaerobic conditions and, following heavy rain, water-logged soil has less air, and therefore less oxygen, in the soil.

b) when straw from the previous crop is ploughed into the soil and is decomposing
Saprobiotic bacteria decomposing the straw use up soil oxygen in respiration, creating the anaerobic conditions in which denitrifying bacteria are more active.

c) when the soil is still warm
Denitrifying bacteria are more active in warmer conditions and diffusion of nitrogen gas from soil air spaces into the atmosphere is more rapid at higher temperatures.

In the UK, the risk of nitrogen loss in autumn is high, so most nitrogen-containing fertiliser is added in spring.

3 Use Figure 2.12 to explain the advantage of applying fertiliser to a winter wheat crop in early spring.
The plants are starting to grow more rapidly again as air temperature and day length increase following the winter, so use phosphate and nitrate ions more rapidly for protein and nucleic acid synthesis, resulting in less leaching of ions from the soil (see page 42)

4 Suggest disadvantages of applying fertiliser in early spring.
Initial growth of the seedlings in autumn does not have the benefit of the fertiliser. In early spring, the soil may still be fairly cold and so root respiration will be slow, so there may not be much ATP being produced for active transport of ions into roots. Earlier growing weeds may also take up some of the ions instead of the crop.

TEST YOURSELF

9 Why is it necessary to use fertilisers when (a) growing crops and (b) on grass when rearing dairy cattle and sheep?

10 Explain why natural fertilisers release the nutrients they contain over a longer period of time than do artificial fertilisers.

11 Explain why the curves in Figure 2.11 do not start at the origin of the graph.

12 From Figure 2.11, what mass of nitrogen-only fertiliser would you recommend a farmer use per hectare for growing maize? Explain your answer.

Fertilisers and the environment

The stream shown in Figure 2.13 runs through an area of farmland. It contains a high concentration of nitrate and phosphate ions. The stream is in a rural area and environmental scientists suspected that these ions came from fertiliser applied to the surrounding farmland.

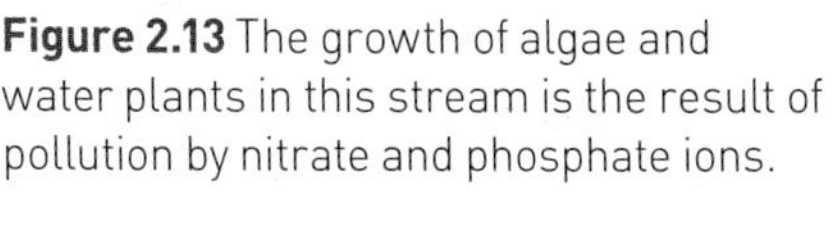

Figure 2.13 The growth of algae and water plants in this stream is the result of pollution by nitrate and phosphate ions.

TIPS

- In an exam you might be asked to recognise the type of data you have been given and to select an appropriate statistical test.
- Look at Chapter 13 on maths skills to find out more about statistical tests and to see a worked example of this test.
- You won't need to do statistical calculations in a written paper.

Table 2.2 Total mass of nitrogen in fertiliser added to surrounding fields and the mean concentration of nitrate in a stream.

Total mass of nitrogen in fertiliser added to surrounding fields/kg $ha^{-1} yr^{-1}$	Mean concentration of nitrate in stream/mg dm^{-3}
41	1.2
41	1.3
51	1.5
56	1.8
63	1.6
69	1.9
72	2.0

Table 2.2 shows data the scientists collected for the same stream over a number of years. It suggests that as the total mass of nitrogen in fertiliser added to fields increases, the mean concentration of nitrate in nearby streams also increases. Because these data consist of two sets of measured variables, the best way to present it would be on a scattergram. The line of best fit would slope upwards, indicating a positive correlation. You would use a Spearman's rank correlation test to find out whether the correlation between these two variables is significant or not.

Eutrophication

Eutrophication The addition of extra nutrients such as nitrate or phosphate ions to aquatic ecosystems.

Leaching The process in which soluble ions dissolved in soil water drain through the soil into aquatic ecosystems.

The addition of extra nutrients to aquatic ecosystems is is called eutrophication. The word is generally used when freshwater streams or lakes, such as the stream in Figure 2.13, are enriched with nitrate and phosphate ions because of leaching. Because nitrate and phosphate ions are so soluble they can be carried in soil water as it drains from fields into ditches.

If fertilisers are used on agricultural land there is a risk of eutrophication in nearby aquatic ecosystems. The risk is higher if the fertilisers are artificial and especially high if the fertiliser is used on bare fields or before heavy rain. This is because artificial fertilisers contain soluble salts such as ammonium phosphate, which readily dissolve in soil water. If they are not quickly absorbed by plant roots, they remain in the soil water and from there they can be leached into ditches and streams by rainwater.

Natural fertiliser such as compost or manure decomposes and releases nutrients slowly. Plants usually absorb them as fast as they are released, so the risk of nutrients leaching into aquatic ecosystems is lower when using natural fertilisers.

If nutrients such as nitrate or phosphate ions reach aquatic ecosystems such as stream or lakes they cause an increase in the growth of aquatic plants and algae. This is because the growth of many aquatic plants and algae is usually limited by nitrate and phosphate ion concentration.

Increased biomass of aquatic plants, and especially algae, reduces the amount of light entering the ecosystem. Dense aquatic plant foliage at the surface can shade the plants beneath them. Large numbers of algae in the water turn it murky green, or turbid. This is sometimes called an algal bloom.

The reduced light available further underwater means that some aquatic plants and algae die. They are decomposed by saprobionts in the water, mostly bacteria. The bacteria grow and reproduce rapidly due to the large amount of dead plant material available. Since the bacteria respire aerobically, they use most or all of the oxygen dissolved in the water.

Aquatic organisms such as stonefly and mayfly larvae, which require a relatively high concentration of oxygen, die. In extreme cases fish may also die. The few animal species that might survive are those tolerant of low oxygen concentrations. However, the diversity of animal species will be much reduced (see *AQA A-level Biology 1 Student's Book* Chapter 13).

Figure 2.14 Stonefly larvae are especially sensitive to depleted oxygen concentration and die if eutrophication occurs in a stream.

Figure 2.15 summarises how eutrophication can affect an ecosystem.

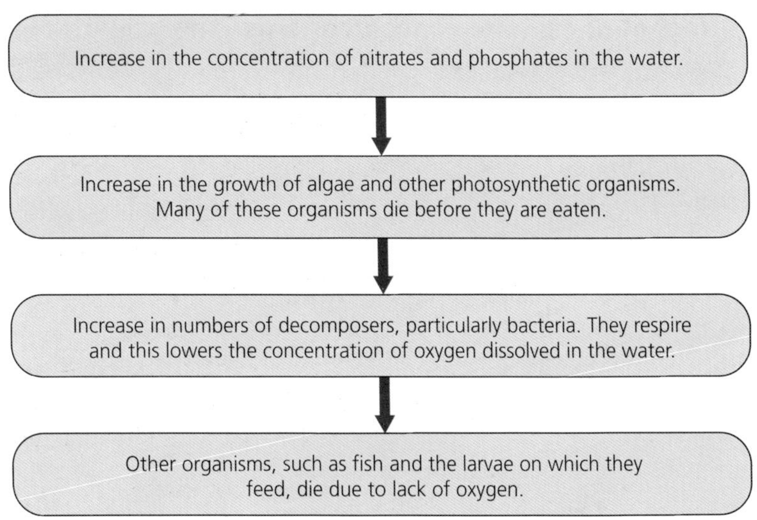

Increase in the concentration of nitrates and phosphates in the water.
Increase in the growth of algae and other photosynthetic organisms. Many of these organisms die before they are eaten.
Increase in numbers of decomposers, particularly bacteria. They respire and this lowers the concentration of oxygen dissolved in the water.
Other organisms, such as fish and the larvae on which they feed, die due to lack of oxygen.

Figure 2.15 The effect of eutrophication on freshwater ecosystems.

Scientists can use two different approaches to study the effects of eutrophication. The first is a fieldwork-based approach. This involves either comparing polluted and unpolluted water, or looking at data collected at a single place over a long period of time. Problems arise with trying to explain the results of investigations such as these, however. It is often very difficult to link an effect to a particular cause because situations are often very complex and many factors may change at the same time.

The second approach is to use an experimental approach, in either the laboratory or the field, to test the effect of changing a specific variable. In this case, scientists have to take care in applying results obtained under carefully managed conditions to natural situations.

Data from fieldwork

The Norfolk Broads are a series of shallow lakes along the lower reaches of three of the main East Anglian rivers. The Broads are surrounded by agricultural land. Some of the land is used for grazing and some for growing crops. Table 2.3 shows figures for the maximum concentration of phosphates, nitrates and algae in the water of seven different lakes in a particular year.

Table 2.3 Maximum concentrations of phosphates, nitrates and algae in seven lakes in the Norfolk Broads.

Lake	Maximum phosphate concentration/g dm^{-3}	Maximum nitrate concentration/ mg dm^{-3}	Maximum concentration of algae given as chlorophyll concentration/µg dm^{-3}
A	340	2	460
B	200	4	230
C	320	3	410
D	240	8	370
E	160	6	220
F	190	8	370
G	190	3	200

ACTIVITY

The effect of phosphate concentration on the growth of algae using data from fieldwork

Start by looking at the data on phosphate concentration and the concentration of algae. Is there a significant correlation between these two variables? If you carry out a Spearman's rank correlation test on these data you should get a value for R_S of 0.85. Look this up in the table of probability values (see Table 13.3 on page 255) and you will see that this value is greater than the critical value for seven pairs of measurements. You can therefore reject your null hypothesis and can conclude that there is a significant correlation between the maximum phosphate concentration and the maximum concentration of algae in the water.

Now look at the relationship between nitrate concentration and the concentration of algae. In this case Spearman's rank correlation test gives an R_S value of −0.28. The minus sign shows us that, in this case, we are looking at a negative correlation. In other words, the greater the concentration of nitrate in the water, the lower the maximum concentration of algae. If you ignore the minus sign and look up this value in the table you will see that it is lower than the critical value. There is a greater than 0.05 probability that this correlation arose by chance. You should therefore accept the null hypothesis and conclude that the correlation between nitrate concentration and concentration of algae is not significant.

The data shown in Table 2.3 confirm what many other scientists have found. In most freshwater ecosystems, the factor limiting growth is very likely to be phosphate concentration. Increasing the concentration of phosphate is often associated with an increase in algal growth. As pointed out earlier, however, scientists must take care in interpreting data like these. There is obviously a strong correlation between phosphate concentration and the concentration of algae. This does not mean that it is the increase in phosphate concentration that causes the rise in the algal population. There may be other ecological factors that vary between these lakes, and any one of them could have affected the algal population.

ACTIVITY

The effect of phosphate concentration on the growth of algae using data from an experiment

A Lund tube is a large rubber tube used to investigate freshwater ecology. It is designed to isolate large volumes of water from the lake outside. The top of the tube is surrounded by a large inflatable ring that floats on the surface, while the bottom of the tube sinks into the mud on the lake floor.

In one experiment, phosphate was added to the water inside a Lund tube in October. The populations of algae in the water in the tube and in the surrounding lake water were measured at regular intervals over the next few months. The results of this investigation are shown in the graph in Figure 2.16.

Look at the data in Figure 2.16 and answer these questions.

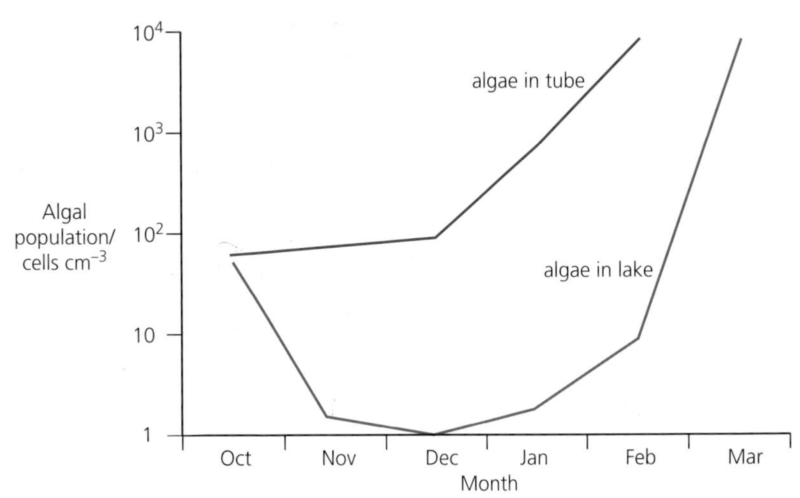

Figure 2.16 The effect of adding phosphate to the population of algae inside a Lund tube.

1 By how many times is the population of algae in the tube greater than the population in the rest of the lake, in the middle of February?

2 a) Describe the difference in the shapes of the two curves shown on the graph.

b) Explain this difference, in terms of phosphate concentration.

3 The Lund tube used in this investigation was approximately 45 m in diameter. Other than the presence of added phosphate, suggest how the water inside the tube might have differed from the surrounding lake water.

4 Use your answers to the questions to explain why scientists have to take care in applying the results of this investigation to natural situations.

TEST YOURSELF

13 What is leaching and why is it a risk to the environment?

14 Why is the risk of leaching often higher if artificial fertilisers are used on farmland rather than natural fertilisers?

15 Describe the effect of eutrophication on a typical freshwater ecosystem.

16 Explain why it was important that all the lakes in Table 2.3 were sampled in the same year.

Practice questions

1 The sequence below illustrates the events after the addition of nitrate and phosphate ions to a stream flowing through farmland.

Increased nitrate and phosphate ion concentration → Increased growth of algae → Death and decay of aquatic plants → Reduced oxygen concentration in the water

a) i) Suggest why the increased growth of algae results in the death of other aquatic plants. (2)

ii) What causes the reduced oxygen concentration in the water? (2)

b) What will be the impact of a reduced oxygen concentration on the animal community in the stream? (2)

2 The diagram shows the phosphorus cycle.

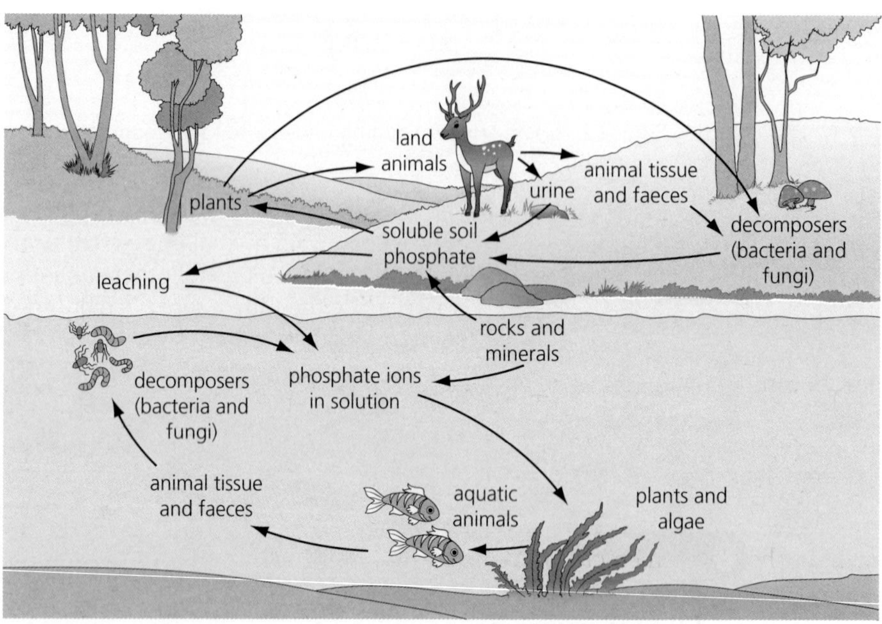

a) Explain how the herbivore obtains phosphorus from the biological molecules in its plant food. (2)

b) Describe two ways that the herbivore would use phosphorus. (2)

c) Explain why the cycle shown in the diagram depends on saprobionts. (2)

3 Sometimes farmers grow mustard plants on land in autumn when there would otherwise be no crop. The mustard takes up nitrate ions, which might be leached from the soil at this time of the year. When it is time to plant seeds of a new crop in the spring, the mustard is ploughed into the soil as a natural fertiliser.

a) Why is the mustard crop used to remove nitrate ions from the soil? *(1)*

b) When the mustard plants are ploughed into the soil, nitrogen contained in biological molecules in their tissues become available for the new crop. Describe the role of saprobionts in this process. (3)

c) Why would using mustard plants as a natural fertiliser reduce the risk of leaching in the spring? (2)

d) Plants also absorb phosphate ions from the soil. Describe why phosphate ions would be needed by a growing crop. (2)

4 Scientists compared the decomposition of beech and fir leaves in leaf litter for one year. The table shows some of their results.

	Beech	Fir
Biomass lost (%)	13.2	23.6
Organic C (%)	45.7	47.9
Total N (%)	0.63	1.57
Lignin (%)	36.1	28.1

a) Explain why it was appropriate that leaf biomass lost was determined as the percentage dry mass lost. (2)

b) Along with cellulose, lignin is a component of plant cell walls. Saprobionts find lignin far less digestible than cellulose. Describe and explain the relationship between the lignin content of the leaves and their decomposition rate. (3)

Organisms that decompose organic material use carbon-containing biological molecules as respiratory substrates and nitrogen-containing biological molecules for protein and nucleic acid synthesis. They need more carbon than nitrogen. Microorganisms require a carbon:nitrogen ratio of about 30 : 1.

c) Explain how the above information relates to the data in the table. (3)

Stretch and challenge

5 Discuss the role of mycorrhizae on the growth of certain plants, especially orchids. Evaluate the recent work by scientists at the University of Maryland in the USA which claims that one of the key factors that determine if certain species of orchid can grow in a particular place is the presence or absence of their mycorrhizal fungi in the soil. Explain possible reasons for this and examine its importance for orchid conservation.

Response

PRIOR KNOWLEDGE

- A nervous system enables humans to react to their surroundings and coordinate their behaviour.
- Receptors detect stimuli (changes in the environment).
- Receptors and the stimuli they detect include receptors in the eyes that are sensitive to light and receptors in the skin that are sensitive to touch, pressure, pain and temperature changes.
- Nerve impulses from receptors pass along neurones in nerves to the brain. The brain coordinates the response. Simple reflex actions are automatic and rapid. They often involve sensory, relay and motor neurones.
- Simple reflex actions involve receptors, sensory neurones, motor neurones, relay neurones, synapses and effectors.
- Plants are sensitive to light, moisture and gravity. Their shoots grow towards light and against the force of gravity; their roots grow towards moisture and in the direction of the force of gravity.
- Plant growth substances (called hormones at GCSE) coordinate and control growth in plants.

TEST YOURSELF ON PRIOR KNOWLEDGE

1 What is a receptor? Give two examples of receptors.
2 Give one advantage of rapid reflex reactions.
3 Describe a synapse.
4 Explain why plant responses to light and water improve their chances of survival.

The idea that some plants respond to music or being talked to is not new but until now has been regarded with scepticism. But recent experiments with a small cabbage-like plant called *Arabidopsis* have shown that they do indeed respond to sound vibrations. More than that, they can detect the difference between different sorts of sounds.

If you think about the possible ecological significance of sound in the life of plants, this is not so unusual as it might seem. Plants are constantly exposed to the feeding of herbivores, from large mammals to a much wider range of smaller animals, especially insects. The chewing, munching and tunnelling of herbivorous animals creates a constant barrage of sound. Since being damaged by feeding animals is disadvantageous to plants, it is actually not surprising that they can detect and respond to the sound of this happening.

Extension

Scientists at the University of Missouri recorded the sounds of feeding caterpillars by shining a laser at a small reflective patch on an *Arabidopsis* leaf and allowing a caterpillar to chew it. The tiny vibrations detected by the laser beam were recorded and then played back to other *Arabidopsis* plants. A second group of plants were kept in silence.

When caterpillars were then allowed to feed on both sets of plants, those that had been exposed to the sound of feeding had more defensive chemicals, substances that caterpillars find distasteful, in their leaves than those that had been kept in silence. Caterpillars respond to the defensive chemicals by moving to another plant.

The advantage of this response to *Arabidopsis* plants is clear. Although plants have other chemical-mediated responses to feeding animals, sound vibrations are a faster way for distant parts of the plant to detect herbivores feeding and produce defensive chemicals. Further experiments showed that other vibrations, such as those caused by wind moving the leaves, did not stimulate this response.

So some plants can detect and respond to sound. Future developments based on this work might include genetic modifications to crop plants that boost the production of defensive chemicals, making them more resistant to damage by pests. These experiments also show that plants respond to more of the same stimuli as animals than you might think, even though their responses may be rather different.

Survival and response

How does a blackbird find an earthworm in a lawn? How do grass leaves grow towards the light? How does a cat catch a blackbird? How does a blackbird recognise a cat as a dangerous predator? How does a worm escape from a blackbird? Getting food, avoiding being eaten and finding favourable conditions to live in are all essential requirements for survival. Any species that does not have the ability to respond to these requirements will die out.

Figure 3.1 Blackbird finding earthworms in a lawn.

Stimulus A change in the internal or external environment.

As we shall see, responses vary in complexity according to circumstances and from plants to animals. Detection requires a **stimulus** that can be detected by **receptors**. Some receptors are cells that secrete substances in response to stimuli such as β cells in the pancreas (see Chapter 6, page 108). Some receptors are specialised cells that produce electrical activity in nerve cells. For example, a sudden movement by a cat may be the stimulus that is detected by receptors in a bird's eyes. Processing involves nerve impulses being conducted to a coordinator, either the brain (or in a worm to a sort of mini-brain) or to the spinal cord, and from there to the parts of the body that will produce the appropriate **response**. The response is carried out by **effectors**. In the bird the effectors will be the muscles that operate the wings. So, the full sequence is:

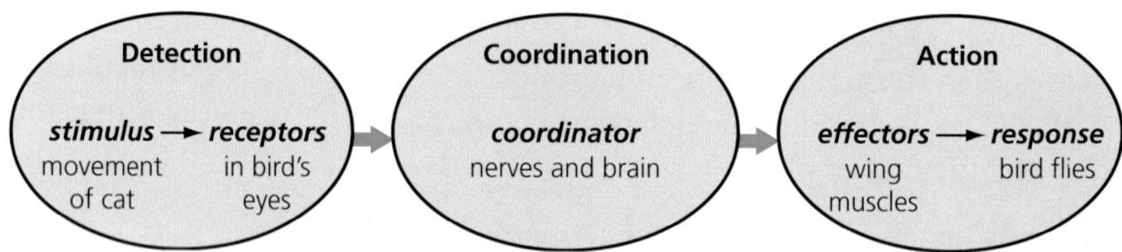

This response obviously has vital survival value to the bird. It is important that the response should be rapid, and a reflex action achieves this. At the same time it cannot be an absolutely fixed and automatic action. A successful escape flight should be controlled so that it carries the bird away from the cat to reach a safe place. After the initial reflex, the bird is able to take control and undertake much more complex behaviour than just flapping its wings.

Simple reflexes

Simple reflex An unlearned, fixed response to a stimulus.

In humans some automatic responses to an external stimuli are called **simple reflexes**. If, for example, you accidentally touch a hotplate on a cooker you will very rapidly pull your hand away. You won't have to think about your action and you won't be able to stop this response if you are not prepared for the heat. This has the obvious advantage that it minimises the damage that might be caused.

Figure 3.2 shows the **neurones** involved in a reflex arc that produces a rapid response. The high temperature of the hotplate stimulates pain receptors close to the surface of the skin. These are actually thin branches at the end of a **sensory neurone** that has a long extension all the way through the arm to the spinal cord. The receptors trigger nerve impulses that are conducted to the opposite end of the sensory neurone. This end of the axon of the sensory neurone has tiny branches that almost touch similar branches on a **relay neurone**. A junction between neurones is called a **synapse** (see page 77). The relay neurone has another synapse linking to a **motor neurone**. The axon of the motor neurone conducts nerve impulses to a muscle in the arm and stimulates it to contract. This produces the response that causes the arm to be quickly pulled away.

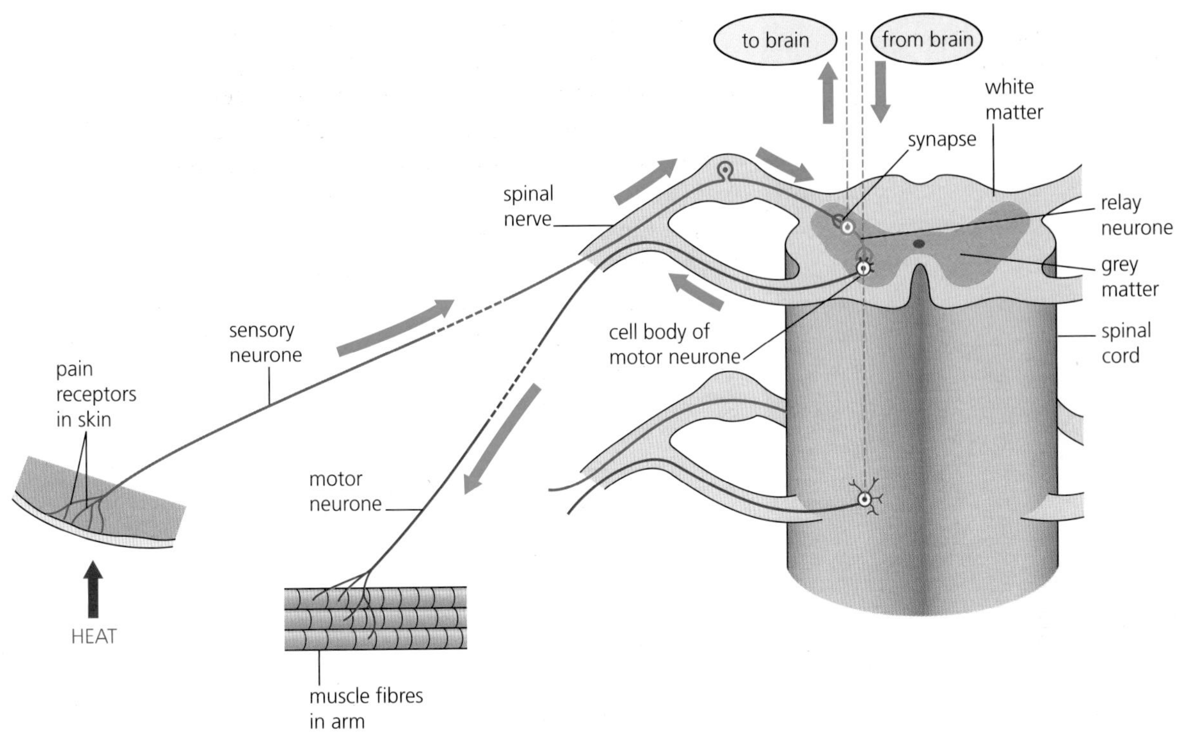

Figure 3.2 A three-neurone reflex arc.

A **three-neurone** reflex arc is an example of a mechanism that results in a totally automatic reflex response. In practice, however, the response of pulling the arm off a hotplate includes much more than a single reflex arc. Touching the hot surface stimulates not just one but a large number of receptors, so many sensory neurones send nerve impulses to the spinal cord. The response involves the coordinated contraction of several muscles in the arm. Therefore the relay neurones despatch nerve impulses not just to one muscle but to several, which is why they have synapses connecting with many other neurones. Nerve impulses also pass to the brain, making us conscious of pain and enabling us to take further action, as well as to shout 'ouch'.

Finding the right environment

Kinesis A non-directional response to a stimulus.

Taxis A directional response to a stimulus.

The ability to respond to environmental stimuli occurs in all living organisms. Even seemingly simple organisms that do not have a complex nervous system can find favourable conditions or get out of trouble. Motile organisms move more or faster in response to a stimulus, or turn less frequently when they experience less favourable conditions. This is called kinesis, which simply means movement. A kinesis is a non-directional response (relative to the direction of the stimulus) in which the rate of movement is affected by the intensity of the stimulus. A second way is to move directly towards or away from a stimulus. This is called a taxis.

Example of kinesis

Planarians are carnivorous flatworms that live in shallow streams and ponds. They have a network of neurones and simple 'eyes' that have light-sensitive cells but no lens.

Planarians are often found clustered on the underside of stones, where they normally remain hidden during daylight. If a stone is turned over,

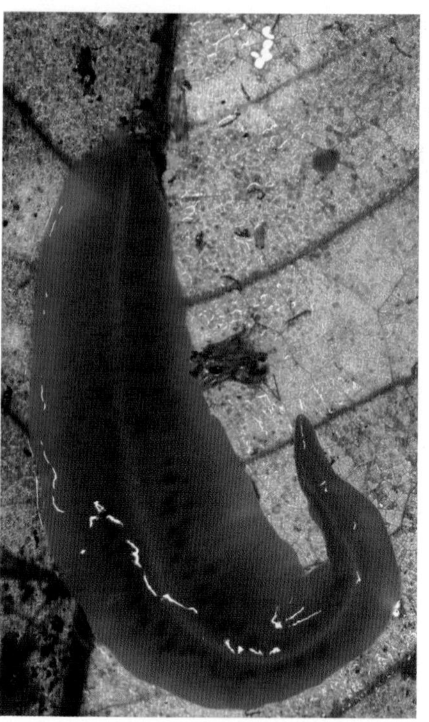

Figure 3.3 The planarian *Dendrocoelum lacteum* is about 15 mm long.

TIP

Bacteria, viruses, plants and small organisms such as flatworms do not think like humans, so avoid using language that suggests that they 'want' or 'prefer' things or 'need to' do something.

the flatworms immediately start moving around in random directions. When their movements bring them back into the darkness they stop moving. This behaviour helps to protect them from predators. Laboratory experiments show that the brighter the light the less frequently they change direction. The flatworms move around randomly until they happen to get to a darker environment, so this is directionless movement. But in the dark they change direction more frequently. This tends to keep them in a darker environment.

Example of taxis

Euglena viridis is a single-celled organism that that lives in small ponds (Figure 3.4). It has chloroplasts and so is able to photosynthesise. It also has a long flagellum that it uses for swimming. There is a receptor near the base of this flagellum that is sensitive to light. You can also see a red spot close to the light-sensitive area.

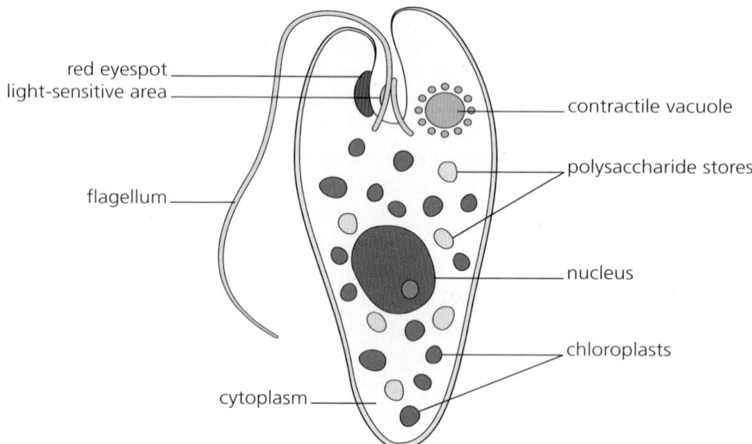

Figure 3.4 Structure of *Euglena*.

Euglena responds to light by swimming towards it. Waves pass along the flagellum from the base to the tip and pull the *Euglena* forward. Because there is only one flagellum the waves also make the cell rotate as it moves forward. If the light is shining from one side the red spot will shade the light-sensitive area (the receptor) each time the cell rotates. When it is moving straight towards the light the area will stay illuminated all the time. The molecules of the light-sensitive pigments are aligned so that when they are illuminated they stimulate the flagellum to beat. This results in the *Euglena* moving directly towards the light. This response is therefore a taxis and because it is movement towards light it is called a phototaxis.

It is amazing that such seemingly complex behaviour can be coordinated by tiny organelles within a single cell. Nevertheless, the ability to move to and away from light is common in single-celled organisms. In shallow seas there is a mass migration of vast numbers of plant photosynthetic plankton towards the surface in the daytime and then at night-time back to deeper levels where there are higher concentrations of mineral nutrients.

Investigating taxes and kineses

It is quite likely that you have already carried out a behaviour experiment using apparatus similar to the choice chamber shown in Figure 3.5. These are suitable for studying taxes and kineses in small animals, such as woodlice and maggots.

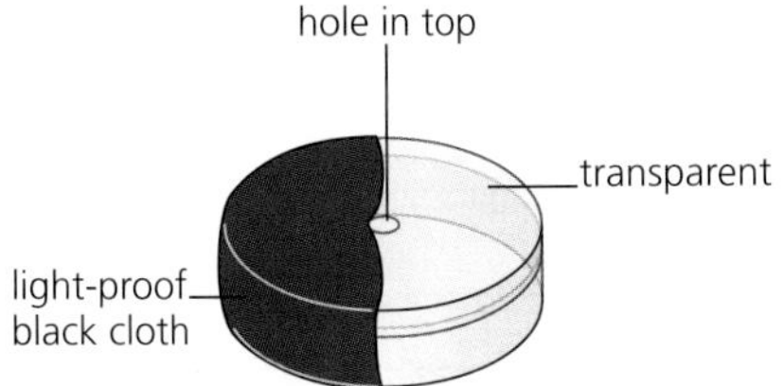

Figure 3.5 A choice chamber is a small container with sub-sections in which different abiotic conditions can be set up, into which small motile animals can be introduced and observed.

A student was investigating the hypothesis that woodlice exhibit negative phototaxis; that is, they move away from the bright light. The student covered one half of a choice chamber with light-proof black cloth (Figure 3.5). The other half of the chamber was illuminated by a light bulb fixed about 20 cm above it. Ten woodlice were put into the chamber through the hole in the centre of the top. Five minutes later the student counted how many woodlice could be seen on the illuminated side (and by simple subtraction worked out how many were on the shaded side). This procedure was repeated 10 times using different woodlice each time. Table 3.1 gives the student's results.

Table 3.1 Results of an experiment to investigate negative phototaxis in woodlice using a choice chamber.

Trial number	Number of woodlice after 5 minutes	
	Bright light	Shade
1	3	7
2	5	5
3	4	6
4	4	6
5	3	7
6	5	5
7	6	4
8	3	7
9	2	8
10	4	6

TIPS

- In an exam you might be asked to recognise the type of data you have been given and to select an appropriate statistical test.
- Look at Chapter 13 on maths skills to find out more about statistical tests and to see a worked example of this test.
- You won't need to do statistical calculations in a written paper.

Looking at these results the student concluded that woodlice do move away from bright light and therefore may show negative phototaxis, since there are clearly more woodlice found in the shade than in the bright light. It fitted with the student's expectations, since woodlice are normally found in damp, dark places, such as under logs or stones. However, it is possible that the response was actually a kinesis. The student only counted the woodlice at the end of 5 minutes. There is no evidence that they were observed during this period and without being able to tell if they showed non-directional responses to end up where they were the experiment is not conclusive.

TIPS

To be able to decide conclusively if a response is a kinesis or a taxis, the animals must be observed during the experiment to see if their rate of movement or turning frequency is different in the different conditions.

Because the student was dealing with two discrete categories, bright light and shade, the best way to present the totals would be on a bar graph. The bars would be different heights indicating that more woodlice moved to the shaded side. You would use a chi-squared (χ^2) test to find out whether the difference in the number moving to each side is significant (see page 255).

REQUIRED PRACTICAL 10

Investigation into the effect of an environmental variable on the movement of an animal using either a choice chamber or a maze

This is just one example of how you might tackle this required practical.

Some of the first experiments on woodlouse behaviour were carried out by J. Cloudsley-Thompson. In one series of experiments he studied the response of woodlice to humidity using large choice chambers similar to the one shown in Figure 3.6. The humidity on each side was controlled. Under the gauze on one side was a dish of a drying agent that absorbed moisture from the air. On the other side was distilled water that maintained a high humidity above it.

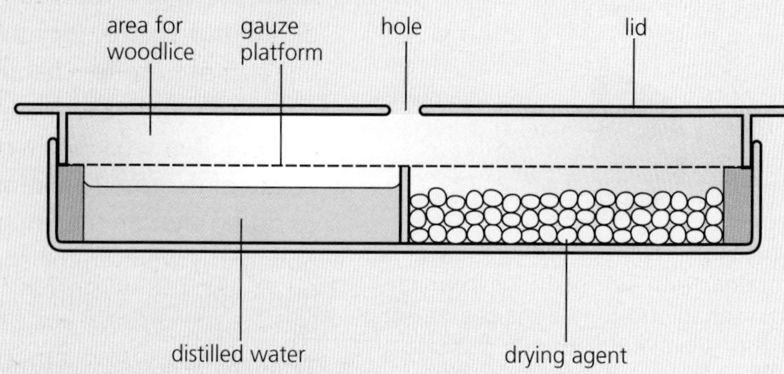

Figure 3.6 Apparatus used by Cloudsley-Thompson to investigate woodlouse behaviour.

Cloudsley-Thompson compared three sets of conditions:

A choice chamber in the light using woodlice that had been in the light for several hours before the experiment
B choice chamber in the dark using woodlice that had been in the light for several hours before the experiment
C choice chamber in the dark using woodlice that had been in the dark for several days before the experiment.

In each experiment, he put five woodlice into the choice chamber and then recorded the positions of the woodlice after 15 minutes (Table 3.2).

Table 3.2 The results of Cloudsley-Thompson's three experiments (A, B and C) in which he investigated three different sets of conditions.

Woodlouse behaviour in choice chamber	Number of woodlice A	B	C
Moving around	113	40	17
Stationary on the dry side	21	35	53
Stationary in the central area between the dry and moist sides	49	47	86
Stationary on the moist side	317	378	344

1 What do the results show about the response to humidity?
2 Suggest how this response might be advantageous to the woodlice.
3 Cloudsley-Thompson noticed that after putting the woodlice into the chamber they usually moved around for a short time before becoming stationary. What does this suggest about the type of behaviour shown by the woodlice?
4 Calculate the percentage change in woodlice still moving around between experiments A and C and show the results on a suitable graph. Suggest a possible explanation for the difference.
5 What does your analysis of the results suggest about the response of woodlice to humidity in the light compared to the dark? Suggest how this might this increase the chances of survival of woodlice.

Plant responses

Plants may just seem to sit around and do nothing, but in fact their survival is as dependent as that of animals on being able to respond to environmental conditions. Whichever way a seed lands in the soil, the shoot will grow upwards and the roots will grow down. Common observations show examples of responses. A pot plant growing on a window sill will grow towards the light unless it is turned frequently. Many flowers close at night and then open again in the morning. You may have come across so-called sensitive plants that respond to a touch. Lightly brushing against the end of a *Mimosa* leaf will stimulate the leaflets to close up like the ripple of a Mexican wave.

Some trees can defend themselves if herbivores damage their leaves. Like the *Arabidopsis* plants in the introduction to this chapter, they respond by producing nasty-tasting or poisonous substances. This ability can pass from affected to unaffected parts of the tree by the use of chemical messengers. Some trees can even pass this chemical message to neighbouring trees, which then produce the same noxious substances before the attackers move in. You might wonder why the trees don't just produce the noxious substances before any herbivores cause some damage. The answer is probably that the production of the poison requires energy and resources, so it is more economical to wait until the threat of attack is real.

Tropism A growth response to a stimulus.

Plants don't have a nervous system, so how are they able to respond to stimuli? Responses such as bending towards or away from light or gravity result from uneven growth. The seedlings in Figure 3.7 are bending towards the light because they have been stimulated to grow slightly faster on the more shady side of the stem. Roots placed horizontally will begin to grow vertically downwards. This sort of growth response to a stimulus from a particular direction is called a tropism. Response to light is referred to as **phototropism** and that to the force of gravity as **gravitropism.** Both may be either positive – towards the stimulus – or negative – away from the stimulus.

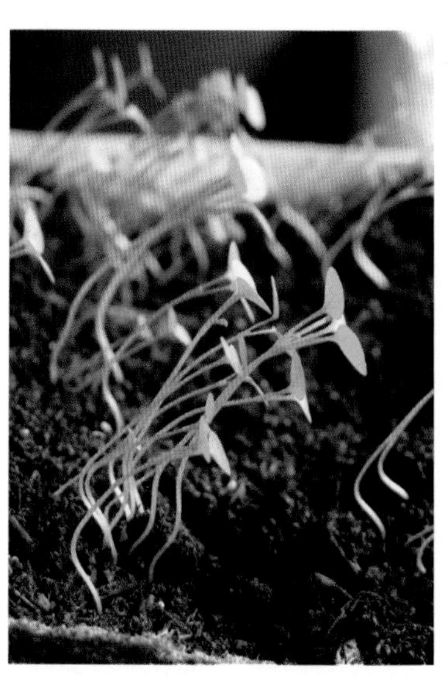

Figure 3.7 Increased growth on the shady side of these seedlings results in a positive phototropic response to a light source.

A growth response depends on chemical substances released in response to a stimulus. These are known as **specific growth factors**, and they act a little like hormones in animals. Although this is much slower than the electrical activity of nerves or the response of some plants to sound vibrations, it can be surprisingly rapid. For example, the phototropic response of plant shoots can be detected within minutes of exposure to light.

The first specific growth factor to be discovered was **indoleacetic acid (IAA)**. Several other substances that affect growth in plants have been discovered since. But there is still uncertainty and disagreement about exactly how these substances work. One of the problems is that the actual concentrations of the substances present in the plant tissues are extremely low.

TIP
You may see specific growth factors called plant hormones. Although they share some similarities with animal hormones, they are different enough that the word 'hormone' is no longer used in the context of plants.

Growth of a root or shoot has two distinct stages (Figure 3.8).

- First, cell division (mitosis) takes place at or near the tip.
- Second, the vacuoles of the new cells expand as they take up water by osmosis, causing the cells to elongate. This starts in the cells a short distance behind the tip in an area called the region of elongation. This is where the most obvious increase in length of a root or shoot occurs.

IAA has its main effect in the elongating region. It is synthesised in the young root or shoot cells at the tip. As it moves into the elongating region it attaches to protein receptors on the membranes of cells. Exactly how it works is still unclear, but one of its effects is to lower the pH by the release of hydrogen ions. These break some of the bonds between the microfibrils in the cellulose walls (see *AQA A-level Biology 1 Student's Book* Chapter 1, page 6), making the microfibrils more easily stretched by the increasing turgor of the cells.

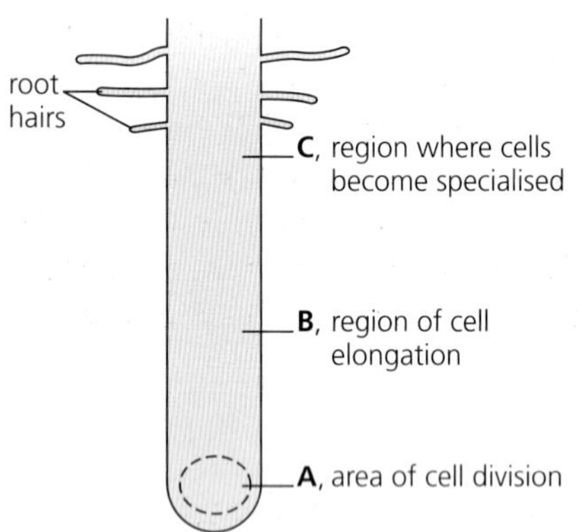

Figure 3.8 Growing regions in a root.

IAA and phototropism

Possible explanations for the phototropic response in plant shoot tips include:

- IAA is destroyed by the light on the illuminated side
- extra IAA is produced on the shaded side
- IAA moves away from the illuminated side.

Figure 3.9 shows results from experiments to investigate these explanations in shoot tips. In each case, the shoot tips were cut off and then placed on thin blocks of agar. In some experiments, very thin slices of mica, which are impermeable to water, were used to separate the two sides of the tips and/or the agar blocks. After 3 hours, the concentration of IAA in each agar block was measured.

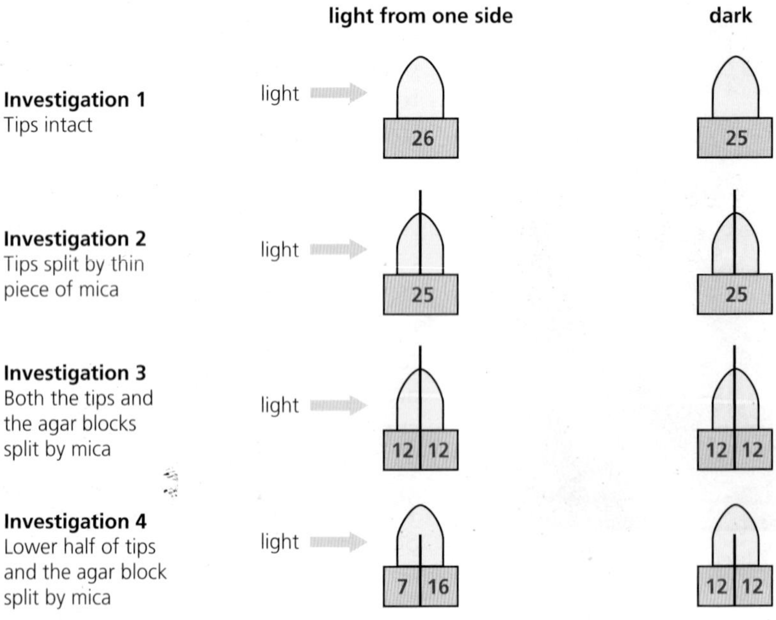

Figure 3.9 Results of four experiments on shoot tips to investigate possible explanations for their phototropic response. The figures show the concentration of IAA in arbitrary units.

In investigations 1 and 2, the concentration of IAA in the agar blocks is about the same whether in the light or dark. If it was destroyed on the illuminated side you might expect an overall decrease in the light. If more was produced on the shaded side you might expect an increase in the dark.

In investigation 3, the total concentration of IAA collected in the agar was about the same as in investigations 1 and 2. Also, equal concentrations of IAA were found in the agar either side of the mica sheets. Again, these observations support the idea that light does not destroy IAA and darkness does not cause more IAA to be produced.

In investigation 4, the total concentrations of IAA collected in the agar blocks were about the same as in the previous investigations. In darkness, equal concentrations of IAA were found in the agar either side of the mica sheets. Under unilateral illumination, however, less IAA was found in the block nearest the light and more was found in the agar block on the shaded side. This suggests that when the tip is illuminated the IAA produced is actively moved towards the shaded side, unless prevented by the mica, as in investigation 3.

The difference in the concentration of IAA moving down each side of the block in the light in investigation 4 explains why the intact shoot would bend towards the light. More IAA would reach the region of elongation on the shaded side. As we have already seen, IAA causes the cell wall to be more easily stretched by the expansion of the cell vacuole due to the uptake of water by osmosis in the cells on this side of the shoot. It elongates faster than the lit side, and because of this uneven growth the shoot would bend like those in Figure 3.7.

IAA and gravitropism

Unequal IAA concentrations either side of the root tip are also why roots bend towards the force of gravity. Cells called columellar cells near the root tip contain dense organelles called **amyloplasts**. These are packed with starch, which makes them heavy, and so sink to the bottom of the cells they are in. When a root is moved from the vertical to the horizontal the amyloplasts fall to what is now the bottom of the columellar cells. This enables these cells to detect the direction of the force of gravity. IAA seems to be actively transported to the side of the root to which the amyloplasts sink (Figure 3.10).

TIP

Remember that the response of roots to higher IAA concentration is the opposite to that of shoots.

As IAA moves from the root tip towards the region of elongation there is more on the lower side. **Unlike shoots, in roots a higher IAA concentration inhibits elongation.** This means that the lower side elongates slower than the upper side, causing the root to bend downwards.

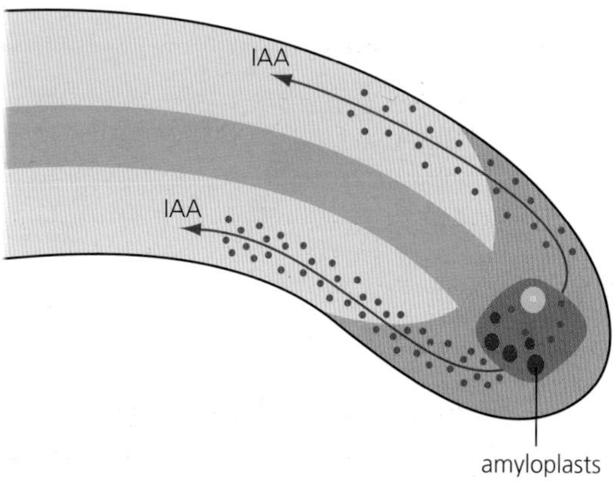

Figure 3.10 The mechanism of the gravitropic response in a root tip.

TEST YOURSELF

1 Describe the components of a three-neurone reflex arc.
2 Maggots that feed on dead animals tend to move away from the light. What type of response is this and why?
3 Would the experiment shown in Table 3.1 distinguish whether the response was a taxis or a kinesis? Explain your answer.
4 If plants have no muscles, how does a plant shoot or root bend towards or away from a stimulus?
5 Describe how amyloplasts allow some root cells to detect a change in the direction of the force of gravity.

Responses to internal stimuli

So far you have seen examples of responses of organisms to external stimuli. Some activities over which we have no conscious control result from responses to internal stimuli. A good example is the control of heart rate.

The muscle of the heart is amazing. For a start, it doesn't get tired. Even in someone who is not very active it can go on steadily contracting and relaxing 70 times a minute, 24 hours a day for 80 years or more. (That works out at about 3 billion contractions in a lifetime, assuming a life with very little exercise or excitement.)

Myogenic Muscle cells that are able to contract without nervous stimulation.

Secondly, it can carry on contracting and relaxing rhythmically without any nerve impulses from the brain. This ability to work on its own is called myogenic. A heart can be removed from the body and, as long as it is given an oxygen and nutrient supply, it will carry on beating.

The sequence of muscle contraction in the heart is initiated by a group of modified heart-muscle cells called the sinoatrial node (**SAN**) near the top of the right atrium. These cells produce regular waves of electrical activity, similar to nerve impulses. The rate at which the SAN produces these waves determines the heart rate because they start off contraction. For this reason the SAN is often called the heart's pacemaker.

During one cardiac cycle, a wave of electrical activity spreads over the walls of both atria, as shown in Figure 3.11. This makes the muscles in the atrial walls contract. Notice that contraction spreads outwards from the top of the atria, squeezing blood towards the ventricles.

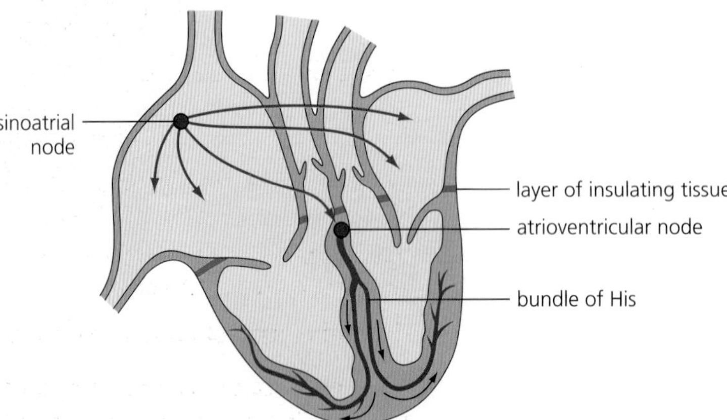

Figure 3.11 The route of the electrical activity that makes the heart beat in a smooth sequence.

The electrical activity cannot pass directly from the walls of the atria to the walls of the ventricles, because it is stopped by a layer of insulating fibrous tissue (Figure 3.11). At the lower end of the wall that separates the atria is another group of specialised cells, the atrioventricular node (**AVN**). These cells can detect the electrical activity passing across the atria.

After a short delay, the AVN triggers electrical activity in specialised muscle cells called **Purkyne tissue**, which in turn conduct the electrical activity rapidly down the wall between the ventricles to the bottom of the heart. The delay allows time for the ventricles to fill completely with blood. Initially, these specialised muscle cells are bunched in a single group, called the **bundle of His**. This then divides into two branches that extend back up the walls of the two ventricles. Electrical activity conducted along these specialised muscle cells stimulates the muscle of the ventricles to contract rapidly from the base of the heart upwards.

However, there are times when the heart rate increases, such as during exercise. For this to happen, the brain is involved, although of course we do not have to *think* about changing our heart rate. In the brain is a special area that controls the heart rate called the cardioregulatory centre and it is situated in the **medulla**. The medulla is tucked away in the base of the brain at the top of the spinal cord (Figure 3.12).

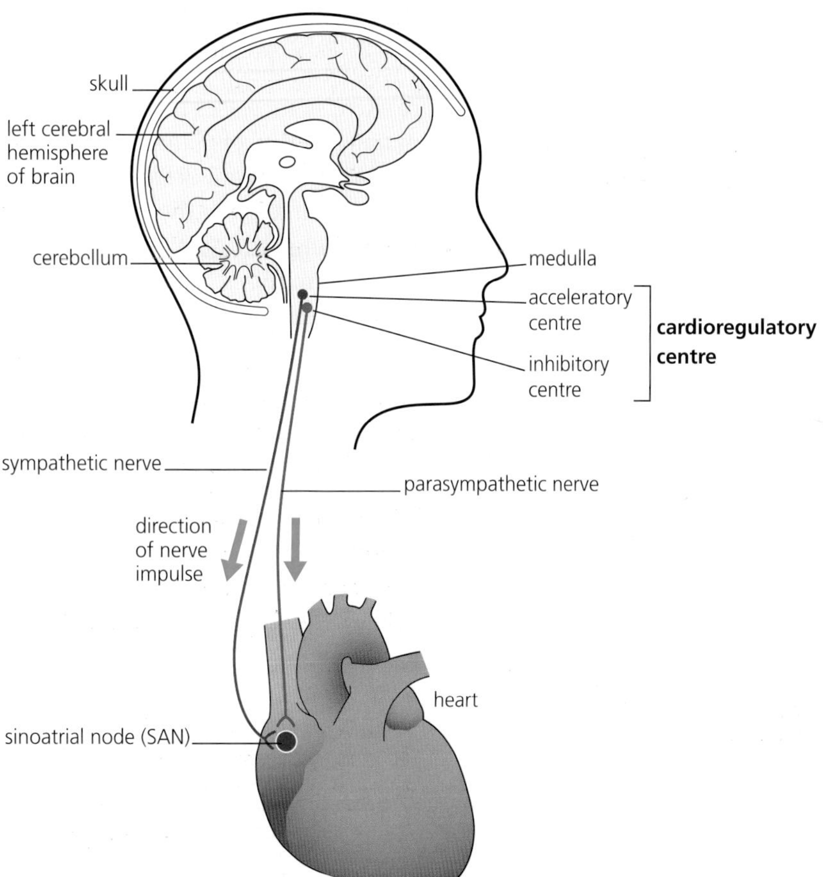

Figure 3.12 The position of the medulla in the brain and its role as a regulatory centre in controlling the heart rate.

The cardioregulatory centre in the medulla actually consists of two discrete parts:

- the acceleratory centre – responsible for speeding up the heartbeat
- the inhibitory centre – responsible for slowing down the heartbeat.

> **TIP**
> You do not need to be able to recall the names of the neurotransmitters involved in the control of heart rate.

A variety of factors can stimulate these centres and we will come back to these shortly. Each of these centres is connected to the SAN via nerves. These nerves are quite separate from the nerves that control our conscious activities. They form part of what is called the **autonomic nervous system**, which means the 'self-controlling' system. When the acceleratory centre is activated, impulses pass along **sympathetic** neurones to the SAN. At the synapse with the SAN, noradrenaline is secreted. It is this substance that causes the SAN to increase the frequency with which it produces waves of electrical activity. Noradrenaline is chemically very similar to adrenaline, which is a hormone produced by the adrenal glands. **Adrenaline** is well known as the hormone that is produced when we are frightened or stressed and which prepares us for a 'fight-or-flight' response. Adrenaline also increases the heart rate and is responsible for the thumping heart when we experience fear. We will find out more about adrenaline in Chapter 6.

Activating the inhibitory centre sends impulses along **parasympathetic** neurones that cause the SAN to decrease the frequency with which it produces waves of electrical activity. This returns the heartbeat to its normal or resting state. Like the sympathetic neurones, the parasympathetic neurones secrete a substance at synapses with the SAN. In this case it is acetylcholine, which has the effect of inhibiting the myogenic activity of the SAN.

So, how do the internal stimuli from exercise lead to activation of the acceleratory centre in the medulla? This question does not have a simple answer and the processes involved are still not fully understood. The onset of exercise increases the concentration of carbon dioxide in the blood and initially causes a fall in blood pressure as muscle arterioles dilate. These internal stimuli are detected by **chemoreceptors** and **pressure receptors**. Both are situated in the aorta close to the heart and in the carotid arteries that pass through the neck to the brain. An increase or decrease in the frequency of nerve impulses from these receptors activates the acceleratory centre and contributes to the increase or decrease in heart rate (Figure 3.13).

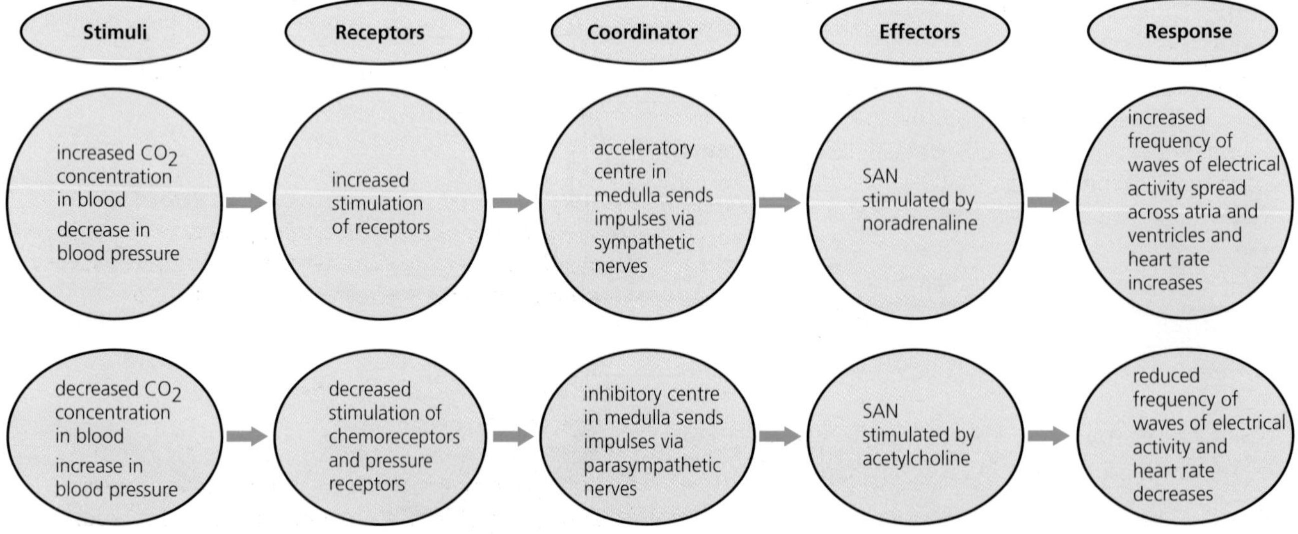

Figure 3.13 Summary of the factors involved in the control of heart rate.

Imagine a hare that suddenly notices a fox closing in on it (Figure 3.14). A rapid response will allow the hare to avoid becoming the fox's prey. The hare immediately leaps into action and escapes at high speed. As long as it can start off quickly it can outrun the fox, which can only reach its maximum speed over short distances. The hare's leg muscles must go to maximum activity as soon as it sees the fox. It is therefore an advantage for the heart rate to increase almost instantaneously, thus speeding up the blood supply and hence the supply of oxygen and glucose for respiration. If the heart responded only *after* the effect of exercise on carbon dioxide production had been detected, vital seconds might be lost. This is the role of the hormone adrenaline, which is released when the hare sees the fox and acts almost immediately on the SAN to increase heart rate.

Figure 3.14 The hare's flight response on seeing the fox must be immediate if it is to get away.

EXAMPLE

Responding to danger

Although we may take part in exercise for sport or pleasure, most other mammals only engage in vigorous exercise to escape danger or to chase food. As well as the increase in the rate of heartbeat there are several other changes that take place in the heart and circulatory system during exercise. Some of the changes are described below and are followed by questions about them. In answering some of the questions you may want to look back to *AQA A-level Biology 1 Student's Book*, and in particular to Chapter 6 about the heart.

1 The cardiac output increases by 100–200% during exercise, or even more in fully trained athletes. Explain how the cardiac output is increased and the advantage of the increase.
Cardiac output is heart rate multiplied by stroke volume (the volume of blood ejected each beat) so it is not just heart rate that increases during exercise. If the volume of blood being pumped each beat increases too, then the increase in cardiac output can be even greater. The increase delivers blood faster to the tissues, especially muscle tissue.

2 The blood pressure increases in the arteries during heart contraction. Explain how this increase is produced.
More forceful contraction of the ventricles results in a stronger pulse.

3 The arterioles carrying blood into skeletal muscles dilate (get wider). Explain the advantage of this.
More blood is supplied to the active muscle tissue, delivering oxygen and glucose faster, and removing carbon dioxide faster, enabling a higher rate of aerobic respiration to continue.

4 The arterioles carrying blood to the abdominal organs and skin constrict (get narrower). What is the advantage of this?
The blood temporarily diverted away from these less critical tissues during exercise can be directed to muscle tissue instead.

5 In the muscles, the higher blood pressure forces open many more of the capillaries. What is the advantage of this?
A larger total surface area becomes available for capillary exchange to occur between the blood and muscle cells.

6 The blood pressure in the major veins returning blood from the body to the heart increases. Explain what causes this increase in blood pressure.
Greater compression of the veins by more muscle contraction taking place in the arms and legs squeezes the blood and increases the pressure.

7 The saturation of haemoglobin with oxygen in the arteries to the muscles is higher than in the veins. During exercise this difference in saturation of haemoglobin with oxygen is much increased. Explain what causes this.
Actively respiring muscle cells use more oxygen so the haemoglobin unloads more oxygen as it passes through muscles, lowering saturation of haemoglobin with oxygen by more than at rest.

Receptors

We have **receptors** that can respond to a wide variety of stimuli, such as light, temperature, chemicals and mechanical effects including pressure, stretching and vibration. Each type of receptor normally only responds to one particular sort of stimulus. This is obviously important, as it enables us to distinguish between a large number of different environmental conditions, both outside the body (external) and inside the body (internal).

Although we often speak of having five senses, we are in fact able to recognise many more different stimuli than this. In the olfactory area of the nose, for example, there are several hundred slightly different receptors that enable us to distinguish as many as 10000 different smells. Some of these receptors are incredibly sensitive, being able to respond to just a few dissolved molecules of a particular substance.

TIP
A receptor responds only to a specific stimulus.

Consider the sense of touch, which is what we usually think of as the sense to which our skin responds. You will easily be able to distinguish between a very light touch and pressure that pushes your skin in a little. Different degrees of pressure feel quite different. This is not just due to different areas of your skin being touched. You can also tell whether your skin is surrounded by warm or cold air. You will be able to detect a slight movement of a hair, and you will certainly get a very different sensation from a jab with a sharp pin – the sensation of pain. All these different sensations are detected by different types of receptors in the skin (Figure 3.15).

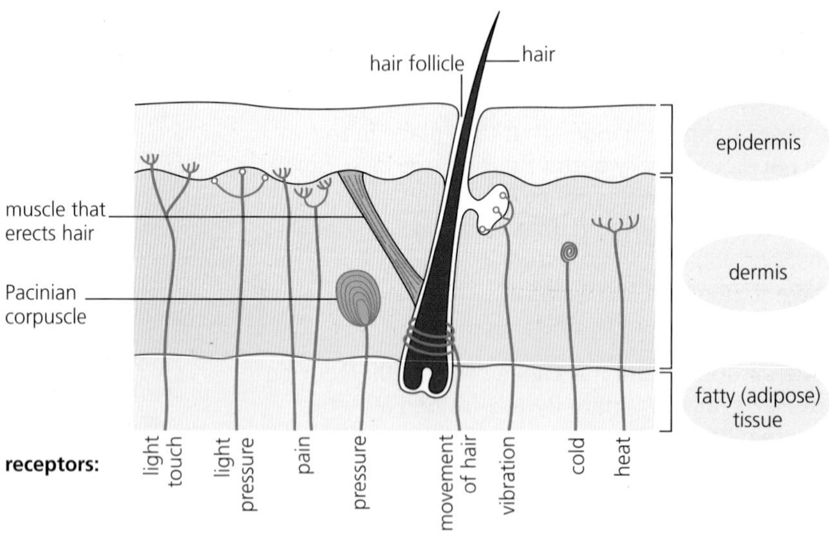

Figure 3.15 Pacinian corpuscles are one of a range of types of receptors in the skin. In an actual area of skin, there are many of each type mixed together. The numbers of receptors vary in different areas; for example there are many more touch receptors in the fingertips than in the middle of the back. The precise structure and functions of some types of receptors is still not understood.

TIP
You do not need to be able to recall details of the structure of the skin.

You will see from Figure 3.15 that receptors in the skin have different shapes and positions. Some of the receptors, such as pain receptors, are not separate cells but just the fine branches at the ends of the fibres of sensory neurones. Other receptors, such as the **Pacinian corpuscles** that respond to changes in pressure on the skin, are more complex structures (Figure 3.16) but are still not separate cells.

Pacinian corpuscles

Pacinian corpuscles are quite large compared with most receptors, so it has been possible to study the way in which these work. You can see from Figure 3.16 that a Pacinian corpuscle consists of many layers of membrane, rather like a tiny onion.

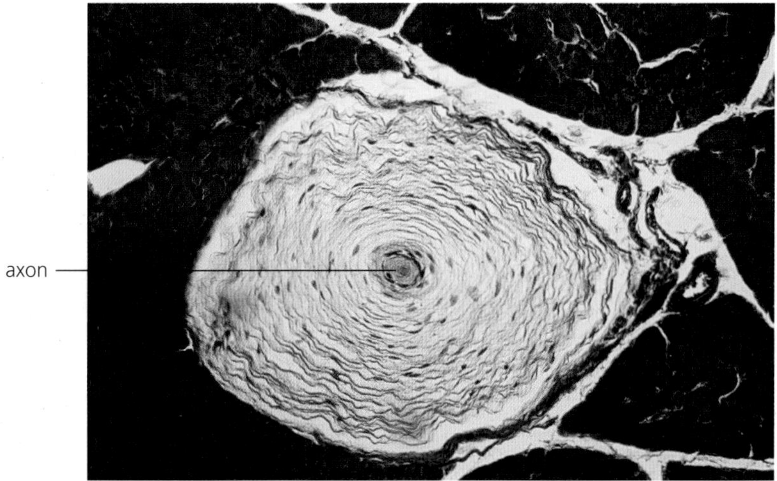

Figure 3.16 Coloured photomicrograph of a Pacinian corpuscle.

These layers surround the end of the axon of a sensory neurone. Normally there is an excess of positive sodium ions (Na^+) outside the axon. But when pressure on the Pacinian corpuscle is increased the layers are distorted and proteins called sodium channels in the membrane of the axon are opened. It is why these particular sodium channels are called **stretch-mediated sodium channels**. Previously you learned about the ways that ions can pass through **carrier proteins** in the membrane. It may be helpful to refer to Chapter 3, page 46 in *AQA A-level Biology 1 Student's Book*.

This allows sodium ions to move into the axon by facilitated diffusion. This changes the electrical potential difference across the membrane, as shown in Figure 3.17. It is this that triggers impulses that can pass along the axon of the neurone to the central nervous system. For this reason it is called a generator potential. We shall see in Chapter 4 how nerve impulses pass along the axons of neurones.

Generator potential The change in electrical potential in a receptor when it is stimulated.

TIP

Remember that you learned about facilitated diffusion and different types of carrier protein during the first year of your course.

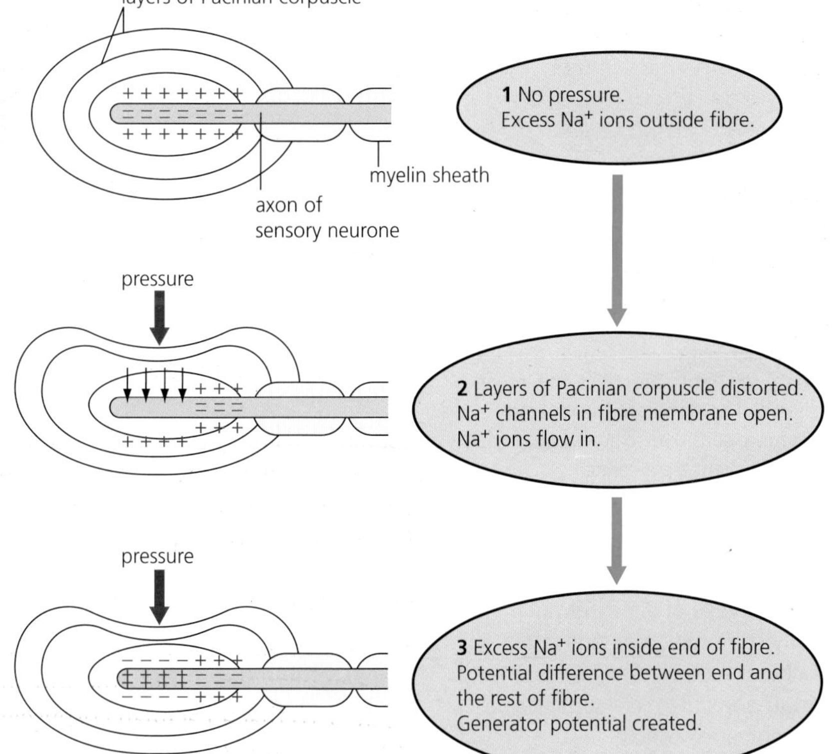

Figure 3.17 How a generator potential is set up in a Pacinian corpuscle.

> **TEST YOURSELF**
> 6 What is a generator potential?
> 7 What happens to stretch-mediated sodium channels in response to pressure?
> 8 What does myogenic mean in relation to the sinoatrial node?
> 9 What is the role of the AVN in coordinating heart activity?
> 10 Where are the chemoreceptors that are involved in the control of heart rate?

Vision

You have probably studied the eye in an earlier stage of your science education. You may remember that images are focused on to the retina at the back of the eye. Figure 3.18 will remind you of the structure of the eye.

> **TIP**
> You do not need to be able to recall the detailed structure of the eye, but this is a useful diagram to remind you of the location of the retina.

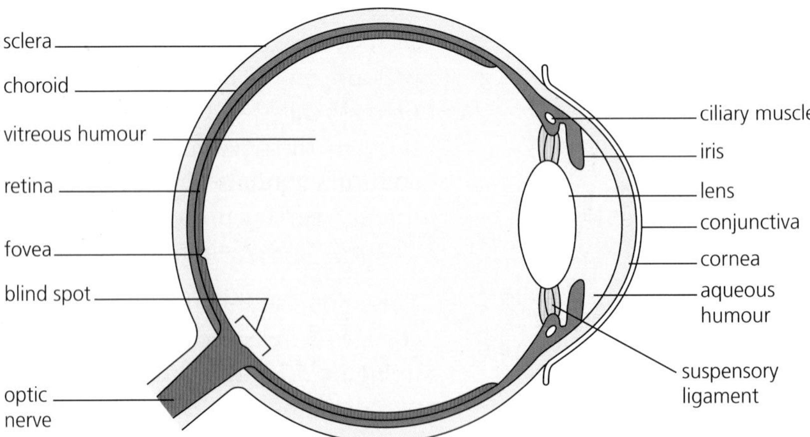

Figure 3.18 Structure of the eye.

The receptors that respond to light are situated in the **retina**. We are going to concentrate on how the different types of receptor cell in the retina helps us to see in different conditions. We have two different types of light-sensitive receptor cell in the retina, the rods and the cones.

Sensitivity

The rods and cones (Figure 3.19) both contain optical pigments, which absorb light (the stimulus) and are broken down. This results in the production of a generator potential. The pigment within rods is broken down in dim light whereas the pigments within cones are only broken down in bright light.

Sensitivity (to light) Describes how much light is needed to stimulate the receptor. Rods are more sensitive so are stimulated in much dimmer light than cones.

- **Rods** are sensitive to a very low intensity of light and therefore enable us to distinguish light from dark in very dim light. They are so sensitive that they can produce a generator potential in response to just one photon. (A photon is a 'particle' of light that is emitted when electrons are excited and jump from one orbit to another in an atom.) Rods do not permit us to distinguish different colours.
- **Cones** are not as sensitive to light as rods. They are, however, sensitive to light of different wavelengths. Human eyes have three types of cone each containing one of three different optical pigments. One pigment is sensitive to the wavelengths of light corresponding to red light, one to green and one to blue. When combinations of the three types of cone are stimulated we perceive the range of other colours in the visible spectrum.

> **TIP**
> The three types of cone are not actually coloured so do not call them blue, red or green cones; they are blue-sensitive, red-sensitive or green-sensitive cones.

Visual acuity

Unlike the receptors in the skin, rods and cones are not directly connected to the central nervous system. As you can see from Figure 3.19, they have synapses connecting them to bipolar neurones that are also in the retina. These neurones have synapses connecting to ganglion cells that have axons extending via the **optic nerve** all the way to the brain.

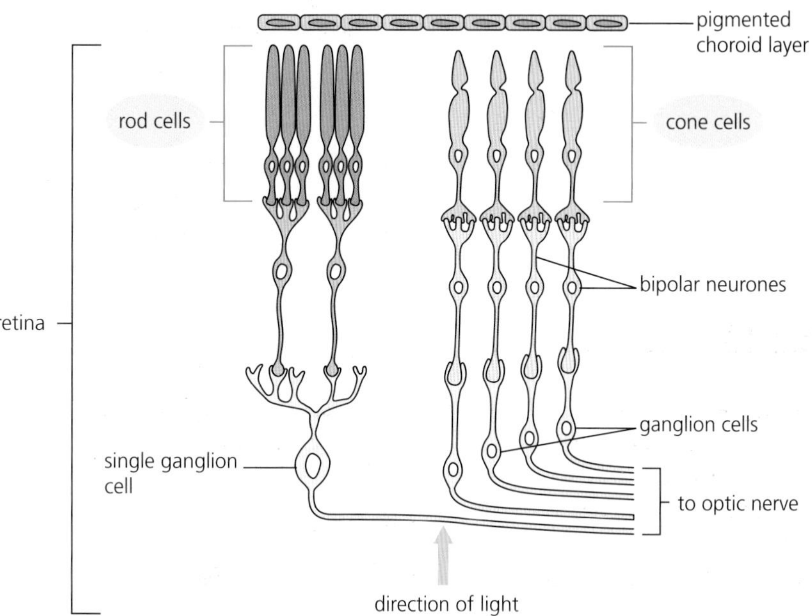

Figure 3.19 How the rods and cones are connected to other cells in the retina.

Each of our eyes has about 125 million rods and 7 million cones. However, they are not evenly distributed in the retina. Most of the cones are concentrated in the centre of the retina, at the position called the **fovea** (look at Figure 3.18). This is where the main part of an image is formed when we look straight at an object. The rods are spread much more widely and they cover most of the back part of the eye. One exception is the area where the optic nerve leaves the eye, the so-called blind spot.

The retina contains about 132 million receptors, each of which can be stimulated by a ray of light falling on it. You can imagine, therefore, that when an image falls on the retina it will have an effect rather like a computer screen, which consist of a mass of tiny dots, called pixels.

Where light falls, the receptors will be stimulated, and where there is a dark patch there will be no stimulation. Each receptor that is stimulated can pass impulses to the brain and the brain can interpret the pattern. At the fovea all of the receptors are cones. Each cone in the fovea connects through a single bipolar cell to one ganglion cell in the optic nerve.

The optic nerve has about 1.2 million ganglion cells. Since there are 132 million receptors there obviously cannot be individual connections to the brain for all the receptors. You can see from Figure 3.19 that several rod cells synapse with a single bipolar cell. In turn, more than one of these bipolar cells synapses with a single ganglion cell. Although the simplified diagram shows only a few synapses, in practice there must be an average of about 100 rod cells with synaptic connections to each ganglion cell. In

Visual acuity How far apart two spots of light must be to be seen separately.

contrast, each cone shown in Figure 3.19 synapses with a single bipolar cell which, in turn, synapses with a single ganglion cell. This affects the **visual acuity**, the amount of detail, or resolution, that can be perceived in an image on the retina.

A ray of light that falls on just one cone in the fovea will show up as a spot of light, as long as it is bright enough to stimulate the cone. This is because the cone is connected to a single ganglion cell in the optic nerve via one bipolar cell (Figure 3.20a). If another ray of light falls on another cone, as in Figure 3.20b, the brain will interpret this as two separate spots.

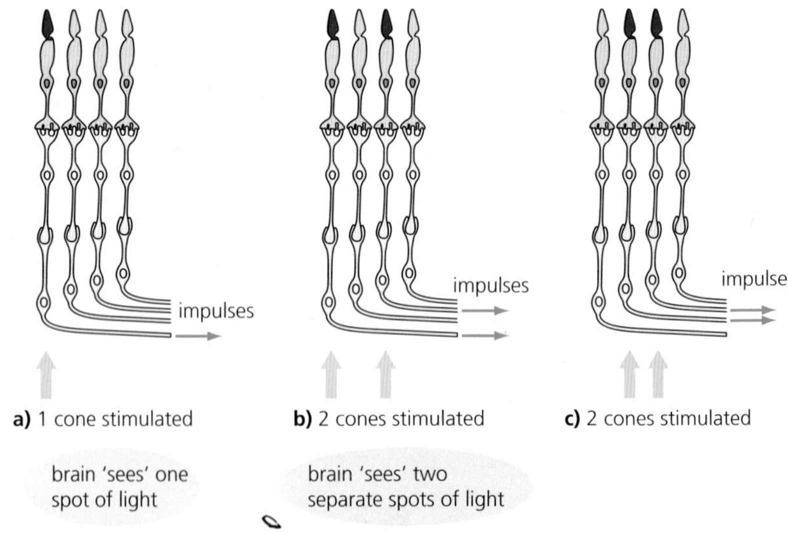

Figure 3.20 How the cones provide higher visual acuity.

TIP

No animal can see using their eyes in complete darkness. Rods enable vision in very low light intensities, not in the dark.

Now consider what will be seen when a ray of light falls on a single rod. Figure 3.19 shows three rods with synapses to each bipolar cell and six rods that connect via the bipolar cells to a single ganglion cell. Assuming that the stimulation of one rod cell is sufficient to send an impulse to the brain, there is no way in which the brain can interpret which of the six rods had been stimulated. In reality there may be many more than six rods with synapses to a single ganglion cell. The rods therefore provide lower visual acuity than the cones.

TEST YOURSELF

11 Give one difference in the structure and one difference in the distribution of rod and cone cells.

12 Which are more sensitive to light, rods or cones? Explain your answer.

13 Explain what is meant by visual acuity.

14 Describe how the way that rods and cones are connected to ganglion cells affects visual acuity.

15 How do humans detect colour?

So, what is the advantage of having many rods connecting to a single ganglion cell in the optic nerve? As we shall see in Chapter 4, synapses can act as barriers to the transmission of impulses. Although the rod is very sensitive to light, a single rod is unlikely to produce a sufficiently large generator potential to be able to stimulate the bipolar cell to conduct nerve impulses. However, if a group of rods is stimulated by dim light at the same time, the combined generator potentials will reach the threshold required to cause nerve impulses to be conducted in the bipolar cell and the ganglion cell in the optic nerve. This process is called **summation**, because the effect of several cells is added together. Although the image is less sharp, we are able to see in much dimmer light than would be possible with cones alone. The eyes of most nocturnal mammals have only, or mostly, rods. For an animal that is hunting or hunted at night it is more useful to see a slightly fuzzy image than to be unable to see at all.

Practice questions

1 Brook lampreys are aquatic animals. Their larvae live buried in mud at the bottom of rivers. Brook lamprey larvae were kept in tanks with different light intensities. They were each observed for 20-minute periods and the time they spent moving around was recorded. The results are shown on the graph as the mean percentage of the observation period when the lampreys were moving.

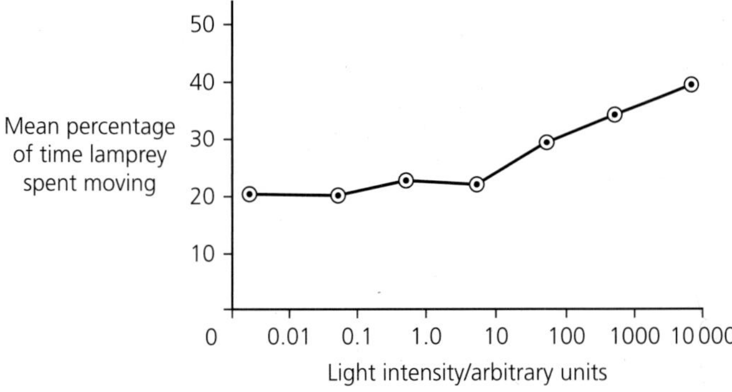

a) i) How many minutes did the lamprey larvae spend moving in the brightest light intensity? *(1)*

ii) What kind of scale is used on the *x*-axis? *(1)*

iii) Give one advantage of using this kind of scale. *(1)*

b) i) What type of response do lamprey larvae show to light? *(1)*

ii) Give a reason for your answer to b(i). *(1)*

c) Suggest how this response might be an advantage to the lamprey larvae. *(2)*

2 The diagrams show the results of four different treatments of shoot tips. The tips were placed on blocks of agar and the mean concentration of indoleacetic acid (IAA) that diffused into the agar is shown by the figures (arbitrary units).

agar block

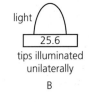

shoot tip

25.8

tips kept dark

A

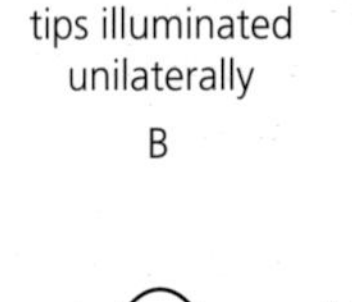

tips illuminated unilaterally

B

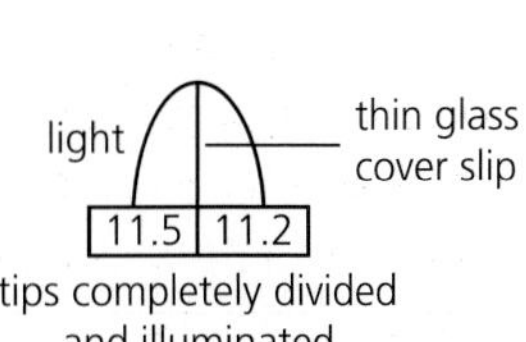

tips completely divided and illuminated unilaterally

C

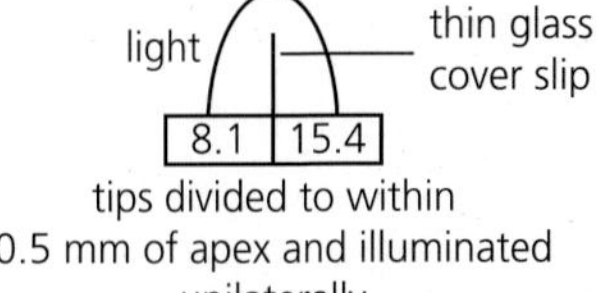

tips divided to within 0.5 mm of apex and illuminated unilaterally

D

a) What can you conclude from comparing the results of treatments A and B? *(1)*

b) Suggest the purpose of the thin glass cover slip. *(1)*

c) Using the results from treatments C and D, describe the response of the shoot tip to unilateral light. *(2)*

d) If the shoot tip from treatment D were on an intact shoot, describe and explain the response you would see. *(3)*

3 Describe the role of the SAN, AVN and Purkyne tissue in coordinating the heartbeat. (6)

4 The number of rods and cones in the field of view of an optical microscope was counted at frequent intervals along a horizontal line across the retina of a human eye. The graph shows the results.

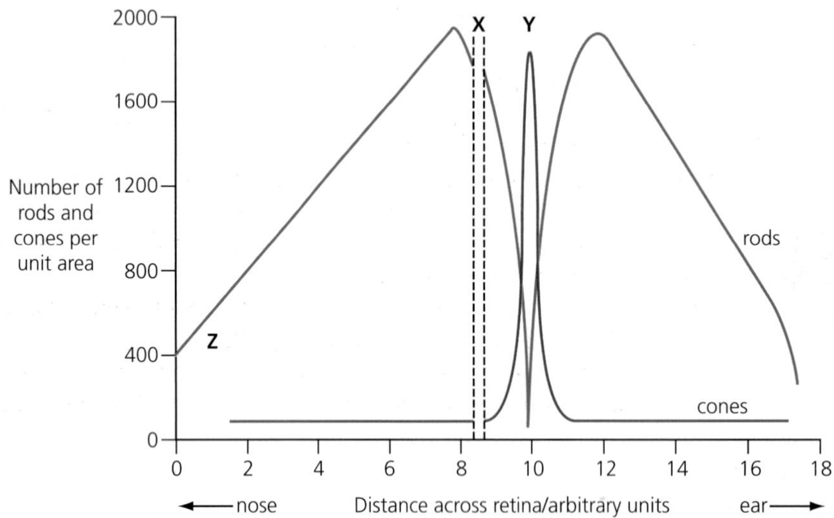

a) Identify the parts of the retina at points X and Y, giving reasons for each answer. *(2)*

b) Use the graph to find the ratio of rods to cones at 6 arbitrary units across the retina. *(1)*

c) At which distance across the retina would sensitivity to colour be highest? Explain your answer. *(3)*

d) i) Describe the sensitivity to light of receptors at Y compared with those at Z. *(1)*

ii) What is the reason for the difference in sensitivity of the receptors at Y and Z? *(2)*

e) Describe the connections made in the optic nerve by receptors at Z and explain how this would affect the visual acuity at this point on the retina compared to Y. *(3)*

Stretch and challenge

5 Evaluate the main theories to explain how the human eye is able to distinguish between so many different colours when cones are sensitive to just three specific wavelengths of light. To what extent are these theories supported by evidence?

6 To what extent are plant growth regulators important in horticulture? You may wish to discuss their use in selective weedkillers, producing seedless fruit and in plant propagation, among other applications.

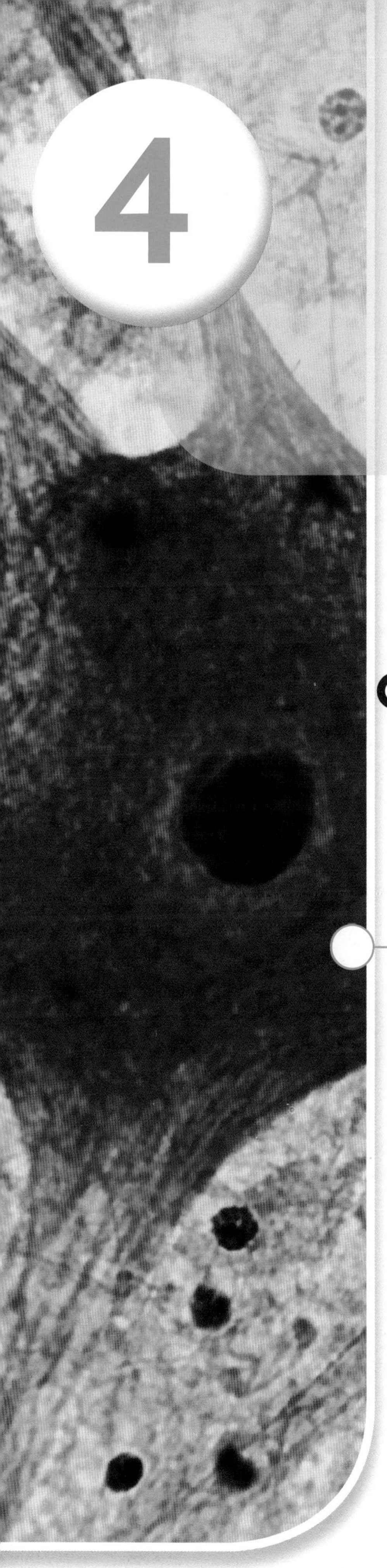

4

Nervous coordination

PRIOR KNOWLEDGE

- Nerve impulses from receptors pass along neurones in nerves to the spinal cord or brain.
- The role of receptors, sensory neurones, motor neurones, relay neurones, synapses and effectors.
- At a junction or synapse, a chemical neurotransmitter is released that causes impulses to be generated in the next neurone.
- Carrier proteins are involved in both facilitated diffusion and active transport across membranes.
- Dissolved substances such as ions can move into and out of cells by facilitated diffusion through channel proteins.
- Active transport is a process by which a cell can transport substances such as ions against a concentration gradient by hydrolysing ATP.

TEST YOURSELF ON PRIOR KNOWLEDGE

1 Give an example of a receptor and the stimulus it detects.
2 Explain what is meant by an effector.
3 Ions cross membranes by facilitated diffusion rather than by diffusion. Explain why.
4 Where do cells make most of the ATP they use in active transport? Explain your answer.

Introduction

The human brain is estimated to have about 100 billion (10^{11}) neurones. If we gave equal portions of a single brain to every person in England, he or she would get about 2000 cells. Moreover, each neurone has synapses connecting it to as many as 10000 other neurones. This creates a neural network that makes the average computer seem like a child's toy.

The brain is the organ responsible for coordinating the great majority of activities in the body. It synchronises most of the automatic activities such as heartbeat and breathing, as well as such complex movements as walking or playing the piano. The brain receives and processes a constant flow of information from sensory receptors. This information may stimulate appropriate responses, or be stored or ignored. The brain is also responsible for what we consider to be the higher human activities, such as thinking, emotions, memory and consciousness.

So, how is it possible that tiny pulses of electrical activity and the transfer of chemicals across synapses and membranes can result in the complex mass of thoughts, feelings, actions, memories and emotions of which humans are capable? Neuroscientists are still a long way from full understanding, but slowly they are discovering the functions of different regions of the brain and how they interact. In this chapter we shall study the basic processes necessary to understand how the brain works and how the nervous system coordinates our activities. This will enable us to explain how chemicals such as alcohol and other drugs can disrupt the system.

Extension

Different areas of the brain are responsible for different functions, but neuroscientists have discovered that these subdivisions are not sharply defined and that the key to the brain's functioning lies in its network of connections. It may seem impossible to understand exactly how such a complex organ works, and our progress has been slow. Much of our understanding has come from evidence based on what happens when things go wrong, for example as a result of damage caused by accidents, blood clots and the effects of drugs.

However, developments in technology, especially brain imaging, mean that scientists are able to examine our brains in new ways. Functional imaging in particular has opened up ways to map the parts of the brain that become more active during different mental processes. Functional magnetic resonance imaging (fMRI) is one technique that confirms the information from previous studies by showing that different brain areas are involved in specific tasks.

Using an MRI scanner, scientists can detect the change in flow of blood to different regions of the brain as their subjects respond to a specific stimulus: sounds, images or even touch. When areas of the brain light up on the image as the result of increased blood flow, it is suggested that this indicates increased neural activity. However, it is not quite mind reading yet. Some scientists have argued that increased blood flow does not necessarily mean increased activity in that area of the brain. Others worry that the method can only reveal large groups of neurones working together. It is a long way from detecting what someone is thinking. But MRI is allowing scientists to begin to recognise the patterns in brain activity that occur with simple tasks such as recognising faces.

Neurones

First we need to look at the structure of the nerve cells, or neurones, that form the conducting tissue in the nervous system (Figures 4.1 and 4.2).

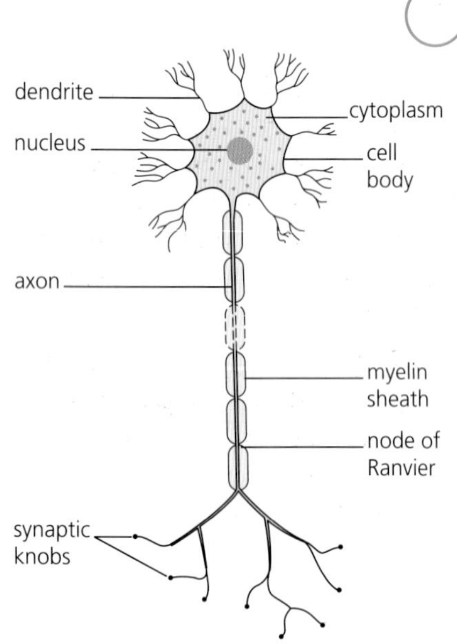

Figure 4.1 The structure of a myelinated motor neurone.

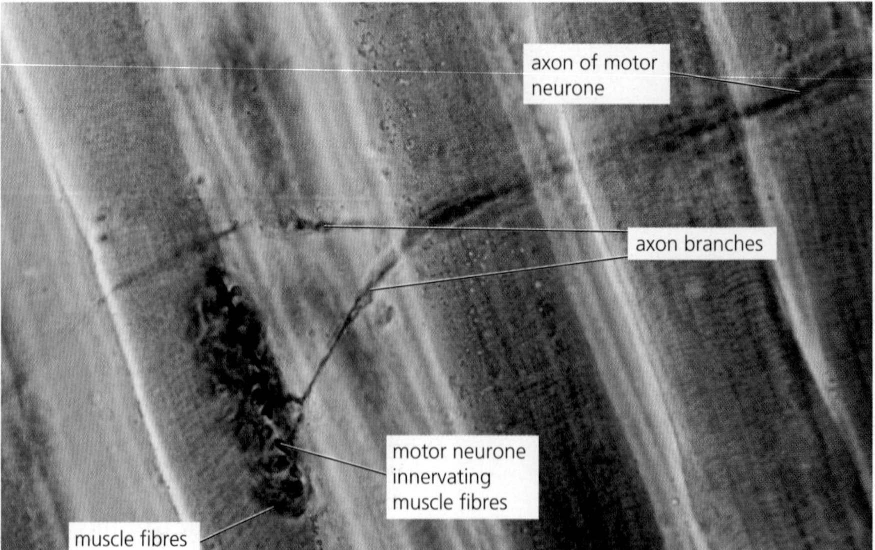

Figure 4.2 Light micrograph of a motor neurone and its effector, a group of muscle cells.

In Chapter 3 you learned about a simple reflex arc that involved three different types of neurone.

- A **sensory neurone** conducts impulses from a receptor to the spinal cord.
- One or more **relay neurones** act as links between the sensory and motor neurones.
- A **motor neurone** conducts impulses from the spinal cord to a muscle.

Look at Figure 4.1, which shows the structure of a motor neurone. You can see that the cell body contains cytoplasm and a nucleus and looks similar in structure to other animal cells. The cytoplasm also contains mitochondria and ribosomes.

Axon The long fibre in a neurone that conducts impulses away from the cell body.

In other respects a motor neurone is highly **specialised**. Its cell body is situated inside the spinal cord. It has large numbers of short branches, called **dendrites**. A motor neurone may have over a hundred of these branches, each with one or more synapses to a relay neurone. There is also one very long branch that extends from the spinal cord to a muscle. This axon can be as much as a metre in length and less than a micrometre in diameter.

Neuromuscular junction A synapse between a neurone and a muscle cell.

At the far end of the axon are short branches that terminate in tiny **synaptic knobs**. These knobs are located close to the cell-surface membrane of muscle cells at the neuromuscular junction. Large numbers of neurones are bundled together into **nerves**. The sciatic nerve, for example, originates from the spinal cord in the lower back. Branches from it pass all the way down the leg to muscles in the foot. The sciatic nerve also contains sensory neurones conducting impulses in the opposite direction. Damage to the lower back can compress this nerve where it passes between the vertebrae, causing a painful condition called sciatica.

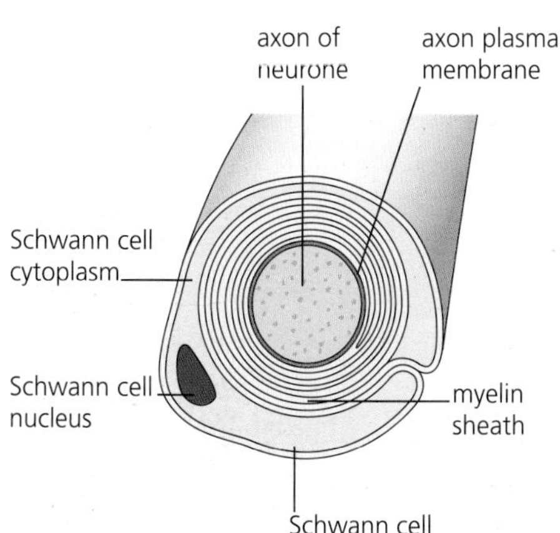

Figure 4.3 A section across an axon and Schwann cell.

You can see from Figure 4.3 that the axon is surrounded by the **myelin sheath**. This sheath is not strictly part of the neurone. It is made from highly specialised cells, called **Schwann cells**, that lie alongside the axon. As the axon grows these cells wrap round and round the axon until there may be up to a hundred layers of lipid and protein membranes. This makes a fatty 'bandage' around the axon, shielding it from surrounding tissue fluid and electrically insulating it from other neurones.

Between each Schwann cell is a tiny gap where the axon is exposed, called a **node of Ranvier** (see Figure 4.1). These nodes are the only places that ions can pass between the tissue fluid and the axon through the cell-surface membrane. The nodes play an important part in speeding up the conduction of impulses along the axon, as we shall see later.

Impulses

Impulses are waves of electrical activity passing along a neurone. The process is not the same as the conduction of electricity through a wire. It is much slower, although it is still quite fast. When a neurone is stimulated at a synapse, there is a brief change in the cell-surface membrane of the neurone. This allows ions to pass through rapidly. As soon as this happens it causes the next small section of the membrane to change, so ions can pass through here. Meanwhile the first section of neurone changes back as the distribution of ions is restored. This process carries on down the neurone. It

TIP

Impulses pass along all parts of the cell-surface membrane of a neurone in the same way. This includes the surface of the cell body as well as the dendrites and axon.

is rather like a long line of standing dominoes that are knocked down one after another once the first one is pushed over, except that in this case the domino is immediately set upright again as soon as the next one falls over.

So, what causes the movement of ions? Before we explain, it might be a good idea to remind yourself of the ways in which ions can pass across a cell-surface membrane. Previously you learned about the ways that ions can pass through **channel proteins** and **carrier proteins** in the membrane. It may be helpful to refer to Chapter 3, page 46 in *AQA A-level Biology 1 Student's Book*.

The resting potential

The cell-surface membranes of neurones contain carrier proteins called **sodium-potassium** pumps. These pumps are special carrier proteins that move sodium ions out of the axon and potassium ions in. The pumps use ATP to actively transport three sodium ions out for every two potassium ions they actively transport in.

There are also proteins in the membrane that allow ions to move back down their concentration gradients by facilitated diffusion. These proteins are called channels and are always open. However, the membrane is more permeable to potassium ions than to sodium ions. This means that the potassium ions diffuse back down their concentration gradient more rapidly than the sodium ions. This is called **differential permeability**.

The result of these two factors is that the outside of the axon membrane always has a slight excess of positive ions. This gives a **potential difference** of about 70 mV (millivolts) between the inside and the outside of the membrane. Because the inside of the axon is more negative and an electric current is a flow of negative electrons from a more negative to a more positive potential this is written as –70 mV. This is called the **resting potential**.

EXAMPLE

Investigating the action potential

Look at Table 4.1, which shows the concentration of sodium and potassium ions inside and outside the axon of a motor neurone.

	Concentration/mmol dm^{-3}	
Ion	Inside axon of motor neurone	Outside axon of motor neurone
Sodium (Na^+)	18.0	145.0
Potassium (K^+)	135.0	3.0

Table 4.1 The concentration of sodium and potassium inside and outside the axon of a motor neurone.

1 Describe the differences in concentration inside and outside the axon for each of the ions in Table 4.1.
The concentration of sodium ions outside the axon is much greater than their concentration inside, whereas the opposite is true for potassium ions.

2 There are channel proteins in the membrane that allow ions to diffuse through. These channels are always open. From Table 4.1, what would you expect to happen to the concentrations of the sodium and potassium ions inside and outside the axon?
You would expect the concentrations to be equalised by sodium ions diffusing in and potassium ions diffusing out.

3 Sodium and potassium ions continuously diffuse through the open channel proteins in the axon membrane. However, the concentrations are not equalised. Explain why.
The sodium-potassium pumps operate all the time actively transporting sodium ions out of the axon and potassium ions in. For every two potassium ions moved in, three sodium ions are moved out.

The action potential

So far we have only considered an axon that is not conducting an impulse. What happens when an impulse passes along an axon?

You may recall from Chapter 3 in *AQA A-level Biology 1 Student's Book* that some channel proteins in membranes have **gates** that prevent facilitated diffusion when they are closed. Gated **ion channels** also play an important role in the action potential.

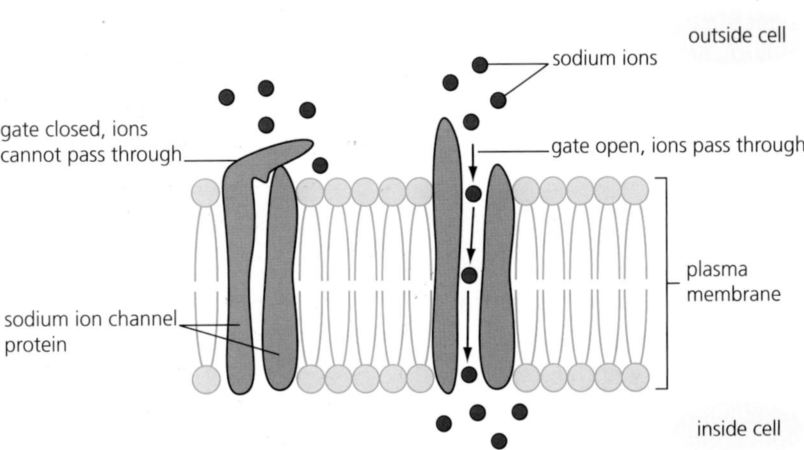

Figure 4.4 Gated sodium ion channels in the cell-surface membrane.

When an axon is inactive, the charges are in an unbalanced state all the time, with an excess of negative charges inside the axon and positive charges outside. As we have explained previously, this is the resting potential. ATP is used to maintain this uneven balance. However, this resting potential means that the axon is poised to conduct an impulse at any time, because of the steep concentration gradient of ions across the membrane.

Depolarisation A rapid temporary reversal of the resting membrane potential.

Stimulation of a neurone at a synapse causes a small change in the potential difference in the membrane close to the synapse. When the change in the potential difference is sufficient (this is called the **threshold potential**), the gates in the sodium ion channels change their shape to 'open' for a brief period. This allows sodium ions to diffuse in. It is estimated that each open channel allows about 20 000 Na^+ ions to diffuse through. This causes a sudden increase in positively charged ions inside the neurone. Instead of having a slight excess of negative ions, the inner surface of the membrane now acquires a positive charge of about 30 mV compared with the outside (so is represented as +30 mV). The change in the potential difference across the membrane is called depolarisation.

The sodium ion channels close as soon as the membrane potential reaches about +30 mV. At this point depolarisation is complete and the diffusion of sodium ions stops. Depolarisation causes potassium ion channels in the membrane to open, allowing potassium ions to diffuse out. They, however, diffuse out more slowly than the sodium ions diffuse in. The membrane **repolarises** because potassium ions continue to diffuse out. In practice, rather more potassium ions diffuse out so the drop in potential difference overshoots a bit, as you can see from Figure 4.6. This is called **hyperpolarisation**.

Repolarisation A return to the resting membrane potential.

Repolarisation does not immediately restore the concentration of ions inside the axon to their original state. The sodium-potassium pumps restore the balance by active transport. This maintains the resting potential and keeps the axon ready for another impulse.

TIP

The actual proportion of ions that move in and out during the passage of an impulse is tiny. Don't get the impression that all the sodium ions diffuse into the axon, or all the potassium ions diffuse out.

TEST YOURSELF

1 In the motor neurone in Figure 4.1, in which direction are impulses conducted?
2 What maintains the resting potential in an inactive neurone?
3 Describe the sequence of events during depolarisation.
4 How does repolarisation occur?

ACTIVITY

Analysing an action potential

Figure 4.5 shows the changes in the permeability of the membrane of an axon to sodium and potassium ions as an impulse passes a particular point. The graph also shows the changes in potential difference compared with the resting potential. These results were obtained from the neurone of a squid, which has large, unmyelinated axons where it is easier to take measurements than in a mammalian neurone.

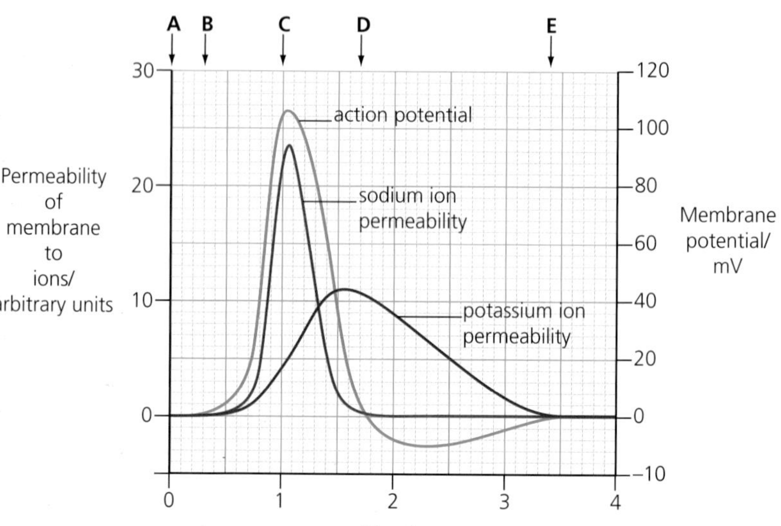

Figure 4.5 Graph of changes in permeability to ions of a tiny part of an axon membrane during an action potential.

1 Use the letters shown in Figure 4.5 to identify each of the following stages: resting potential, depolarisation, repolarisation, hyperpolarisation.
2 Explain how the change in permeability to sodium ions results in an action potential.
3 Why does the increase in sodium ion permeability start later than the change from the resting potential?
4 Explain the effect of the change in permeability to potassium ions.
5 Look at the curves on the graph showing the changes in the permeability of the membrane to sodium and potassium ions. How do these changes compare?
6 Suggest the advantage of differences in the permeability to sodium and potassium ions.
7 The resting potential in this axon was –70 mV. What was the maximum value of the potential inside the axon?
8 For how long was the potential above –70 mV?
9 For how long did hyperpolarisation last?

TIP

Remember that 1 ms is one-thousandth of a second.

How do impulses move along a neurone?

The depolarisation in one small section of the membrane sets off depolarisation of the next section of the neurone because the change in voltage in the membrane stimulates adjacent sodium ion channels to open. Therefore the action potential travels like a wave along the neurone Figures 4.6 and 4.7. This wave is what we commonly call a **nerve impulse**. Once started the impulse travels all the way to the next synapse.

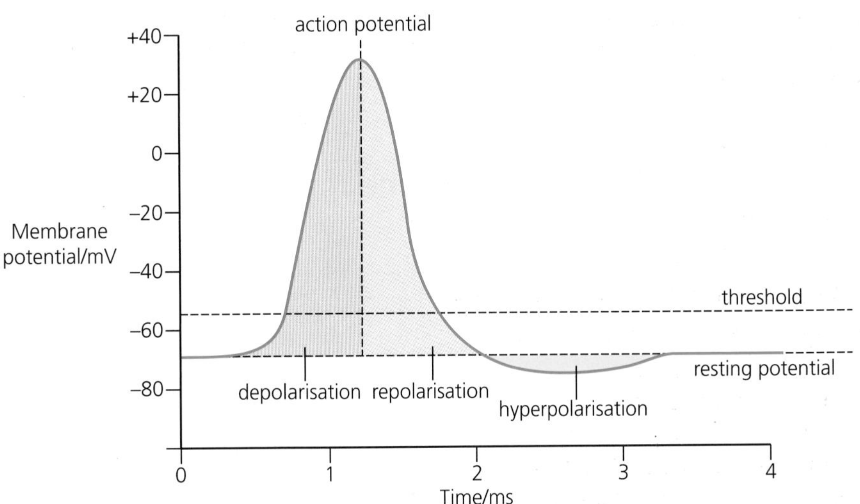

Figure 4.6 Changes in membrane potential as an impulse passes a particular point in a neurone.

The strength of the stimulus does not affect the impulse, as long as the stimulus is strong enough to get above the threshold value. This is known as the **all-or-nothing principle**. It is rather like squeezing the trigger on a rifle. As long as the trigger is pulled back far enough the bullet is fired. Pulling harder will not make the bullet go any further or faster.

However, a stronger stimulus does result in an increased frequency of impulses. This is how nerve impulses carry information. If receptors are stimulated more, they establish a larger generator potential (page 63), which in turn triggers more rapid impulses in the sensory neurone to which they are connected.

Refractory period The time following an action potential during which another action potential cannot take place, regardless of the strength of the stimulus.

There is always a time gap between impulses. This means that impulses are always **discrete**, in other words they never merge together. This is because after the action potential has reached its peak the ionic balance has to be restored during repolarisation. Only after repolarisation can another action potential be generated. The minimum interval between action potentials, and therefore between impulses, is the refractory period. The refractory period means that there is a maximum frequency at which nerve impulses can be conducted along the axon. This means that there is a limit to the strength of a stimulus that can be detected.

When the sodium ion channels close at the peak of the action potential there is a short period of about 0.5 ms when it is impossible for the channels to re-open. Therefore no stimulus can generate an impulse during this period. The full refractory period lasts until repolarisation is complete. In practice most neurones can only conduct about 300 impulses per second.

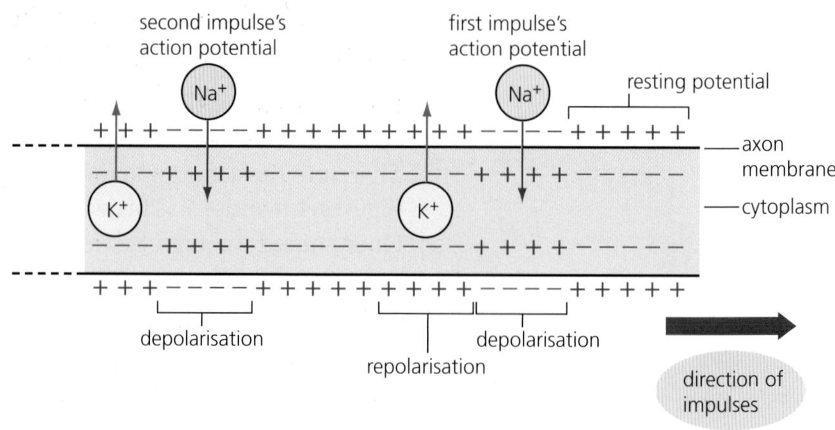

Figure 4.7 A change in charge at an axon membrane during the passage of impulses.

The myelin sheath

You may have noticed that so far we have talked about an axon as though the cell-surface membrane has a continuous and unobstructed outer boundary. But if you look back to Figure 4.1 you will recall that the axon of a motor neurone is myelinated. It is surrounded by a fatty myelin sheath that only has gaps at the nodes of Ranvier. Not all neurones in humans have myelin sheaths. **Non-myelinated neurones** conduct impulses in the way described above.

Myelinated neurones Neurones with axons surrounded by a series of Schwann cells.

In myelinated neurones, ions can only pass through the cell-surface membrane at the nodes of Ranvier. Therefore action potentials can only occur at these nodes. The effects of the depolarisation cause almost immediate depolarisation/action potentials at the next node. In effect action potentials jump rapidly from node to node, as shown in Figure 4.8. This is called **saltatory conduction**.

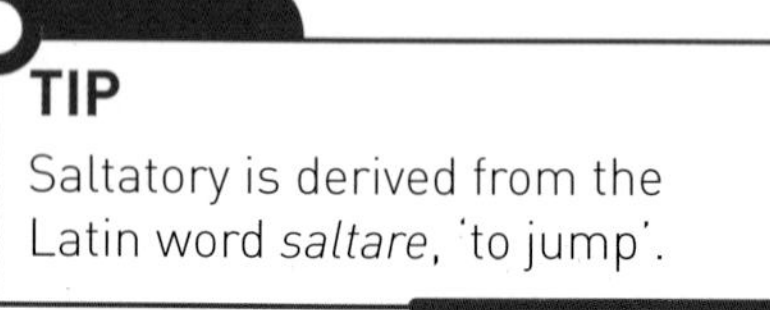

TIP

Saltatory is derived from the Latin word *saltare*, 'to jump'.

At each node there are large numbers of sodium and potassium channels. Since the nodes are about 1 mm apart, saltatory conduction has the great advantage of being much faster.

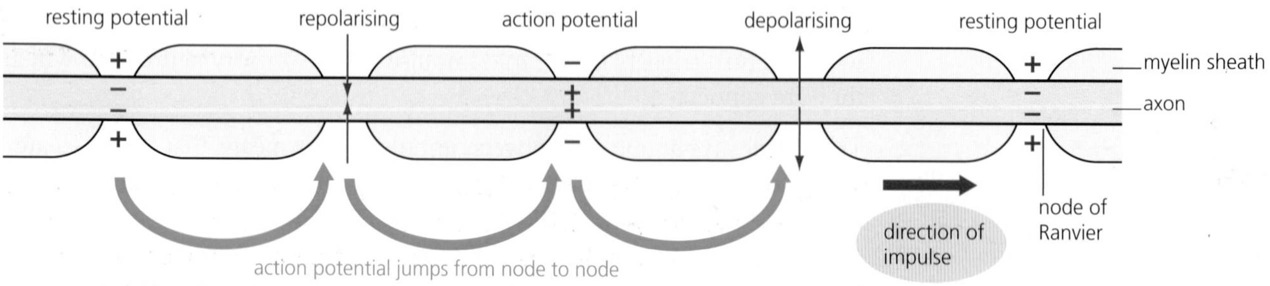

Figure 4.8 Saltatory conduction in myelinated neurones. Conduction of impulses along these neurones is approximately 50 times faster than in non-myelinated neurones.

Two other factors affect the rate at which neurones conduct impulses – temperature and the diameter of the axon. The rate of conduction is slower at low temperatures and in narrow axons.

In mammals, temperature rarely makes a significant difference since they normally maintain a fairly stable temperature. In animals whose temperature varies with the environmental temperature, cold conditions can considerably slow reaction times because the facilitated diffusion of sodium and potassium ions during an action potential will be slower.

The diameter of an axon makes a difference because the surface area to volume ratio of an axon is related to its diameter. Axons with a greater diameter have a larger volume of cytoplasm, containing more ions, which reduces their electrical resistance. This means that an action potential in one part of the axon pushes the next section to threshold more quickly. Nerve impulses are therefore conducted faster in larger diameter axons.

Apart from vertebrates, many animal species possess only non-myelinated neurones. These usually have a small diameter and are relatively slow conductors. Some animal species have evolved exceptionally large-diameter axons. Much of the early research on nerve conduction was done using giant axons from squid, marine animals closely related to octopuses. We saw some of the results of these investigations in Figure 4.7. These giant axons have a diameter of nearly 1 mm, over a hundred times greater than other axons in the squid. They can conduct impulses about 10 times as fast as normal motor neurones.

TIP

You do not need to recall details of the giant axons in squid and earthworms.

The giant axons extend the full length of the main part of the body, with branches to the muscles along the way. When the squid is startled, impulses are conducted at about 35 m per second along the giant axons. The muscles contract almost instantly, providing the sudden force that squirts water out backwards and jet-propels the squid away from danger. Earthworms also have large-diameter axons running the length of their body. These allow them to contract muscles rapidly and withdraw quickly into a burrow when threatened by a bird.

TEST YOURSELF

5 Explain the importance of the refractory period during impulse conduction.

6 Describe the all-or-nothing principle.

7 Give **three** factors that affect the speed of impulse conduction.

8 Explain what is meant by saltatory conduction.

Synapses

Synapses are the junctions between neurones. At a synapse, there is a small gap called the **synaptic cleft**, which is usually about 20 nm wide. This gap prevents electrical impulses passing directly from one neurone to another. Communication between neurones is by chemical **neurotransmitters** that diffuse across the cleft from one neurone to another. This is a slower process than the passage of an impulse along a neurone. Figure 4.9 shows a synaptic cleft.

TIP

Remember that 1 μm is 1000 nm.

TIPS

- Note that you should be careful not to refer to impulses or action potentials crossing synapses. It is information that crosses a synapse.
- Communication between neurones is by chemical neurotransmitter.

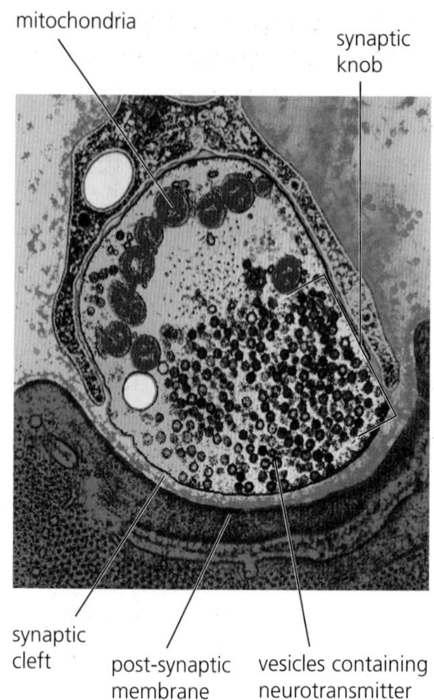

Figure 4.9 Coloured scanning electron micrograph of a synaptic cleft. The synaptic knob containing vesicles of neurotransmitter is clearly visible.

A synapse is a gap between the tiny branching end of a neurone and either a dendrite or the cell body of the next neurone. The branching end is always slightly swollen into a **synaptic knob** (Figure 4.10). A motor neurone may have approximately 8000 synapses on its dendrites and another 2000 directly on the cell body. It has been estimated that neurones in the brain have an average of approximately 40 000 synapses and that some have as many as 200 000. The number of possible pathways between neurones is phenomenal, which helps to explain the amazing complexity of the brain (see the introduction to this chapter).

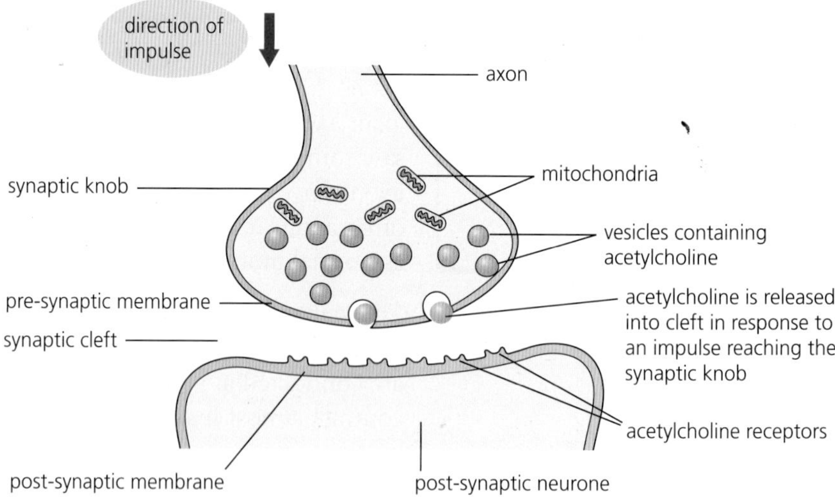

Figure 4.10 Structure of a synapse.

Cholinergic synapse A synapse that uses acetylcholine as a neurotransmitter.

So, what happens when an impulse reaches a synapse at the end of an axon? We will describe here what happens in a cholinergic synapse, which uses **acetylcholine** as a neurotransmitter. Acetylcholine (ACh) is the main neurotransmitter in synapses in the nerves outside the central nervous system, as well as in some parts of the brain. It is also used at the junctions between motor neurones and muscles. These **neuromuscular junctions** work in much the same way as synapses, as we shall see in Chapter 5. Many other neurotransmitters are used in the nervous system, but the principles of synaptic transmission are similar in all cases.

When an action potential arrives at the **pre-synaptic membrane of an axon** at the synaptic knob it causes calcium ion channels to open. The concentration of calcium ions (Ca^{2+}) in the fluid of the synaptic cleft is higher than the concentration inside the synaptic knob. Therefore calcium ions rapidly move by facilitated diffusion into the knob.

TIP

Don't confuse receptor molecules with receptor cells, such as the rods and cones that we studied in Chapter 3.

As you can see from Figure 4.10, inside the synaptic knob there are many vesicles containing acetylcholine. These vesicles are tiny droplets of acetylcholine surrounded by a membrane. When calcium ions enter, some of these vesicles move to the pre-synaptic membrane and fuse with it. As a result, the acetylcholine is secreted out into the cleft and diffuses across a very short distance to the **post-synaptic membrane** on the other side. This post-synaptic membrane has receptor proteins on its surface to which the acetylcholine molecules attach. Since the diffusion distance is so short, the acetylcholine diffuses quickly across the synaptic cleft (see *AQA A-level Biology 1 Student's Book* Chapter 3).

TIP

Don't refer to the binding site of a receptor protein as an active site. Although it has a complementary shape to the neurotransmitter, it is not an enzyme.

The receptor proteins have **binding sites** that are complementary in shape to acetylcholine molecules. Each receptor protein is located next to a sodium channel. When acetylcholine binds to a receptor protein, it changes shape and pushes the neighbouring sodium channel open. You will recall that the excess of sodium ions entering the neurone depolarises the membrane. If enough sodium ions enter, and the threshold is exceeded, an action potential is triggered (page 73). Once an action potential does start it travels all the way along the neurone as an impulse until it reaches the synapses at the other end. Figure 4.11 illustrates the sequence of the events that occur at a synapse.

1 Action potential opens gated channels. Ca^{2+} ions flow in.
2 ACh vesicles move to membrane and fuse with it.
3 ACh released into synaptic cleft and diffuses across it.
pre-synaptic membrane
Ca^{2+}
receptors
molecules of ACh
post-synaptic membrane
Na^+
K^+
4 ACh attaches to receptors and gated Na^+ and K^+ channels open.
5 Na^+ ions diffuse in faster than K^+ ions diffuse out. Action potential created in post-synaptic membrane.

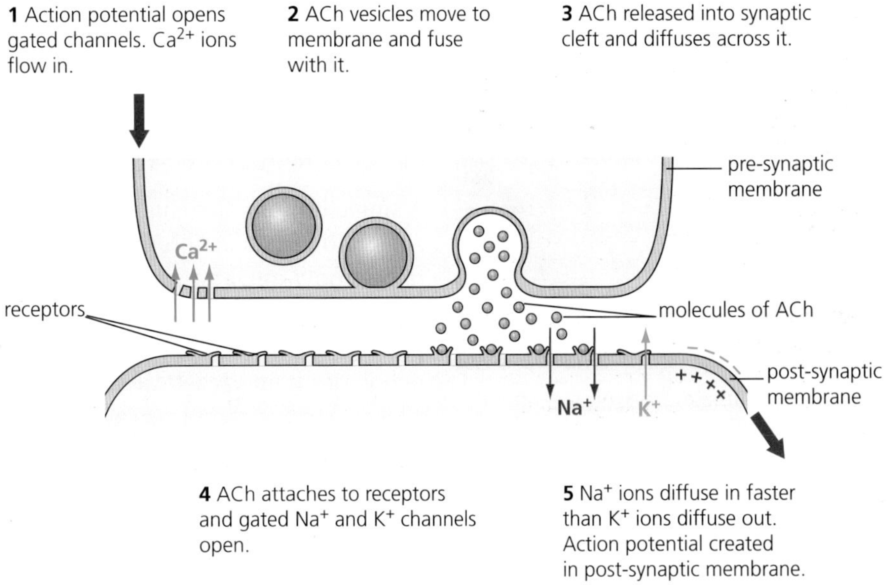

Figure 4.11 Sequence of events during an impulse transmission at a cholinergic synapse.

You will no doubt have realised that during transmission across a synapse changes take place that must be reversed if it is to keep on transmitting. Calcium ions are **actively transported back into the synaptic cleft**, ensuring that the concentration outside the membrane is always higher. The acetylcholine is removed from the receptors by an enzyme called **acetylcholinesterase** that hydrolyses it to acetate and choline. These are actively transported back into the synaptic knob where they are synthesised into acetylcholine once again. Vesicles are refilled, so the acetylcholine is recycled continuously.

The role of synapses

It takes much longer to describe what happens at a synapse than for the transmission to occur. But synapses do slow down the rate at which impulses pass through the nervous system. There is a delay of about 0.5 ms at each synapse. This is because the rate of diffusion is much slower than the rate at which a nerve impulse is propagated along a neurone, especially by saltatory conduction.

This does not mean that synapses are a bad thing. Synapses are vitally important as the points in the nervous system where the passage of impulses is controlled. Remember the all-or-nothing principle. If impulses could cross straight from one neurone to another without any control, stimulation at any point in the system would spread to all neurones.

Specialisation of functions in different parts would be impossible. Some of the control mechanisms involved are complex and there is still much to learn about them, especially about the processes involved in the brain.

One simple control method is that transmission across a synapse can only be in one direction. Only the synaptic knob contains the vesicles of neurotransmitter. Therefore neurotransmitter can only be released into the cleft from the pre-synaptic membrane. Once an action potential is established in the post-synaptic neurone it can only travel as an impulse in one direction because depolarisation always occurs in front of the action potential. This is called **unidirectionality**.

Summation

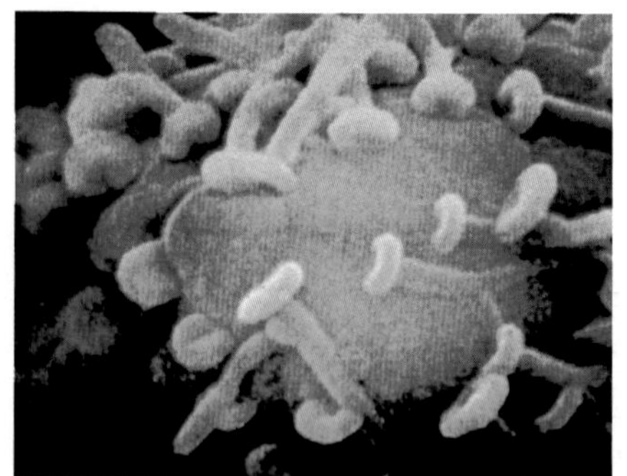

Figure 4.12 An electron micrograph showing a mass of synaptic knobs spread over the surface of the cell body of a neurone.

Impulses arriving at a synapse do not always result in impulses being generated in the next neurone. The electron micrograph in Figure 4.12 shows the synapses on the cell body of a motor neurone. As you can see there are many synaptic knobs all linking to this motor neurone.

A single impulse arriving at one synaptic knob is not likely to lead to the generation of an action potential in the post-synaptic neurone. It may release only one or a small number of vesicles of acetylcholine. If, as a result, only a few of the gated ion channels in the post-synaptic membrane are opened, not enough sodium ions will pass through the membrane to depolarise it and reach the threshold value. The acetylcholine only remains attached to the receptors for a very brief time since the acetylcholinesterase breaks it down within a couple of milliseconds. If several impulses reach the synaptic knob in quick succession, enough acetylcholine is released for the membrane potential to reach the threshold value. This causes the sodium ion channels to open and produces an action potential. Because the effects of several impulses are added together in a short time this process is called **temporal** (meaning 'time') **summation** (Figure 4.13).

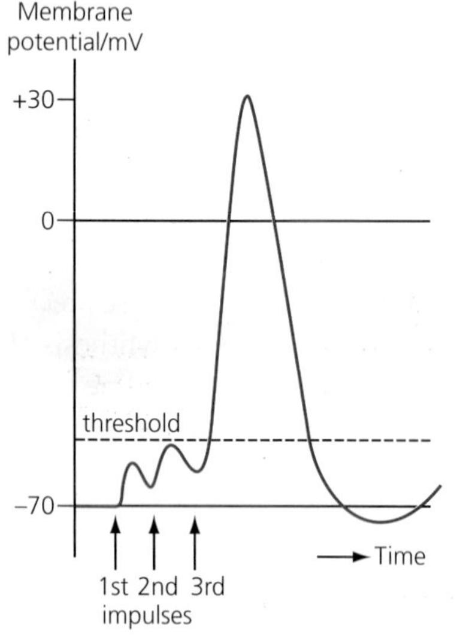

Figure 4.13 Temporal summation. You can see from the graph that the first impulse reaching the synapse fails to raise the membrane potential above the threshold for an action potential. The three impulses arriving in quick succession do produce an action potential.

A second way in which an action potential can be generated is when several impulses arrive simultaneously at different synaptic knobs stimulating the same cell body. The cell body may then be depolarised enough for an action potential to be generated in the axon. This is called **spatial summation**.

One advantage of summation is that the effect of a stimulus can be magnified. We have, for example, already come across summation in the retina (see page 66). It is also a way for different stimuli, or a combination of different stimuli, to trigger a response. Another effect of summation is that it avoids the system being swamped by impulses. A synapse acts as a barrier that only allows impulses to pass on if there is a significant input from receptors or other neurones, which may relate to the strength of the stimulus.

Inhibition

So far we have only described situations in which transmission across a synapse results in an action potential being generated in the post-synaptic neurone. Just as important are synapses where impulses are stopped altogether. Some neurotransmitters prevent action potentials being generated in the post-synaptic neurone. This is called **inhibition**. One way in which this occurs is that the inhibitory neurotransmitter stimulates gated potassium ion channels to open. Potassium ions therefore diffuse out of the cell body. If a cell body is affected by both excitatory and inhibitory synapses, the effect of sodium ions entering following stimulation by an excitatory synapse may be cancelled out by potassium ions leaving following stimulation by an inhibitory synapse. Therefore the membrane potential does not reach the threshold for an action potential to be produced.

Inhibition is very important in nervous circuits. It enables specific pathways to be stimulated, while preventing random impulses all over the body. For example in reflex actions it is desirable for neural pathways to produce a response only where it is needed. If your hand is on a hotplate you need your arm muscles to pull it away, but not your leg muscles to contract. The development of inhibitory pathways can be very important in learning specific skills, such as drawing and writing. If you watch a young child learning to draw you will see that at first the movements are uncontrolled. As the skill develops the movements become much more coordinated as the appropriate pathways are refined by inhibitory circuits.

Look at Figure 4.14. This shows in simplified form the cell body of a neurone and an axon with two synapses on it. Both are excitatory synapses.

If an action potential arrives at synapse A the membrane potential becomes more positive but not enough to reach the threshold value for an action potential. If an action potential arrives at synapse B at more or less the same time, the two effects are added together and the combined effect is now above the threshold value for an action potential to be generated in the axon.

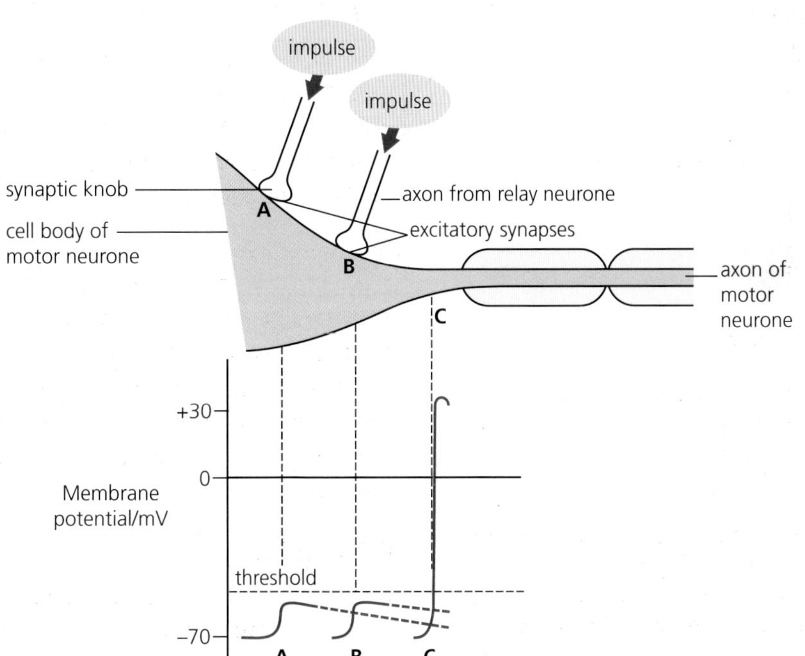

Figure 4.14 The cell body and an axon of a neurone with two excitatory synapses on it and a graph of the membrane potential that results.

Figure 4.15 shows a similar arrangement, but in this case synapse B is an inhibitory synapse. At an inhibitory synapse the effect is that the membrane potential actually decreases. In this example the two effects balance out the EPSP so the threshold value for an action potential is not reached.

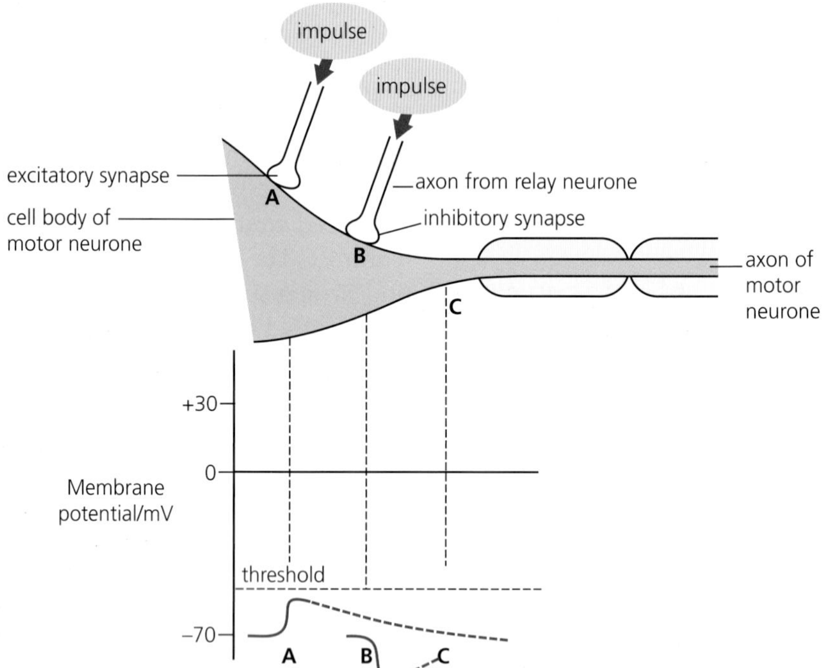

Figure 4.15 The cell body and an axon of a neurone with one inhibitory synapse and one excitatory synapse on it. The threshold value for an action potential is not reached.

TEST YOURSELF

9 Explain why synaptic transmission can only work in one direction.

10 How would inhibition of acetylcholinesterase affect synaptic transmission?

11 Explain the difference between temporal summation and spatial summation.

12 What does an inhibitory neurotransmitter do?

The effects of drugs

We mentioned at the beginning of this chapter that chemicals may have major effects on the working of the brain and other parts of the nervous system. Many of these substances are chemically quite simple. We often refer to these substances as **drugs**.

The term drugs, however, is quite vague and is applied both to substances that are used as cures and to mind-altering substances that may be damaging or potential poisons. Studies of drugs and poisons have revealed that many have quite specific effects on transmission at synapses or on neural circuits within the brain. In some cases, research has helped scientists to understand more about how the nervous system works and to offer ways of using certain drugs to treat nervous disorders.

TIP

You do not need to recall the names or modes of action of any specific drugs on synpases. The curare example illustrates the sort of information you might be asked to explain.

An example of a poison that affects synaptic transmission is curare. Curare is a naturally occurring poison that is found in the bark of trees in the Amazon forest. It has been used for centuries by South American Indians to coat the tips of arrows used for hunting. Curare molecules attach to the acetylcholine receptors on post-synaptic membranes, especially the membranes of neuromuscular junctions at the end of motor neurones. The curare molecules block the receptors and prevent sodium channels opening. Therefore impulses that pass down the motor neurones fail to stimulate muscle contraction, so any animals shot with such an arrow are paralysed.

EXAMPLE

The effect of drugs on synapses

Read the following passage carefully.

The brain has many different neurotransmitters. Over 50 have been identified. Many work in a similar way to acetylcholine. Many occur only in certain parts of the brain and are therefore restricted to particular functions. Some are inhibitory and some excitatory. Each binds to receptor molecules that are specific to the neurotransmitter. For example dopamine is a neurotransmitter that is important in parts of the brain concerned with control of muscle action. Parkinson's disease is a fairly common condition in elderly people in whom the neurones that produce the dopamine degenerate. The muscles become too tense, making delicate muscular movements difficult so that, for example, it is hard to pour a cup of tea without trembling. As the disease develops, walking and other everyday activities become harder. Two types of drug can be used in treatment. One is a substance from which dopamine is synthesised in the neurones. Another is an agonist, something that mimics the effect of dopamine by binding to the same receptors.

Another condition in which research suggests that dopamine plays a part is schizophrenia. This is a condition that affects about 1% of people at some time in their lives but most frequently as young adults. Schizophrenia causes people to behave untypically and often to suffer from delusions and hallucinations. It appears to have a variety of causes but one feature that has been discovered is an excess of dopamine receptors in the brain.

It is not easy to develop drugs that have specific effects in the brain because many substances are blocked from entering the brain by the blood–brain barrier. However, a number of drugs have been produced that help in schizophrenia by blocking the dopamine receptors. Dosage is very important and there is, as with many drugs, a danger of side effects.

1 Two types of drug can be used to treat Parkinson's disease. Suggest how each type of drug mentioned in the passage helps to treat Parkinson's disease.

Providing the brain with more of the precursor for dopamine helps the neurones that have not degenerated to manufacture more dopamine. The agonist attaches to the same receptors and has the same effect as dopamine.

2 Use information in the passage to suggest:

a) one type of drug that might help to reduce the symptoms of schizophrenia

Drugs that block the dopamine receptors should help.

b) one side effect that could result from the use of this drug.

A possible side effect is loss of muscle control, as found in Parkinson's disease.

Practice questions

1 The graph shows some changes in the permeability of an axon membrane during an action potential.

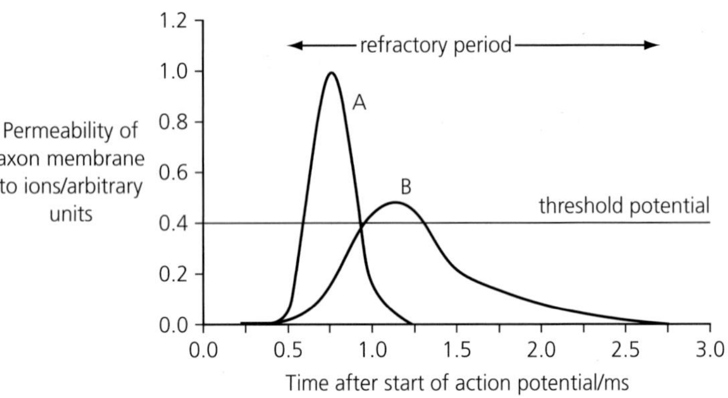

a) Name

i) ion A

ii) ion B. (1)

b) Explain the changes in permeability to ion A that take place between 0.5 and 1.0 ms. (2)

c) Explain how changes in permeability to ion B contribute to repolarisation of the axon membrane. (2)

d) Use the graph to explain what is meant by

i) the refractory period

ii) the all-or-nothing principle. (2)

2 The drawing shows a cholinergic synapse.

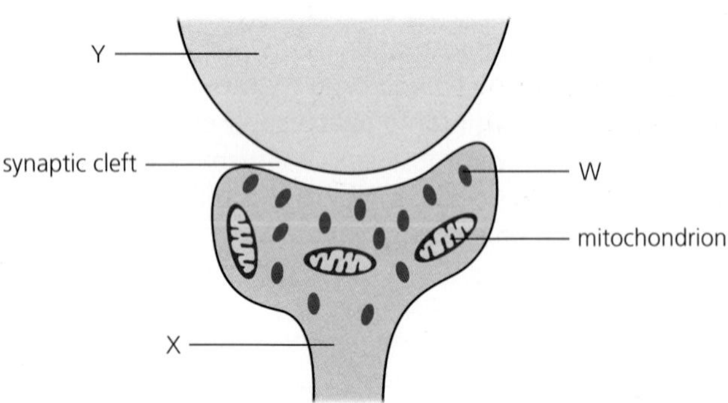

a) Name

i) organelle W

ii) the neurotransmitter used in this synapse. (1)

b) In which direction does synaptic transmission take place, X–Y or Y–X? Give a reason for your answer. (1)

c) Describe the sequence of events that occur at this synapse during synaptic transmission. (4)

d) Atropine is a chemical substance found in the deadly nightshade plant (*Atropa belladona*). It binds to and blocks acetylcholine receptors. Describe the effect of atropine on the sequence of events at a cholinergic synapse. (2)

3 Explain how a resting potential is established and maintained in a neurone. (5)

4 The diagram shows three different neurones synapsing with a motor neurone.

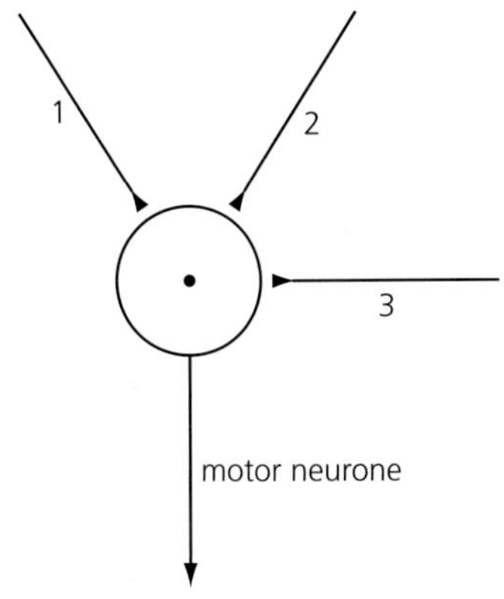

a) The motor neurone is myelinated. Describe and explain how this affects the conduction of action potentials. (3)

b) If an action potential arrives at synapse 1, no action potential occurs in the motor neurone. If action potentials arrive together at synapses 1 and 2, an action potential occurs in the motor neurone. Explain why. (3)

c) Synapse 3 releases an inhibitory neurotransmitter. Suggest what would happen if action potentials arrived at all three synapses at the same time. (2)

Stretch and challenge

5 Botox is a widely available but controversial cosmetic treatment. Discuss its mode of action and evaluate its use as a cosmetic treatment.

6 Examine the role of myelin in nerve tissue, by discussing the symptoms of multiple sclerosis. To what extent may it be regarded as an autoimmune condition?

5 Muscles and movement

PRIOR KNOWLEDGE

- Skeletal muscles work in antagonistic pairs.
- Muscles and skeleton interact to produce movement.
- Muscles are examples of effectors, and are stimulated by nerve impulses from motor neurones.
- An action potential results from depolarisation of the cell-surface membrane of a neurone.
- Action potentials spread across the cell-surface membrane of a neurone.
- Cholinergic synapses are those that use acetylcholine as a neurotransmitter.
- ATP hydrolase hydrolyses ATP to ADP and inorganic phosphate.

TEST YOURSELF ON PRIOR KNOWLEDGE

1 What is an antagonistic pair of muscles?
2 Explain why effective movement requires both muscle and an incompressible skeleton.
3 Describe the structure of a myelinated motor neurone.

Have you ever wondered why the meat of chicken breast is almost white and seemingly more or less bloodless, whereas the meat on a chicken leg is much darker and looks more like meat from mammals? After all, both are muscle. The obvious suggestion might be that farmed chickens can't fly, so they don't use their breast muscles for flying as a wild bird does. This, however, is not the complete answer. Chickens have been bred from wild junglefowl that live in the forests of South-east Asia. These birds spend most of their time scratching around for food on the ground, but when danger threatens they quickly fly to safety in a tree. Unlike most wild birds they never fly long distances. The breast muscles that flap their wings are only used in short bursts but have to produce enough force to lift quite a heavy bird off the ground at speed. On the other hand their leg muscles are used more or less all the time for wandering about the forest floor, scratching away at the leaf litter to find food and stopping them falling off their perch.

The junglefowl is adapted to its way of life by having different proportions of two types of muscle fibre in its muscles. The wing muscles mainly consist of fast fibres. These fibres contract rapidly but quickly become tired. They are well suited to a fast and short escape flight into a tree, lasting just a few seconds, but would not be suitable for sustained flight over a long distance. The muscle looks pale for two reasons. It has fewer capillaries containing blood with red blood cells than red muscle and therefore obtains its supplies of oxygen and glucose quite slowly. It also has very little of the red pigment myoglobin. Myoglobin is similar to haemoglobin and acts as a store of oxygen in muscles. Myoglobin also speeds up the absorption of oxygen from the capillaries.

The junglefowl's leg muscles contain mostly slow fibres. These have larger amounts of myoglobin and more mitochondria than the fast fibres. The muscles also have a denser network of capillaries. The combination of larger quantities of haemoglobin and myoglobin make the muscles appear a much darker red. Slow fibres contract more slowly than fast fibres but they do not fatigue anything like as quickly. They are therefore well suited for the walking, scratching and perching activities required of the legs.

It is not just junglefowl that have both fast and slow fibres. The skeletal muscles of vertebrates in general include a mixture of both muscle types. The proportions in different muscles vary according to the activity for which the muscles are adapted. For example, cheetahs and wolves pursue their prey in quite different ways (Figure 5.1).

Figure 5.1 a) Cheetahs depend on taking their prey by surprise with bursts of very high speed over short distances. They have a high proportion of fast fibres in their leg muscles. These muscles fatigue very quickly, so if a cheetah does not catch its prey within a few metres it has to give up.

b) Wolves work as a pack and may follow their prey, which is often much larger than themselves, over long distances before singling out a weak member of a herd. Their leg muscles are adapted for endurance by having a high proportion of slow fibres.

In this chapter we will find out how muscles contract and how they use the ATP that is needed for contraction. We will also see how their activity is coordinated by the nervous system.

Skeletal muscles

The human body has over 600 skeletal muscles. About 40% of the mass of an average man is skeletal muscle. Humans have both fast and slow fibres in their muscles. Some of our muscles have particularly high proportions of one type. For example the muscles that move our eyes and eyelids have mostly fast fibres. These operate rapidly but do not need to keep up their movements for long periods. The muscles in our back have a large proportion of slow fibres. They are required to remain contracted for long periods so that we can stand or sit upright but they rarely have to move quickly. Most people have roughly equal proportions of each type of fibre in their main leg and arm muscles, although some individuals have distinctly more of one sort. Studies of highly trained sportspeople and athletes have shown that the most successful often have a bias that benefits their type of activity. The table shows some results from one study of trained athletes.

Table 5.1 Results of a study of muscle fibre type in the leg/arm muscles of athletes trained in different disciplines.

Sport/discipline	Mean percentage of slow fibres in leg or arm muscles	Mean maximum oxygen consumption/cm^3 per kg body mass per minute
Cross-country skiing	79	82
Marathon running	65	79
Swimming	56	63
Fit students	53	55
Weightlifting	51	48
Running 800 m	48	65
Sprinting 100–200 m	42	55

You can see from Table 5.1 that athletes in endurance events, such as cross-country skiing and marathon running, have high percentages of slow fibres. These enable them to maintain activity for long periods without fatigue. These muscle fibres are well supplied with blood capillaries that provide a good oxygen supply. In contrast, athletes trained for short bursts of activity, such as weightlifting and sprinting, have lower proportions of slow fibres, which, of course, means that they have more fast fibres. These contract rapidly and can carry out intense activity for a few seconds. However, the rate at which the muscles are re-supplied with glucose and oxygen is low, because there are relatively few capillaries, so it is only possible to keep up this rate of muscle contraction for a short time. The average running speed of an Olympic 100 m sprinter is almost double that of a marathon runner.

An obvious question is whether athletes are born or bred. There is clear evidence that some people are born with unusually high proportions of one type of muscle fibre and that this tendency is inherited. However, training can make one type of fibre develop much more than the other. For example, weightlifting routines can increase the size of fast fibres. The number of fibres in a muscle does not increase but the number and length of the contractile units inside the fibres can increase considerably. Training can also increase the numbers of mitochondria and capillaries in the muscles. So, although a person may be born with muscles that are more suited for sprinting or for long-distance running, appropriate training can make a big difference to their chances of success (Figure 5.2).

Figure 5.2 Inheritance of desired muscle fibre types and training methods can make an athlete more successful at their chosen discipline. a) For a short race at maximum speed the sprinter benefits from a high proportion of fast fibres in his leg muscles. b) It is an advantage for a long-distance runner to have a high proportion of slow muscle fibres that fatigue less easily.

Muscle mechanics

The muscular system is very complex: many different muscles cross each other in many directions. Figure 5.3 shows some of the muscles on the front (anterior) of the body.

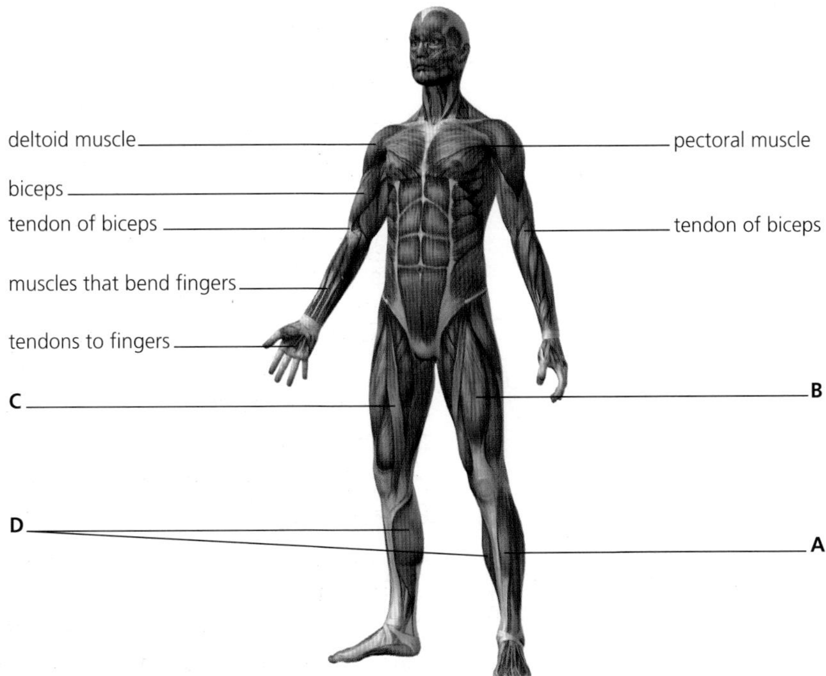

Figure 5.3 Some of the anterior muscles in the human body.

You will probably be familiar with the biceps and the triceps muscles in the arm from your GCSE course. These are used to flex and extend the arm at the elbow, respectively. But as you can see in Figure 5.3 there are several other muscles that enable us to raise and twist our arms as well as making delicate movements with our hands. The biceps and triceps are attached to

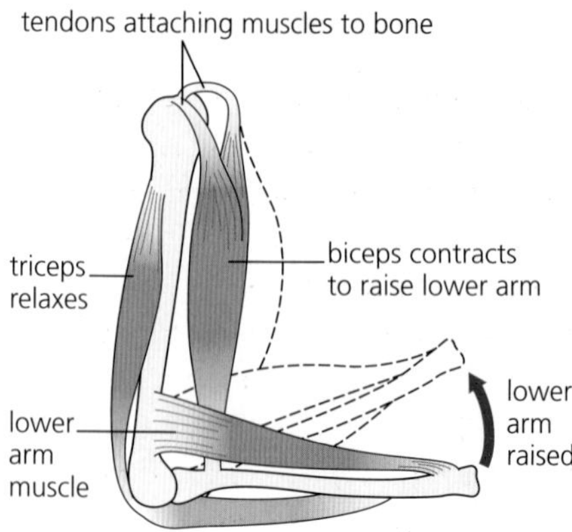

Figure 5.4 The triceps and biceps in the arm work antagonistically; here the lower arm is being raised.

bones by thick tendons made of strong fibrous connective tissue. A tendon has to be flexible, but not elastic. It must not stretch when a muscle is contracting and pulling on a bone. You can see in Figure 5.4 the tendon that attaches the lower end of the biceps to the radius bone in the lower arm. The upper end of the biceps has two tendons attaching it to the shoulder blade. These cannot be seen in Figure 5.3 as the deltoid muscle hides them.

Some muscles have very long tendons. The muscles that bend the fingers are situated in the arms just below the elbows. Their tendons stretch across the wrists and the palms of the hands. There are also muscles that attach directly to a bone instead of having tendons. For example, if you look at the pectoral muscle in Figure 5.3 you will see that it is attached all the way down the breastbone. The muscles generate force as they contract and pull on the bones of the skeleton, and this produces movement.

Muscles can only contract to move parts of the body. They can only *pull*; they cannot *push*. For this reason muscles in general operate as pairs, with one pulling in one direction at a joint and a partner that pulls the opposite way. This is called **antagonistic muscle action**. The biceps and triceps act as an antagonistic pair of muscles (Figure 5.4). When the biceps contracts the triceps relaxes so the lower arm is pulled upwards. When the biceps contracts it also pulls the triceps back out to its starting length so that it is ready to contract again. To straighten the arm again the triceps contracts and the biceps relaxes. When the biceps or triceps contracts, it pulls against the bones in the lower arm and because these bones can't be stretched, the arm is flexed around the elbow joint.

Extension

In practice, when you lift a heavy object in your hand, many muscles are involved. For example, another muscle connects from the lower end of the bone in the upper arm to the lower arm near the wrist. Several muscles are used to enable the hand to grip and the wrist to rotate. Others help to keep the shoulder steady. As the biceps contracts, the triceps slowly relaxes, so the movement takes place smoothly. This involves the brain, and in particular the cerebellum, in a highly complex process of coordination. Receptors in the muscles, tendons and joints constantly send impulses to the brain indicating the degree of stretching of muscles and the positions of tendons and joints. Reflex adjustments ensure that the operation is a continuously smooth process. Damage to the cerebellum, for example a stroke or tumour, can result in clumsy and jerky movements.

TEST YOURSELF

1 Which of the other muscles in the leg labelled B to D is antagonistic to muscle A in Figure 5.3? Explain your answer.
2 What are tendons and what is their function?
3 What sort of muscle fibres are mainly found in the eye muscles? What is the advantage of having this type of fibre here?
4 Describe and explain the relationship between mean percentage of slow fibres and mean maximum oxygen concentration in Table 5.1.
5 Why might people with a higher proportion of slow fibres in their muscles be more suited to long-distance running?

The structure of a muscle

The 'cells' that make up skeletal muscles are so different from most cells that they are referred to as **muscle fibres**. Each fibre is long, on average about 30 mm, but in some cases they can be as much as 300 mm. Part of a fibre is shown in Figure 5.5. You can see that it is surrounded by a cell-surface membrane, called the sarcolemma. Beneath the sarcolemma is cytoplasm, containing many mitochondria. There is also a specialised network of tubules, the sarcoplasmic reticulum, that store calcium ions. We shall see later how the sarcoplasmic reticulum is important in muscle contraction. Dotted within the sarcoplasm along the length of the fibre are many nuclei. Most of the fibre is filled with much thinner fibres, called **myofibrils** (Figure 5.5).

TIP

The names of many parts of muscles begin with the prefix 'sarco'. Examples include sarcolemma, sarcoplasm and sarcoplasmic reticulum.

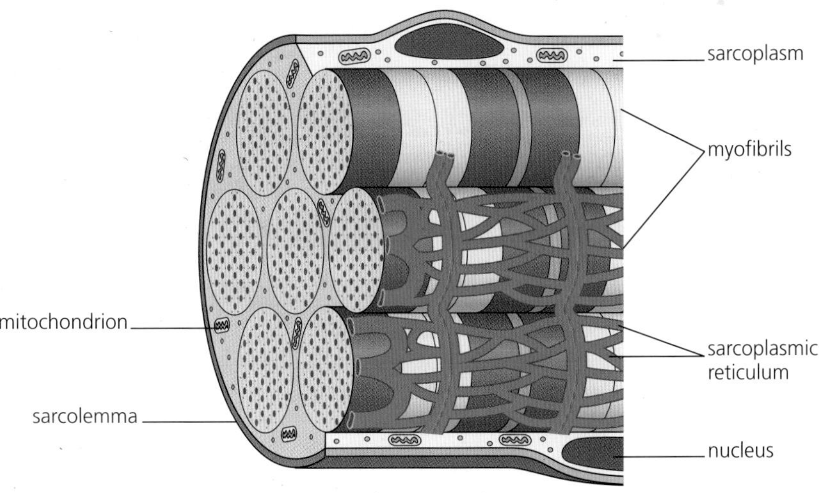

Figure 5.5 Section through a skeletal muscle fibre.

The muscle fibres are packed together in bundles surrounded by a sheath of connective tissue. It is these bundles that can get stuck in your teeth when eating tough meat. Each bundle is well supplied with blood capillaries and branches of motor neurones. In larger muscles, such as the biceps, many bundles are wrapped together in a thick and tough connective tissue layer. This connective tissue is continuous with the tendon that attaches it to the bone (Figure 5.6).

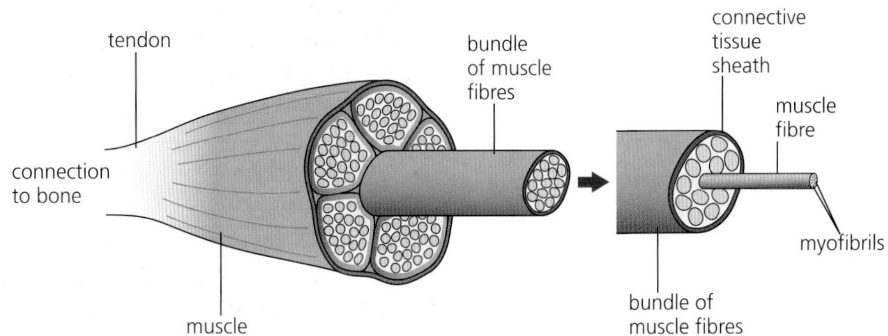

Figure 5.6 The arrangement of skeletal muscle fibres in a muscle.

Myofibrils

With an optical microscope very little detail of muscle fibres can be seen. Dark bands are just visible, as you can see in Figure 5.7a, and for this reason skeletal muscle is often called striated muscle. An electron microscope enables the structure of the myofibrils inside a muscle fibre to be seen. Figure 5.7b shows an electron micrograph of part of a muscle fibre showing short sections of five myofibrils.

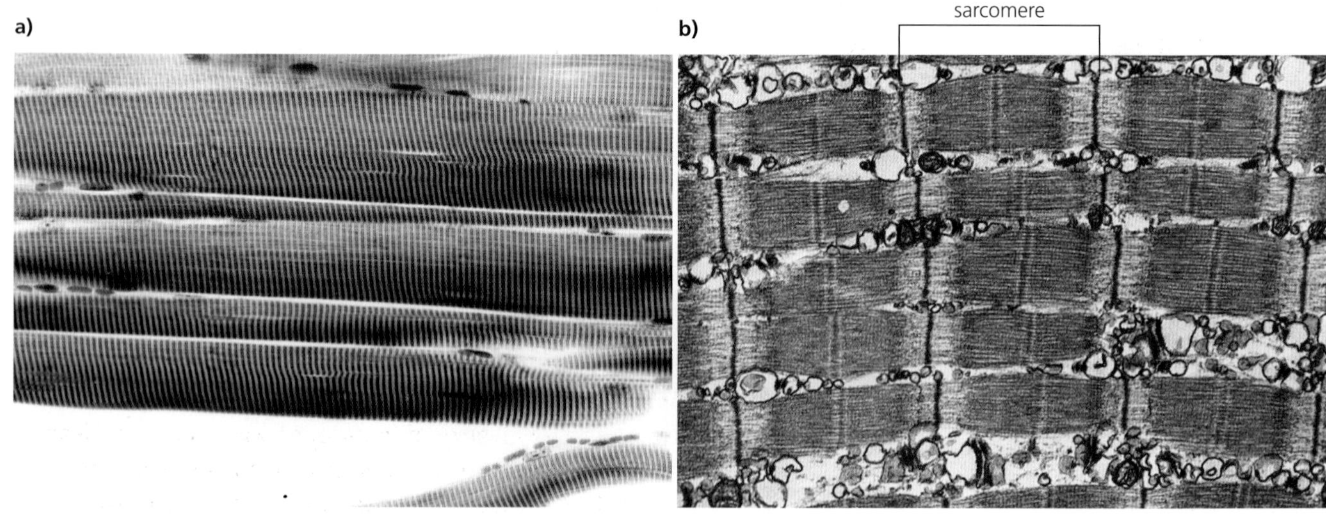

Figure 5.7 a) Skeletal muscle fibres viewed through an optical microscope, magnification approximately ×600; b) myofibrils in a skeletal muscle fibre viewed with an electron microscope with magnification approximately ×10 000.

TIP

Electron micrographs are black and white images. Figure 5.7b is a false-colour electron micrograph, where colour has been added using image-processing software.

Look carefully at Figure 5.7b. You will see that each myofibril has dark bands across it. The section between a pair of dark bands is called a sarcomere. In the centre of each sarcomere is a paler (blue) band, and to either side of this are broad darkish (green) areas. You will notice that there appear to be many thin lines running along the length of the sarcomere. The explanation for this appearance is shown in Figure 5.8.

Each sarcomere contains thin filaments of **actin** and thick filaments of **myosin**. The different bands and zones are known by the letters shown in Figure 5.8.

Sarcomere The basic unit of contraction in a myofibril. The sarcomere is defined by the distance between two Z lines.

TIP

The decreasing size of these structures follows their alphabetical sequence: fibre, fibril, filament.

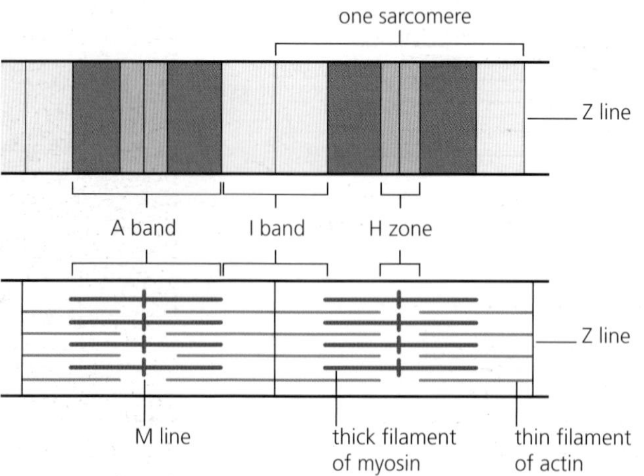

Figure 5.8 The arrangement of actin and myosin filaments in two sarcomeres.

You can see that the thin actin filaments are attached to the dark Z lines at the ends of each sarcomere. The thick myosin filaments are attached at the centre of the sarcomere to the M line and the tails of myosin molecules facing in opposite directions are bound to each other. Where there are myosin filaments the sarcomere looks darker than at the ends where there are only actin filaments. The darkest zones are where the actin and myosin filaments overlap. If you were to cut across a fibril at different points, you would see the mixtures of thin and thick filaments shown in Figure 5.9.

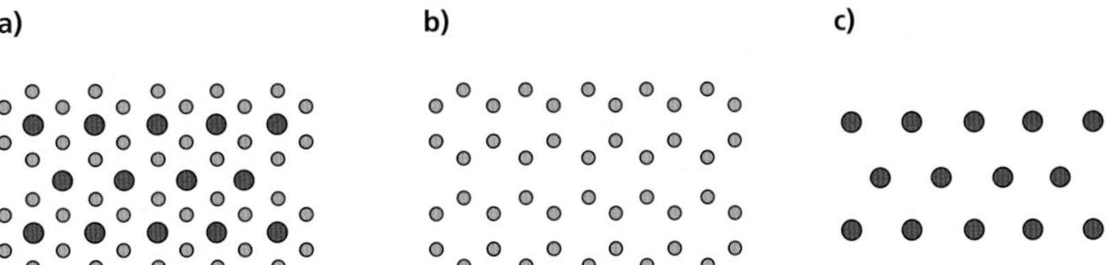

Figure 5.9 Actin and myosin filaments in section at different points along a sarcomere: a) where actin and myosin overlap within the A band, b) in the I band and c) in the H zone.

How do the muscle fibres contract?

Not until the development of electron microscopes was there any clear idea of how muscle contraction works. Then two scientists at London University, called Jean Hanson and Hugh Huxley, came up with the sliding filament hypothesis. They suggested that when a muscle fibre contracts the thin actin filaments slide between the myosin filaments. This is shown in Figure 5.10. They obtained evidence for their hypothesis by examining electron micrographs of the light and dark bands in contracted and relaxed myofibrils. If the actin filaments are sliding between the myosin filaments you would expect the light I bands to get smaller as the sarcomere contracts. This is exactly what Hanson and Huxley found. One feature that can be seen by electron microscopes is that there are 'cross-bridges' connecting the myosin and actin filaments. These can be seen in the scanning electron micrograph in Figure 5.11.

TIP

During contraction, neither the actin nor the myosin filaments get shorter. Instead, the actin slides between the myosin and the region of overlap becomes larger.

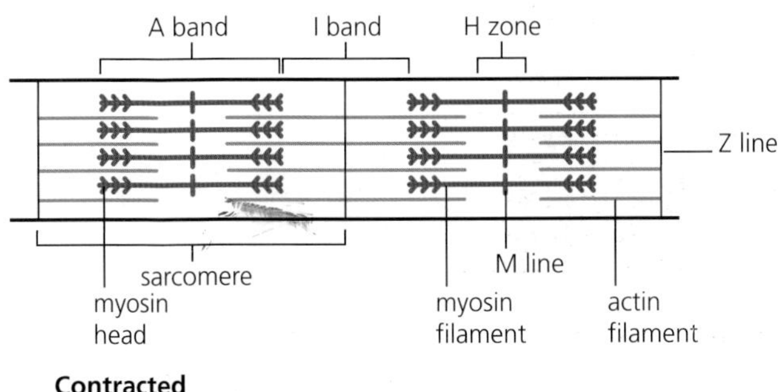

Figure 5.10 Uncontracted (relaxed) and contracted sarcomeres.

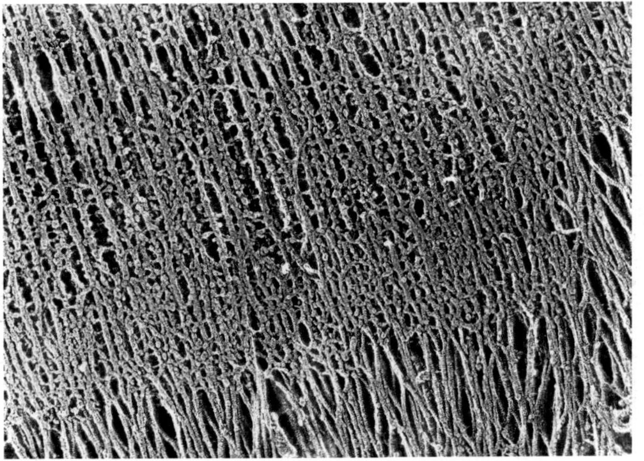

Figure 5.11 Electron micrograph of a myofibril showing the cross-bridges between myosin and actin filaments.

In simple terms the bridges are like tiny hands extending from the myosin towards the actin. These 'hands' can attach to the actin and then bend over. As a hand bends the actin is pulled along for a short distance before the hand lets go and straightens up again. It is then able to attach to the actin filament a bit further along and give it another pull. As you can see from the diagram in Figure 5.12, these 'hands' are referred to as myosin heads. The summary below describes the process in rather more scientific terms.

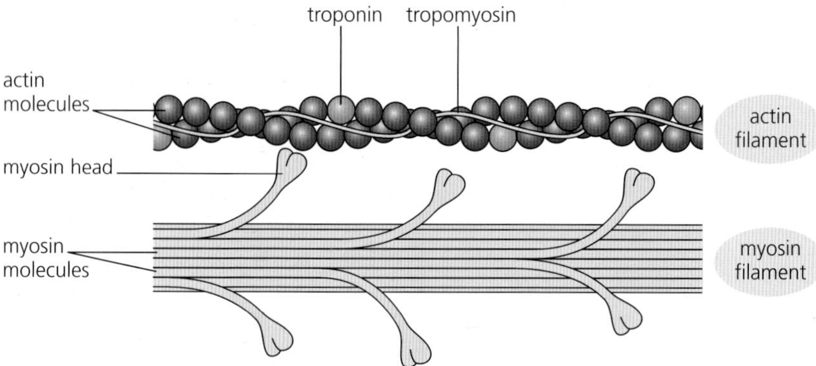

Figure 5.12 Actin and myosin filaments are quaternary proteins, each consisting of many individual actin and myosin molecules.

Actin and myosin filaments are quaternary proteins (see *AQA A-level Biology 1 Student's Book*, Chapter 1, page 13) which means that they each consist of many individual actin and myosin molecules held together (Figure 5.12). This is why the myosin filaments shown in Figure 5.10 have many heads and why actin filaments have many binding sites. You can see that the heads at each end of the myosin filaments in Figure 5.10 point in opposite directions. Figure 5.12 only shows one end of a myosin filament close up. When a myofibril is in a relaxed state the myosin heads protruding from the myosin filaments are detached from the actin filaments, although still very close, as shown in Figure 5.12. As well as molecules of actin the actin filaments have two other proteins, **tropomyosin** and **troponin**. In the relaxed state the tropomyosin covers the sites on the actin filaments to which the myosin heads can attach.

When nerve impulses reach a neuromuscular junction, the following steps occur, resulting in contraction of the muscle. These are illustrated in Figure 5.13.

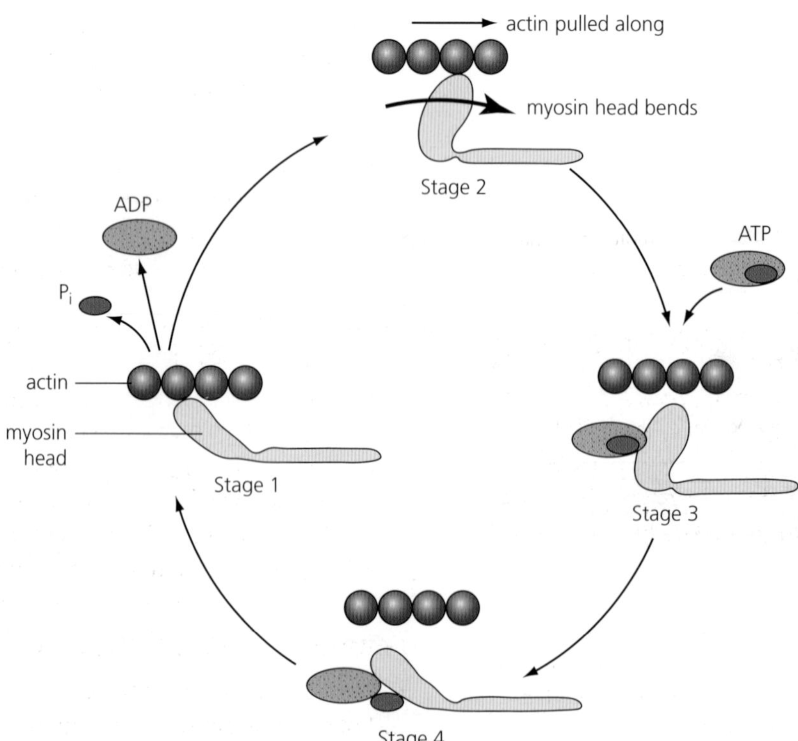

Figure 5.13 Stages in the cycle of actinomyosin bridge formation.

Stages in the cycle of actinomyosin bridge formation.

- A wave of depolarisation travels across the muscle cell-surface membrane, or sarcolemma, and calcium ion channels open.
- Calcium ions (Ca^{2+}) diffuse from the sarcoplasmic reticulum and also diffuse into the muscle cell across the sarcolemma.
- Calcium ions attach to the troponin and cause a change of shape, making the tropomyosin twist away from myosin-binding sites on the actin.
- The myosin heads attach to the binding sites forming **actinomyosin bridges**.
- Actinomyosin bridge formation causes the myosin molecules to spontaneously bend, releasing ADP and inorganic phosphate (P_i) and pulling the actin molecule along for short distance.
- Another ATP molecule attaches to each myosin head, leading to a change in shape, and it separates from the actin (but always in the same direction).
- ATP hydrolase hydrolyses the ATP to ADP and inorganic phosphate (P_i), resulting in straightening of the myosin molecule. This is called the recovery stroke.
- Each head is now able to repeat the process by attaching to another binding site and sliding the actin along a bit more.

TIP

You would think that ATP hydrolysis would be used to drive the power stroke. However, it is actually used to drive the recovery stroke and allow the myosin head to reposition for the next cycle.

The speed of this process is hard to imagine. Each myosin head can bind and detach up to 100 times a second. Each myosin filament is made up of many myosin molecules held together by their tails, and during contraction their many heads are all flicking to and fro. They do not all bind to actin filaments at the same time, which would make the sliding action jerky. However, the most important thing to understand is that the myosin heads at each end of the myosin filaments point in opposite directions, so when they bind they pull both sets of actin filaments in the sarcomere towards one another at the same time. This explains why it is the actin filaments and not the myosin filaments that slide, and why sarcomeres shorten when muscle contracts (see Figure 5.10). The combined effect of all the sarcomeres and therefore all the myofibrils contracting together can produce remarkably large forces in muscles at considerable speed.

While the calcium ions remain in the cytoplasm, the myosin heads can continue to bind to and detach from the same site, so the fibre stays contracted. This process requires a continuous supply of ATP. When nervous stimulation ceases the calcium ions are reabsorbed into the sarcoplasmic reticulum by active transport. The tropomyosin moves back, covering the binding sites, and the myofibrils relax. However, the sarcomeres remain the same length until an outside force such as an antagonistic muscle pulls the actin filaments back out from between the myosin filaments.

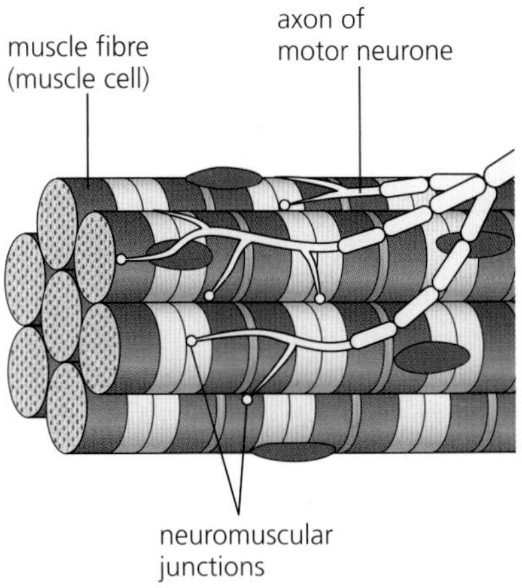

TEST YOURSELF

6 Explain why the myofibrils of skeletal muscle appear striated.
7 Use Figure 5.13 to describe the structure of an actin filament.
8 What do you expect to happen to the A band and the I band during contraction? Explain your answer.
9 What is ATP used for during the cycle of actinomyosin bridge formation?
10 What is the role of the sarcoplasmic reticulum in a muscle fibre?

Figure 5.14 Neuromuscular junctions between a motor neurone and a group of muscle fibres.

How are skeletal muscle fibres stimulated to contract?

A skeletal muscle contracts when stimulated by nerve impulses reaching the end of a motor neurones. Figure 5.14 shows the branches of a motor neurone connecting with muscle fibres. The branches spread across the surface of the muscle fibres and at various points terminate at a flattened structure called an endplate close to the sarcolemma of a muscle fibre. These points are called **neuromuscular junctions**. Neuromuscular junctions are very similar to synapses and work in much the same way. It might be useful to look back at the section on synapses in Chapter 4, particularly at Figure 4.10.

As you can see in Figure 5.15, the synaptic knob contains vesicles of acetylcholine. All neuromuscular junctions in skeletal muscles are cholinergic. When action potentials reach the pre-synaptic membrane, calcium ion channels open and calcium ions diffuse in across the membrane. This leads to vesicles containing acetylcholine fusing with the membrane. The acetylcholine diffuses into the cleft. This is stimulated by calcium ions that enter the pre-synaptic membrane from the cleft through channel proteins. The acetylcholine diffuses across the cleft to the sarcolemma of the muscle fibre, where it binds to protein receptors. This causes the sarcolemma to be depolarised in exactly the same way as a post-synaptic membrane is depolarised (see Chapter 4 page 73). Assuming the threshold value is reached, action potentials spread over the muscle cell-surface membrane and into the sarcoplasmic reticulum. This causes calcium ions to be released from the sarcoplasmic reticulum and to enter the muscle cell across the sarcolemma. As described earlier, the calcium ions attach to troponin and allow the binding sites on actin to be exposed.

Figure 5.15 A neuromuscular junction.

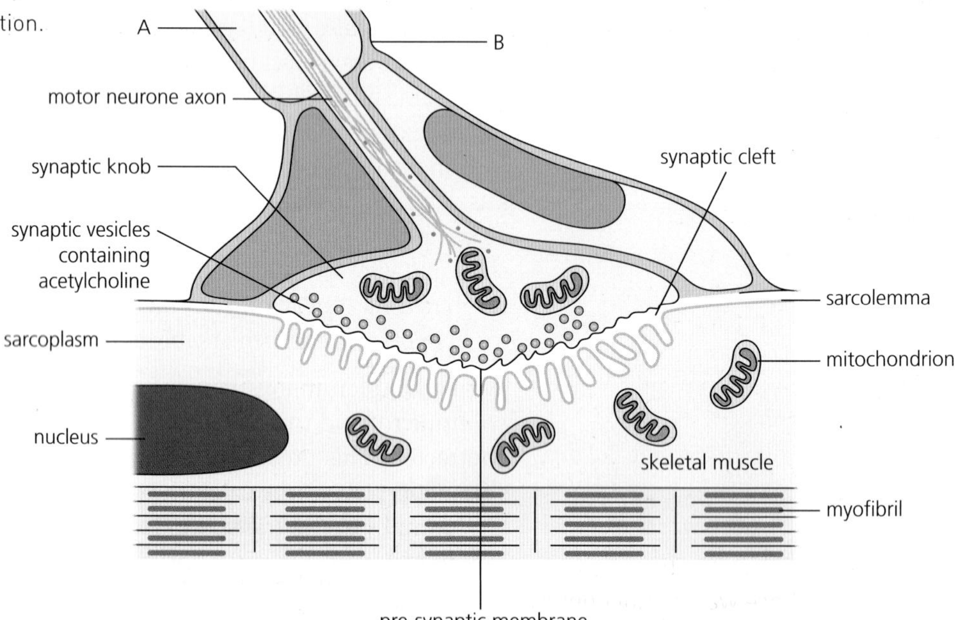

ATP hydrolysis is required for muscle contraction

A supply of ATP is required both for the return movements of the myosin heads that make the actin filaments slide and for the return of calcium ions into the reticulum by active transport. Resting muscle fibres only store enough ATP to maintain contraction for a very short time. The supply will only last for 3 or 4 seconds of intensive exercise. The muscle fibres contain mitochondria that can generate ATP by respiration of glucose. However, full aerobic respiration is relatively slow and even anaerobic respiration takes about 10 seconds to begin to produce some ATP.

But muscle fibres also contain stores of **phosphocreatine**. This can be used to produce ATP very rapidly by transferring a phosphate ion to ADP and thus replacing the ATP that has been hydrolysed.

$$\text{ADP} + \text{phosphocreatine} \longrightarrow \text{ATP} + \text{creatine}$$

> **TIP**
>
> Summation can occur at neuromuscular junctions as well as at ordinary synapses.

The amount of phosphocreatine available is limited and depends on the type of fibre but enables the muscles to keep going until more ATP can be supplied from the mitochondria. Nevertheless this means that really intense muscle activity can only be kept up for quite a short time. This may be enough for a trained athlete to sprint 100 m in 10 seconds, but such a level of activity could not be sustained for a much longer event. For prolonged activity the rate of muscle contraction has to be matched by the rate at which ATP can be provided from a combination of aerobic and anaerobic respiration.

Extension

How are muscle movements controlled?

When you pick up a heavy bag of books you may use much the same muscles as when you pick up a pen. But obviously you do not need to pick up the pen with the same force as the bag. If you carry the bag home you will have to hold it in the same position by keeping your muscles contracted. Individual muscle fibres are like neurones in that they work on an all-or-nothing principle. They either contract fully or, if the stimulus is below the threshold value, they do not contract at all. The strength of a muscle contraction therefore depends on how many fibres contract and how often the stimulus is repeated. In skeletal muscles each motor neurone has branches to a group of muscle fibres called a **motor group.** This group may have several hundred fibres. When an action potential is conducted along the neurone, all the muscle fibres in the group contract. Different motor neurones in a motor nerve have slightly different thresholds, so how many groups of fibres are activated depends on the strength of the initial stimulus.

ACTIVITY

Muscle contraction

Figure 5.16 shows the effect of applying a *single* stimulus to a motor nerve connecting to a small muscle. The stimulus produced single action potentials in a few of the motor neurones in the nerve.

1 What fraction of a second is 100 ms?
2 For how long was the muscle contracting?
3 Before the muscle started to contract there was a short delay. Suggest what was happening during this period.
4 What stops the muscle contracting?

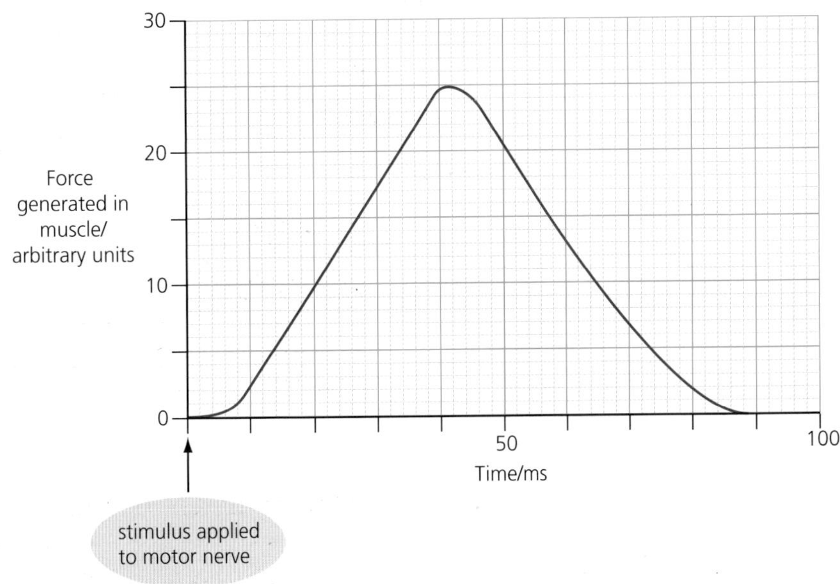

Figure 5.16 Graph showing the effect of a single stimulus on a small skeletal muscle.

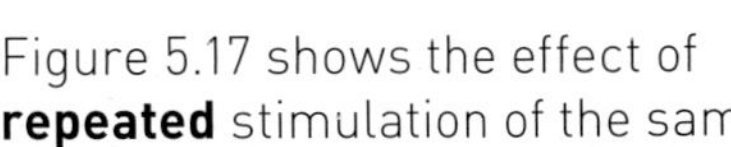

Figure 5.17 shows the effect of **repeated** stimulation of the same motor nerve. The nerve is stimulated at the times shown by the arrows A, B, C and D.

5 Describe what happens when the muscle is stimulated several times in quick succession. Suggest a name for this effect.
6 Suggest how the repeated stimulation causes the muscle to contract with a greater force.
7 After the sarcolemma of a skeletal muscle fibre has been depolarised and an action potential has been produced, there is a delay of about 5 milliseconds before it can be depolarised again. This is similar to the period of delay in a neurone. What is this period called?

8 Skeletal muscle can start to contract again when relaxation is only about one-third complete. Cardiac muscle has a much longer delay of about 300 milliseconds. Use your knowledge of the cardiac cycle from *AQA A-level Biology 1 Student's Book* to suggest the advantage of this longer delay.

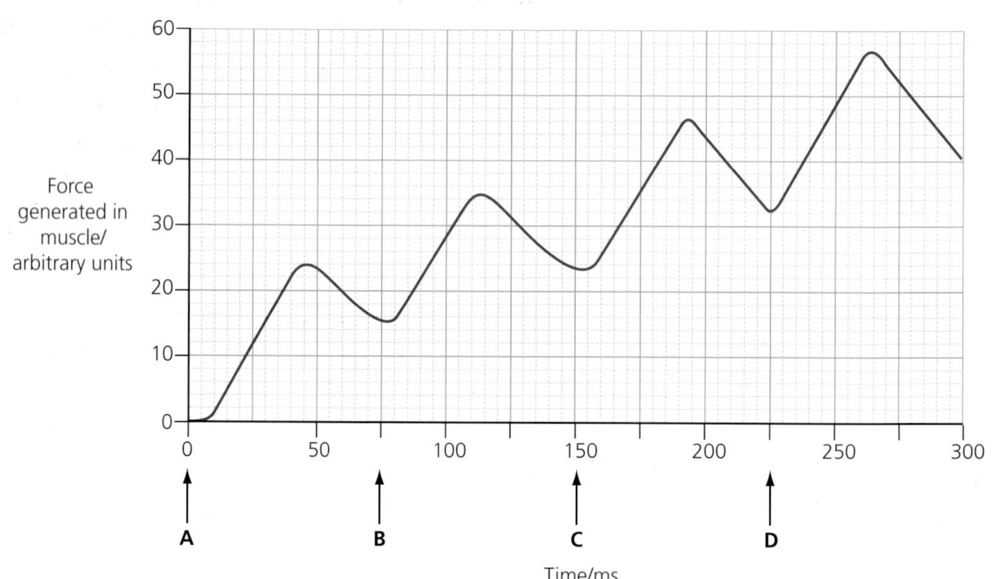

Figure 5.17 Graph showing the effect of a repeated stimulus on a small skeletal muscle.

Fast and slow skeletal muscle fibres

The chapter introduction about junglefowl explained that muscle fibres are not all the same. As the name implies, fast fibres contract much more quickly than slow fibres. During contraction their myosin heads connect and disconnect from the binding sites on the actin up to five times as fast. Now that we've studied the detailed structure of muscles we can explain how the two types of fibre function. Table 5.2 shows the key differences.

Slow muscle fibres	Fast muscle fibres
Long contraction–relaxation cycle	Short contraction–relaxation cycle
Smaller store of calcium ions in sarcoplasmic reticulum	Large store of calcium ions and more sarcoplasmic reticulum
Dense network of blood capillaries around fibres for supply of oxygen and glucose for aerobic respiration	Fewer blood capillaries around fibres
ATP largely obtained from aerobic respiration	ATP largely obtained from anaerobic respiration
Many, large mitochondria, nearer the surface of the fibres	Fewer, smaller mitochondria, more evenly distributed
Small amount of glycogen	Larger amount of glycogen and phosphocreatine
Slower rate of ATP hydrolysis in myosin heads	Higher rate of ATP hydrolysis in myosin heads, so more actinomyosin bridges formed per second
Resistant to fatigue, since less lactate is formed	Quickly become fatigued, since more lactate is formed

Table 5.2 Comparing the properties of slow muscle fibres and fast muscle fibres.

Slow fibres are able to keep working for long periods. They have plenty of mitochondria, so they are able to obtain most of their ATP from aerobic respiration, even though it produces ATP relatively slowly. They have a dense capillary network, and their mitochondria are near the surface, so diffusion distances for oxygen are shorter. This maximises the delivery of oxygen, so slow fibres are more likely to remain aerobic. Because they contract more slowly they require a lower concentration of calcium ions to initiate contraction. They tend not to fatigue easily because lactate is less likely to build up if they are respiring aerobically.

Fast fibres have fewer capillaries to deliver oxygen and glucose and fewer mitochondria, and they obtain much of their ATP from glycolysis, or anaerobic respiration, which is faster. They work in rapid bursts for short periods of time so require a higher concentration of calcium ions to initiate more contraction. Larger amounts of phosphocreatine help to supply ATP for the first few seconds until more can be produced by respiration. Fast fibres have larger glycogen stores to act as an internal source of glucose, or respiratory substrate, for glycolysis. However, they tend to fatigue more quickly than slow fibres because anaerobic respiration produces lactate, which is acidic and lowers their pH.

EXAMPLE

Comparing fast and slow muscle fibres

1 Explain the importance of mitochondria in muscle contraction.

As you learned in Chapter 1, page 17, the mitochondria are where the Krebs cycle takes place during aerobic respiration and where the majority of ATP is produced. This is a significantly slower process than anaerobic respiration but produces much more ATP per molecule of glucose than does anaerobic respiration.

2 Explain the importance of the dense network of capillaries in slow muscle fibres.

Slow fibres depend on aerobic respiration for ATP. This requires a constant supply of oxygen and glucose, which has to be obtained from the haemoglobin in the red cells of the blood. The dense network ensures that all fibres are close to this source of oxygen and glucose.

3 How does the availability of calcium ions in the sarcoplasmic reticulum affect the rate of contraction?

Calcium ions are released from the sarcoplasmic reticulum when a muscle fibre is stimulated. The ions diffuse into the myofibrils and attach to troponin, causing the tropomyosin to move away from the binding sites on actin. The larger network of sarcoplasmic reticulum in fast muscle fibres reduces the distance for diffusion and the higher concentration of calcium ions increases the diffusion gradient. Both factors help to speed up the rate at which the ions reach the troponin in the sarcomeres.

4 Explain the importance of ATP hydrolysis in the myosin heads.

The hydrolysis of ATP into ADP and inorganic phosphate drives the recovery stroke. The myosin molecule straightens and becomes ready to attach to take part in another cycle of actinomyosin bridge formation. The faster hydrolysis occurs the faster the fibre contracts.

5 Explain how glycolysis provides ATP more rapidly than aerobic respiration.

Glycolysis takes place in the cytoplasm whereas the rest of aerobic respiration occurs in the mitochondria. In the absence of oxygen the pyruvate produced by glycolysis is converted to lactate. This process occurs at a much faster rate than the diffusion of pyruvate into the mitochondria, so anaerobic respiration can occur very rapidly for a short period, making ATP available more quickly.

6 Suggest what causes fast muscle fibres to fatigue rapidly.

Anaerobic respiration produces lactate and this increases the acidity inside the fibre. One effect is to disrupt the ATP hydrolase in the myosin head that hydrolyses ATP. In practice, the effects are complex and not fully understood. Other factors are interference with the release of and uptake of calcium ions and depletion of the reserves of phosphocreatine and glucose.

7 A top-class 100 m sprinter can complete a race without taking a breath. Suggest an explanation for this.

All the ATP for muscle contraction during the race is derived from phosphocreatine and glycolysis. There is no need to replenish the supplies of oxygen through breathing. However, following the race, deep breathing is required to overcome the oxygen deficit. Oxygen is required for the oxidation of the accumulated lactate.

TEST YOURSELF

11 Name the parts labelled A and B in the neuromuscular junction shown in Figure 5.15.

12 Describe how a sarcolemma is depolarised by acetylcholine.

13 Explain how the resting potential of the sarcolemma is restored.

14 What is the role of phosphocreatine in muscle contraction?

15 Give three differences between fast and slow muscle fibres.

Practice questions

1 The figure shows a diagram of the structure of two filaments in a myofibril.

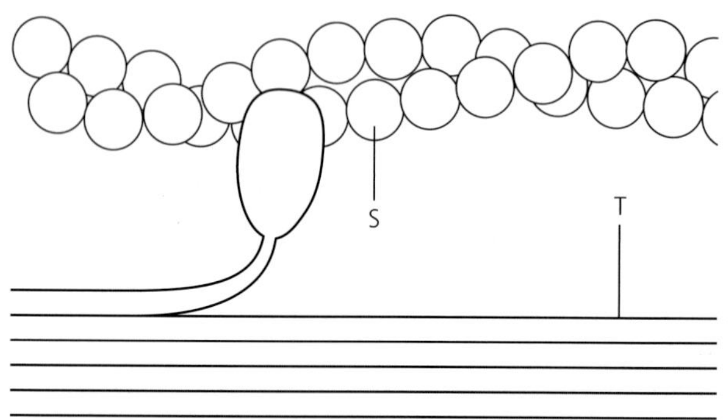

a) i) Identify filament S and filament T. *(1)*

ii) Mark with an X on the diagram the location of ATP hydrolase. *(1)*

iii) Indicate with an arrow on the diagram the direction in which filament S would move during myofibril contraction. *(1)*

b) What is the role of ATP in myofibril contraction? (1)

c) Describe the sequence of events in the cycle of actinomyosin bridge formation. (5)

2 The figure is an electron micrograph of part of a skeletal muscle myofibril showing a sarcomere.

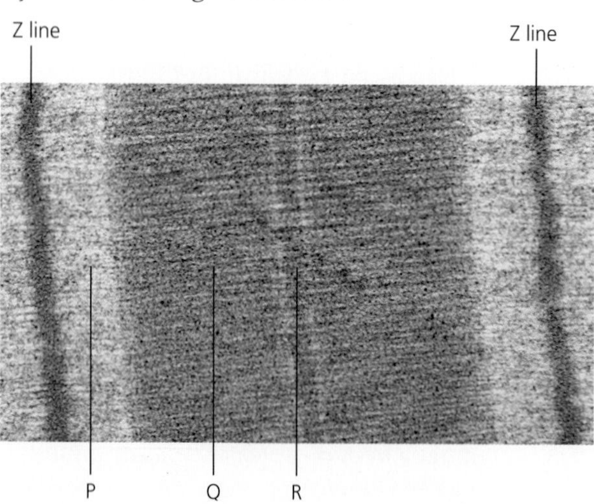

a) Which type of filaments would be present in a cross-section cut at

i) R?

ii) Q? *(1)*

b) If this myofibril were to contract, describe and explain what would happen to the distance between the Z lines. *(2)*

c) What would happen to the zone labelled P during contraction? Explain your answer. (2)

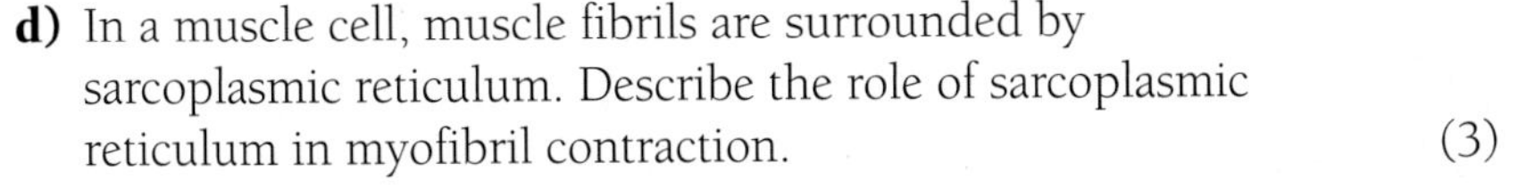

d) In a muscle cell, muscle fibrils are surrounded by sarcoplasmic reticulum. Describe the role of sarcoplasmic reticulum in myofibril contraction. (3)

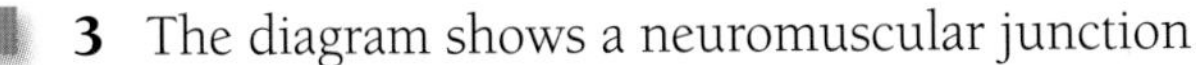

3 The diagram shows a neuromuscular junction.

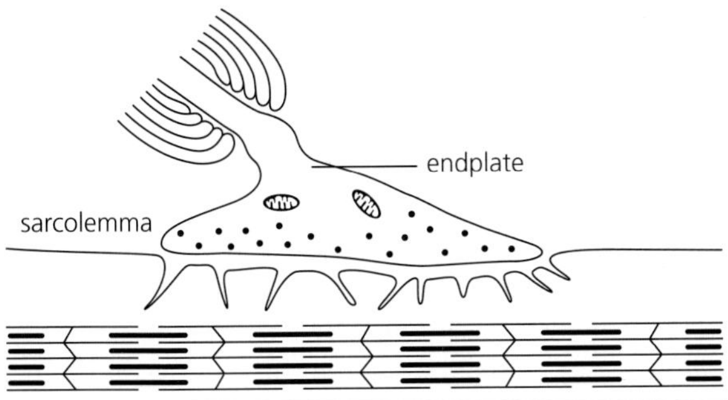

a) Which neurotransmitter would be released at a neuromuscular junction? (1)

b) Suggest the advantage of a neuromuscular junction having a flattened endplate rather than a synaptic knob. (1)

c) Describe the events that take place when action potentials arrive at this neuromuscular junction. (5)

d) Explain how one action potential arriving at this neuromuscular junction may not cause contraction, but several arriving one after another might. (3)

4 The slow loris is a tree-dwelling mammal that shows very slow, deliberate movements along branches and cannot jump from tree to tree. The skeletal muscles in its limbs contain mainly slow fibres.

a) Explain why slow fibres are surrounded by very dense networks of blood capillaries. (2)

b) Give two other differences between slow and fast muscle fibres. (1)

c) The ATP hydrolase on the myosin heads of slow fibres has a slow rate of ATP hydrolysis. Explain why this prevents the slow loris from jumping from one tree to another. (3)

d) The slow loris is only active at night. Suggest why having mainly slow fibres in its limb muscles might be an advantage for the slow loris. (1)

Stretch and challenge

5 Cardiac muscle is similar to skeletal muscle in some ways but very different in others. Compare cardiac muscle fibres with skeletal muscle fibres and discuss reasons for the differences.

6 Examine the role of dystrophin in muscle tissue. Discuss its importance by contrasting muscle tissue in a healthy person with muscle tissue in a person with Duchenne muscular dystrophy.

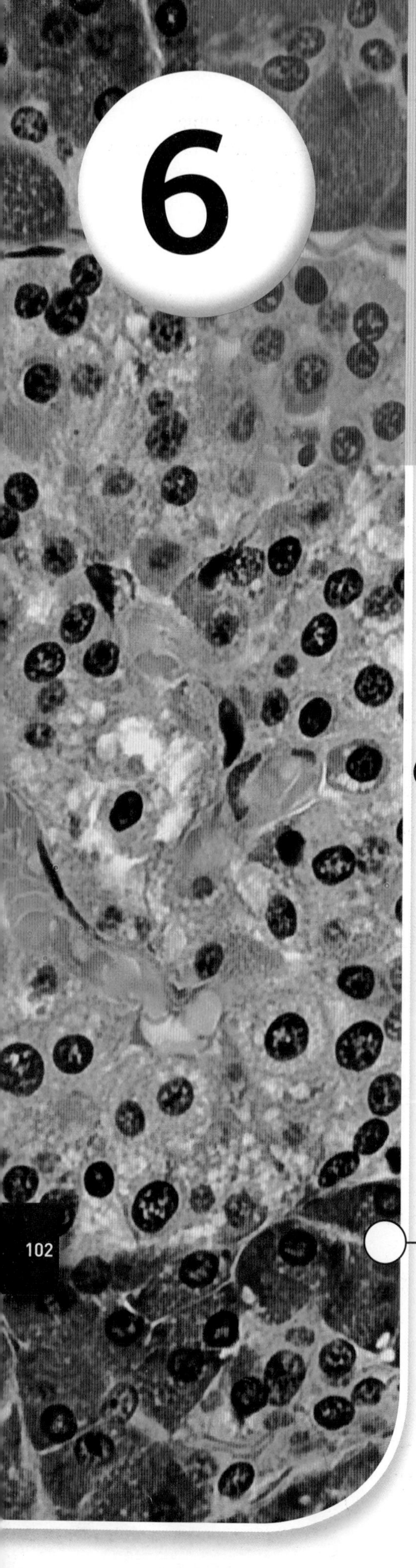

6

Internal control

PRIOR KNOWLEDGE

- Many internal conditions in our body are controlled, including water content, ion content, temperature and blood sugar concentration.
- Water leaves the body via the lungs when we breathe out, via the skin when we sweat to cool us down, and via the kidneys in the urine.
- Ions are lost via the skin when we sweat and excess ions are lost via the kidneys in the urine.
- Our body's temperature is maintained at the temperature at which enzymes work best.
- Blood sugar concentration is maintained to provide the cells with a constant supply of energy.
- Many processes within the body are coordinated by chemical substances called hormones. Hormones are secreted by endocrine glands and are transported to their target organs by the bloodstream.

TEST YOURSELF ON PRIOR KNOWLEDGE

1 Complete Table 6.1 to explain advantage of controlling the following conditions in the body.

Table 6.1

Condition	Advantage of controlling this condition
Water content of the body	
Ion content of the body	
Temperature	
Blood sugar	

2 Hormones travel to all parts of the body via the blood, but only affect certain target cells. Use your knowledge of cell-surface membranes to suggest how they are able to affect only certain cells.

Introduction

It was only in the 20th century that biologists discovered that chemical 'messengers' or hormones are transported round the body in the blood and that these play an essential part in the body's communication system. Compared to a nervous response, mammalian hormones produce responses that are usually slow, long-lasting and widespread. The term 'hormone' was not coined until 1906.

In this chapter we shall look at some examples of the ways in which hormones control physiological processes in humans and other mammals.

One important role of hormones is maintaining more or less stable conditions in the body. Hormones ensure that the glucose concentration in the blood doesn't rise too high or fall too low. They keep the pH and water potential of the blood fairly constant.

What is a hormone?

A hormone is a substance that is secreted directly into the blood from the tissue where it is made. The blood carries it to all parts of the body but the hormone only has an effect on certain **target organs** or **target cells**. Many hormones work in a similar way to neurotransmitters. They attach to **receptor proteins** on the cell-surface membranes of target cells. This sets off reactions that activate enzymes in the cells, as we shall see later. Some hormones that are soluble in lipids can pass through the cell-surface membrane into the cell. Here they may combine with receptor proteins in the cytoplasm, then enter the nucleus and cause genes to be expressed, leading to the production of enzymes.

Hormones are synthesised in **endocrine** tissues. Endocrine glands are organs whose function is to release hormones. These include the pituitary gland situated below the brain and the adrenal glands, one at the top of each kidney. Other hormones are made by groups of cells in other organs, such as the cells that produce the sex hormones in the testes and ovaries, and the insulin-making cells in the pancreas.

Homeostasis in mammals

Homeostasis in mammals involves physiological control systems that maintain the internal environment within restricted limits. For example, it is obviously advantageous to control

- the temperature and pH of tissue fluid around cells since they both affect the rate of enzyme-controlled reactions
- the concentration of glucose in the blood since this affects both the water potential of the blood and the availability of a respiratory substrate to cells.

Negative and positive feedback

You will remember from Chapter 3 that receptors detect stimuli. Deviations from the normal range are stimuli, detected by receptors. This leads to a corrective mechanism to bring the factor back to the normal range. This is regulated by **negative feedback.** Usually, there is one corrective mechanism when the factor becomes too high and another corrective mechanism when the factor falls too low.

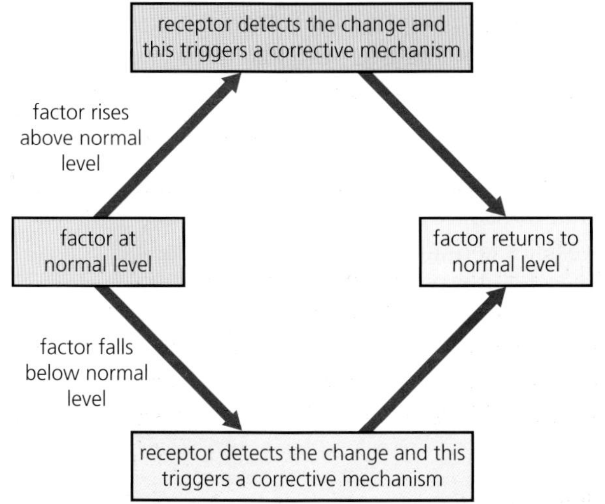

Figure 6.1 Negative feedback.

Figure 6.1 shows how, when a factor changes from its normal level, the body detects the change. It then causes a corrective mechanism to change the factor back to the normal level. This corrective mechanism may involve nervous mechanisms and/or hormones. The amount of correction needed to bring the factor back to the normal level is regulated by **negative feedback.** This makes sure that the corrective mechanism is reduced as the factor returns to its normal level.

As you can see, with negative feedback the corrective mechanism reduces the original effect of the stimulus. In a positive feedback system, the stimulus brings about a response that changes the factor even further from the normal level.

Control of blood glucose concentration

The control of the concentration of glucose in the blood involves two homeostatic mechanisms. The possession of two separate mechanisms, controlling departures in different directions from the norm, provides a greater degree of control than would a single mechanism.

The normal value for the concentration in human blood is approximately 90 mg per 100 cm^3. It is important that there is always some glucose circulating in the blood to enable cell respiration to continue. The brain cells in particular will very soon die if their supply is cut off. On the other hand, if the concentration of glucose rises too high it has a major effect on the water potential of the blood.

Glucose enters the blood from three main sources

- absorption from the gut following the digestion of carbohydrates
- hydrolysis of stored glycogen
- conversion of non-carbohydrates such as lactate, fats and amino acids.

The amount of glucose absorbed into the blood from digestion can fluctuate greatly. The control systems remove excess glucose entering the blood after a carbohydrate-rich meal and release glucose rapidly from storage compounds when muscles are depleting the content of the blood at a fast rate during exercise.

What happens when blood glucose concentration rises?

A rise in the blood glucose concentration above the 'norm' is detected by the **beta cells** (**β cells**) in the **pancreas**. The β cells are situated in little groups of cells dotted around the pancreas called **islets of Langerhans** (Figure 6.2). The β cells synthesise a hormone called **insulin**. You will probably recall that the pancreas secretes digestive enzymes that pour through a duct into the upper part of the small intestine. Since they produce hormones and not enzymes, the islets are like mini-endocrine glands.

When blood with a high concentration of glucose reaches the islets, glucose is absorbed into the β cells. The cell-surface membrane of a β cell contains carrier protein molecules that transport glucose into the cells by facilitated diffusion. This stimulates vesicles containing insulin to move to the cell-surface membrane and release insulin into the surrounding capillaries. This is similar to the process at synapses where acetylcholine is released into the synaptic cleft (see page 78).

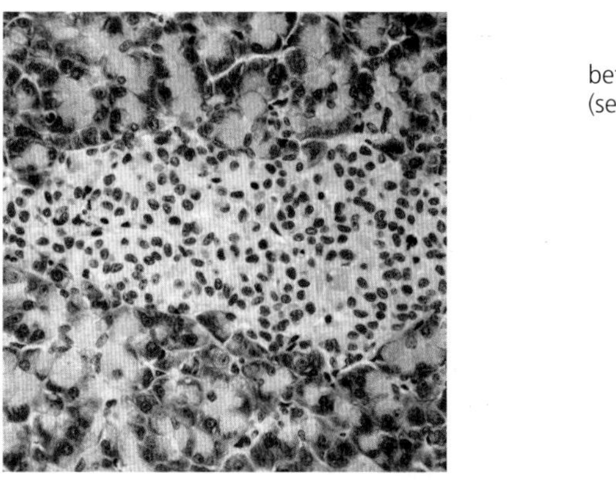

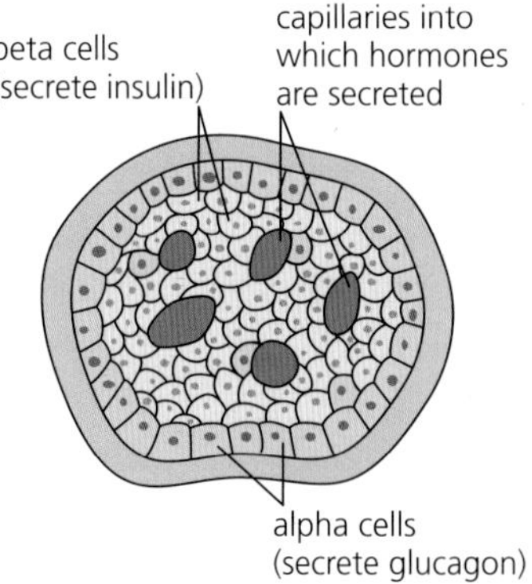

Figure 6.2 a) Coloured photomicrograph of islets of Langerhans in the pancreas, magnification ×100; b) diagram of an islet of Langerhans.

Insulin then circulates round the body in the bloodstream. Its main effects are to stimulate uptake of glucose by cells in muscles, adipose (fat storage) tissue and the liver. Insulin attaches to receptor proteins in the cell-surface membranes of the cells in these tissues. Glucose cannot diffuse into cells through the phospholipid bilayer. Cells have channel proteins, called glucose transporter proteins, that allow glucose to enter by facilitated diffusion; but the rate of uptake is limited by the number of these channel proteins. The glucose transporter molecules in muscles and adipose tissue (called GLUT4) are insulin-sensitive. Insulin causes additional transporter molecules to join the cell-surface membranes of muscle and adipose cells (Figure 6.3). By adding many more channel proteins the rate of uptake of glucose from the blood by facilitated diffusion is greatly increased.

Glycogenesis The formation of glycogen from glucose. This occurs in the liver.

Liver cells already have large numbers of glucose transporter molecules in their membranes. In the liver, insulin leads to an increase in glucose uptake in a different way. After the glucose has entered the liver cells it activates an enzyme that rapidly converts the glucose to glucose phosphate. This lowers the glucose concentration inside the cells and this maintains a steep **diffusion gradient** between the blood and the liver cells. Other enzymes then synthesise the glucose phosphate to glycogen, a process known as glycogenesis. Glycogenesis also occurs in the muscles, replenishing the stores there. In fat storage tissue, insulin activates enzymes that manufacture fatty acids and glycerol, which are then stored as fat.

Figure 6.3 In the cells of muscles and adipose tissue, insulin causes additional glucose transporter molecules to join the cell-surface membrane.

To summarise, insulin stimulates removal of glucose from the blood by:

- increasing the rate of facilitated diffusion into muscle and fat storage cells by stimulating these cells to add more carrier proteins (glucose transporter molecules) to their cell-surface membranes
- increasing the rate of glucose uptake in the liver by stimulating glycogen synthesis
- increasing the rate of glucose uptake in adipose tissue by stimulating fat synthesis.

What happens when blood glucose concentration falls?

Glycogenolysis The breakdown of glycogen to release glucose.

Gluconeogenesis The synthesis of glucose from molecules that are not carbohydrates, such as amino acids and fatty acids.

When the glucose concentration in the blood drops below approximately 90 mg per 100 cm^3, insulin secretion declines. As blood concentration falls, this is detected by the beta cells, which respond by producing less insulin. This is a negative feedback mechanism by which insulin regulates its own secretion. Furthermore, the lower glucose concentration stimulates the **alpha cells** (**α cells**) in the islets of Langerhans to secrete another hormone, called **glucagon**. Glucagon's main effects are in the liver. It activates enzymes that break down the stored glycogen to glucose, which is then released into the blood. This process is called glycogenolysis. It can also activate enzymes that convert other substances to glucose, notably lactate and amino acids. This is a process called gluconeogenesis, which roughly means 'making glucose from new substances'. Both of these effects increase the blood glucose concentration again so that it quickly returns to its normal value. In practice, there are constant adjustments to the amounts of insulin and glucagon being secreted. Much of the time both hormones are secreted in small quantities with the proportions adjusted to maintain the glucose concentration at a fairly constant level.

Figure 6.4 summarises how negative feedback mechanisms operate in the control of the blood glucose concentration.

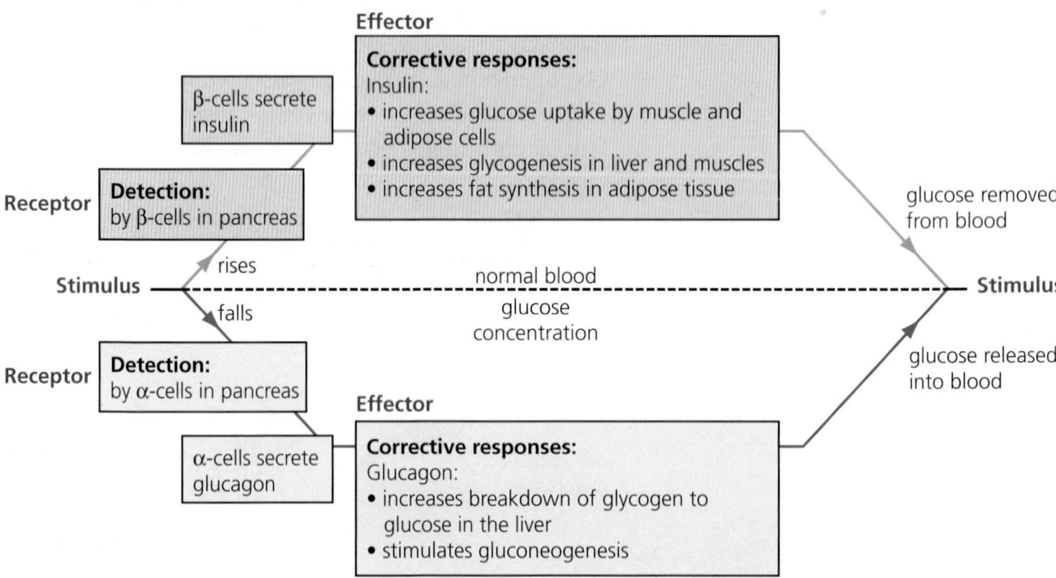

Figure 6.4 Negative feedback in the control of blood glucose concentration, showing the effects of insulin and glucagon.

TIP

Do not confuse glyco**genesis** with glycogen**olysis**. Remember genesis is about creation. Similarly, take care not to confuse **glu**cagon with **gly**cogen.

As often happens in scientific research, continuing investigations have revealed that the simple story does not give the full picture and that systems are more complex than at first thought. The breakdown of glycogen in the liver is not only stimulated by glucagon. **Adrenaline**, the 'fight-or-flight' hormone, has a similar effect. Adrenaline attaches to specific receptor proteins in the cell-surface membranes of liver cells, which, in turn, leads to the activation of enzymes involved in the conversion of glycogen to glucose.

The second messenger model

Unlike insulin, adrenaline and glucagon do not have a direct effect on the liver cells. Instead, when adrenaline or glucagon attaches to a specific receptor protein that spans the cell-surface membrane of a liver cell, it causes a cascade of reactions within the cell. This is called the **second messenger model**, where glucagon is the first messenger.

Figure 6.5 shows what happens in this model, using adrenaline as an example; the process is similar when glucagon attaches to its specific receptor protein. Binding of adrenaline causes a change in the shape of an enzyme also in the cell-surface membrane, called **adenyl cyclase**. This makes the enzyme active. This enzyme acts on ATP, which has three phosphate groups. Instead of producing ADP, adenyl cyclase removes two of the phosphate groups. This makes **cyclic adenosine monophosphate**, usually written as **cAMP**. cAMP is the second messenger. The cAMP then binds to another enzyme in the cytoplasm, called **protein kinase**. This exposes its active site, which leads to the hydrolysis of glycogen to produce glucose phosphate. Hormones, and indeed chemicals in general, often get a bad press. Hormones are blamed for erratic behaviour in teenagers and for a variety of eccentricities in men and women. Hormones are frequently thought to be minor players in the body's communication system compared with nerves. As you have seen in the previous chapters, the nervous system depends on chemicals, for example at synapses and neuromuscular junctions. Many of the neurotransmitters are now known to be chemically very similar to hormones.

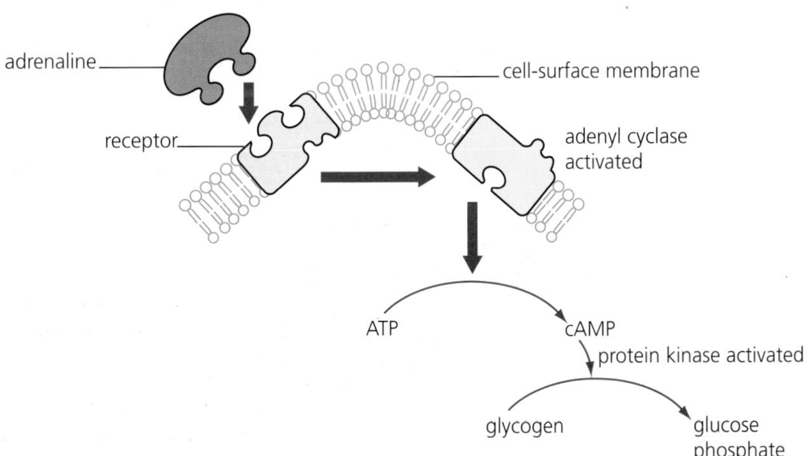

Figure 6.5 The second messenger model of hormone action, which follows stimulation of liver cells by adrenaline or glucagon.

The advantage of this process is that each molecule of hormone can stimulate many molecules of cAMP, and these in turn activate large numbers of enzyme molecules.

TEST YOURSELF

1 Explain the effect of high glucose concentration on the water potential of the blood.

2 What is meant by facilitated diffusion? (If you can't remember you may need to refer back to your first year course; see *AQA A-level Biology 1 Student's Book*).

3 How does facilitated diffusion differ from active transport?

4 After its secretion, glucagon breaks down within a few minutes. Suggest the advantage of this.

5 The veins from the small intestine and the pancreas connect to the hepatic portal vein. This transports blood directly to the liver. Therefore the blood flows through the liver before returning to the heart and entering the general circulation. Suggest the advantages of this arrangement in relation to the control of the blood glucose concentration.

6 Suggest why it is an advantage for adrenaline to have its amplification effect through a second messenger.

Diabetes

Diabetes is a condition in which homeostatic control of the blood glucose concentration breaks down. In 2013, 3 208 014 adults in the UK were living with diabetes, and its incidence is growing. The increase is attributed to the trend towards over-eating and obesity. Insulin function is disrupted and the concentration of glucose in the blood rises above the normal range after meals. One effect is that the kidneys are unable to reabsorb all the glucose into the blood as they normally do. Therefore glucose appears in the urine. In response the kidneys tend to produce excess quantities of urine. This dehydrates the body and makes the sufferer very thirsty.

There are two main types of diabetes, called **Type I** and **Type II**. Table 6.2 shows the proportion of these two types of diabetes in 2011 in the UK.

Table 6.2 The incidence of diagnosed Type I and Type II diabetes in the UK in 2011.

	Percentage of cases	
Age group	**Type I diabetes**	**Type II diabetes**
Children	98	2
Adult	10	90
All	15	85

Type I diabetes

Auto-immune condition A condition in which the immune system destroys some of the body's own cells.

Type I diabetes is caused by an inability to make enough insulin. It usually develops quite early in life, often in childhood (see Table 6.2). It is an auto-immune condition in which the immune system destroys some of the body's own cells. In *AQA A-level Biology 1 Student's Book* we described how T cells attack infected cells by attaching to antigens on their cell-surface membranes.

For some unexplained reason, in Type I diabetes T cells start to attack and destroy the β cells in the islets of Langerhans. The shortage of insulin causes fatigue because not enough glycogen is stored in the liver. The concentration of glucose in the blood can rise to dangerously high levels after a meal. This can lead to serious organ damage.

This type of diabetes can be controlled by eating appropriate food and taking regular blood tests and injections of insulin.

The food recommended for people with diabetes by health authorities and by Diabetes UK differs little from that recommended for everyone wishing to have a healthy diet. The recommendations include the consumption of food containing polysaccharides, rather than monosaccharides and disaccharides, at least five portions of fresh fruit and vegetables each day, and few processed foods.

Before the isolation and use of insulin, Type I diabetes was always fatal. For many years insulin was extracted from pancreas tissue of cattle and pigs. In the 1980s, synthesis of human insulin by recombinant DNA technology was one of the early successes of genetic engineering (see Chapter 12). Treatment has been transformed by the development of fast-acting soluble and slow-release insulin preparations. These allow either for rapid control at a mealtime or control that can last for many hours. Methods of injection have been made much more convenient, for example by the use of injector pens. As a result people with Type I diabetes can live a normal and full life. Nevertheless most have to inject themselves with insulin twice a day in order to maintain a stable blood glucose concentration. There is hope that transplants of islets of Langerhans will provide more permanent treatment, but so far this has been much less successful than, for example, kidney transplantation.

Type II diabetes

As you can see from Table 6.2, this form of diabetes is much more common than Type I. Usually it develops in people over 40 years old but it is increasingly being found at younger ages. It is caused, not by failure to produce insulin, but by reduced sensitivity of the liver and fat storage tissue to insulin. The receptors either fail to respond or are reduced in number so glucose uptake is much reduced. The exact effects are still unclear because they are often associated with a range of other factors, often resulting from obesity and unbalanced diets. One response is that the β cells are stimulated to produce larger quantities of insulin. If the condition persists the β cells are damaged and the condition becomes more like Type I.

In the early stages Type II diabetes can be treated simply by careful attention to the diet. Reducing the intake of refined carbohydrates, such as sugars and forms of starch that are rapidly digested to sugars, avoids large surges in blood sugar concentration. The fat content of the diet also needs to be kept low and exercise helps to improve the body's sugar metabolism.

TEST YOURSELF

7 Insulin is a protein. Explain why it has to be injected instead of being taken orally.

8 Suggest how Type II diabetes could be controlled by exercise and diet.

9 One effect of poorly controlled diabetes is high blood pressure. Use your knowledge of osmosis to suggest why.

EXAMPLE

The glucose tolerance test

TIP

You do not need to recall details of the glucose tolerance test, but learning how to evaluate such material is an important skill, so make sure you can do this.

One of the tests used in diagnosing diabetes is the **glucose tolerance test**. After a period of several hours without food the person being tested has a drink of water containing 75 g of glucose. The concentration of glucose in the blood plasma is measured at 30 minute intervals. The graph in Figure 6.6 shows the results of this test from three adults. One had no symptoms of diabetes. Both of the others had tested positive for glucose in the urine.

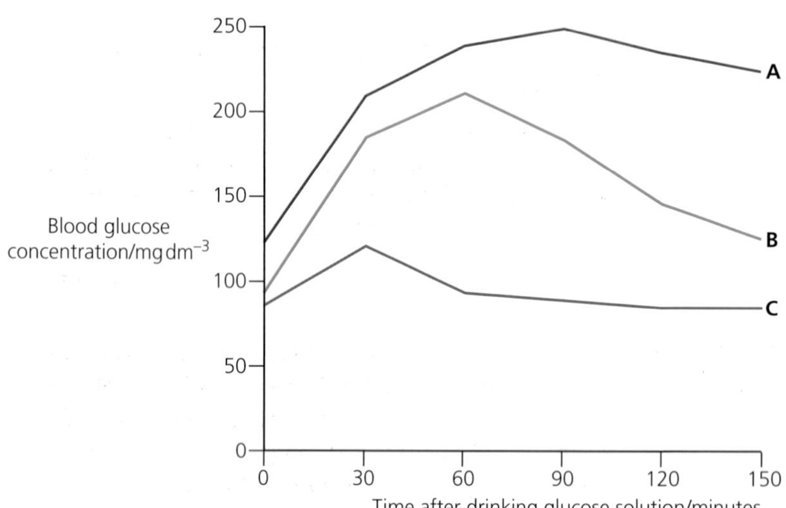

Figure 6.6 Results of a glucose tolerance test taken by three adults, labelled A, B and C.

1 Curve C shows the results for the person without diabetes. Suggest why the glucose concentration rose in the first 30 minutes.
The glucose from the drink was absorbed into the blood.

2 Describe and explain the results for curve C after 30 minutes.
The β cells of the pancreas release insulin, so glucose is removed from the blood and stored as glycogen in liver and muscle cells.

3 Describe the differences between curves A and B.
Both curves show a large increase in glucose concentration in the blood, but while curve B peaks and starts to fall after an hour, curve A does not peak until 90 minutes and then falls more slowly than curve B.

4 Of the people A and B, one was diagnosed as having Type I diabetes and the other Type II. Which was which? Explain your reasoning.
A has Type I diabetes as the glucose concentration rises in the blood for 90 minutes, which indicates that insulin is not produced. The concentration of glucose falls very slowly, which again indicates that there is no insulin produced and the glucose only gradually decreases as it is used in respiration. B has Type II diabetes as the curve peaks earlier than A and starts to fall more rapidly. This implies that insulin is produced and some of the gucose is converted to glycogen, but the body is not as responsive to insulin as in a healthy person.

REQUIRED PRACTICAL 11

Production of a dilution series of a glucose solution and use of colorimetric techniques to produce a calibration curve with which to identify the concentration of glucose in an unknown 'urine' sample

This is just one example of how you might tackle this required practical.

A colorimeter is a piece of equipment that passes light
of a particular wavelength through a sample and measures the amount of light transmitted through the sample. You learned about this in *AQA A-level Biology 1 Student's Book (see Chapter 1 page 16)*.

In using a colorimeter, you need to produce a calibration curve. This is done using solutions of a known concentration.

TIP

Refresh your memory of colorimetry by reading the section about it in *AQA A-level Biology 1 Student's Book* Chapter 15.

SAFETY

Wear eye protection when carrying out this practical.

A student was given a 1 mol dm^{-3} solution of glucose and then made a series of dilutions. She made 20 cm^3 of each dilution.

1 Complete the table to give the volumes of water and 1 mol dm^{-3} sucrose needed for each solution.

Table 6.3

Concentration of sucrose/ mol dm^{-3}	Volume of 1 mol dm^{-3} required/cm^3	Volume of distilled water required/cm^3
0.9		
0.8		
0.7		
0.6		
0.5		
0.4		
0.3		
0.2		
0.1		

An alternative version of Benedict's reagent for quantitative testing contains potassium thiocyanate and does not form red copper oxide. Instead the presence of reducing sugar is measured by the loss of the blue colour of copper sulfate and a white precipitate is formed. This will settle out or can be removed by filtering. Then the **filtrate** is placed in a cuvette in a colorimeter. The intensity of the blue colour is measured by the amount of light that is able to pass through the solution. This method can give an accurate measurement of the concentration of reducing sugar in a solution, and it is much more sensitive that the qualitative Benedict's test.

Filtrate The precipitate that is left in the filter paper after filtering has taken place.

The student put 4 cm^3 of each solution into separate labelled test tubes.

Next she added 2 cm^3 of quantitative Benedict's reagent to each tube and placed the tubes in a boiling water bath for 5 minutes. After this time she filtered each solution to remove the precipitate.

The student set the wavelength on the colorimeter to red.

First of all she filled a cuvette with distilled water and put it into the colorimeter. This is called a 'blank'. She set the transmission of light through the tube to 100%. This means that she could compare the transmission of light through the test solutions to the blank.

The student put a sample of each test solution into cuvettes in turn, and measured the percentage transmission of light through each tube. Next she plotted a graph with concentration of glucose on the *x*-axis and percentage transmission of light through the solution on the *y*-axis.

After this, she used the same method to identify the concentration of glucose in two urine samples.

2 Why did the student use a red light in the colorimeter?

3 What is the purpose of the blank?

4 How could the student use her graph to find the concentration of glucose in an unknown solution?

Control of blood water potential (osmoregulation)

The body cells of mammals are bathed by tissue fluid. Exchange of substances, including water, occurs constantly between this tissue fluid and the cells. Unless this tissue fluid has the same water potential as their cytoplasm, the cells will lose or gain water by osmosis, which could prove fatal. You learned in *AQA A-level Biology 1 Student's Book* how tissue fluid is formed from capillary blood. Control of the water potential of tissue fluid is achieved by controlling the water potential of the blood from which it is formed.

TIP

Remember to use water potential terminology when discussing this topic.

The homeostatic control of water potential is called **osmoregulation**. Like the homeostatic control of blood glucose concentration, osmoregulation involves negative feedback loops. Unlike the homeostatic control of blood glucose concentration, the control of blood water potential involves nervous as well as hormonal coordination. In this section we will look at how the hypothalamus, the posterior lobe of the pituitary gland and antidiuretic hormone (ADH) regulate the amount of water that is lost from the body in urine. Look at Figure 6.7 to remind yourself where the hypothalamus and pituitary gland are located. Mammalian urine is produced by the kidneys. The two kidneys form part of the human urinary system shown in Figure 6.8. They are supplied with blood via renal arteries. The kidneys filter the blood brought in these arteries. Initially, they remove some ions and molecules including amino acids, glucose, urea and water. Later, they reabsorb ions and molecules that are useful but not molecules like urea, which is a breakdown product of amino acids that is excreted in the urine. As we shall see later, it is the control of water reabsorption that is at the heart of osmoregulation. The functional unit of the kidney that removes ions and small molecules from the blood and then selectively reabsorbs them is the kidney tubule, or **nephron**.

TIP

You will soon realise that two of the processes involved in urine production are fundamentally similar to the production of tissue fluid and the countercurrent flow of blood that you learned about in your first year of study.

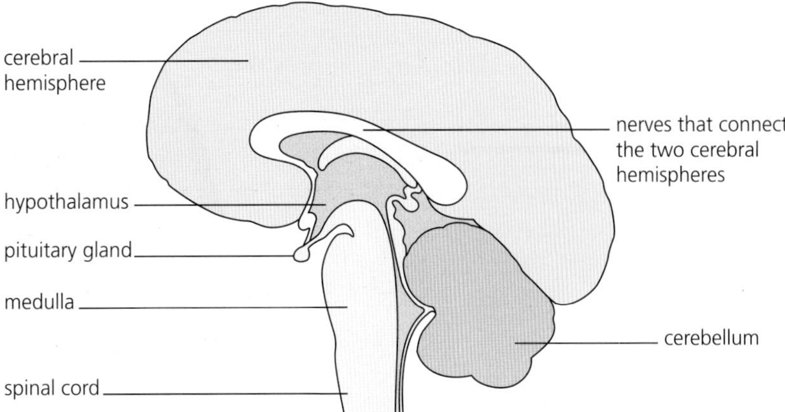

Figure 6.7 This shows a section through the centre of the brain, separating the two cerebral hemispheres. The pituitary gland and the hypothalamus have important roles in several different control systems. Although you do not need to learn the parts of the brain shown here, we shall be referring to them several times in this chapter so it will be useful to know where in the brain they are situated.

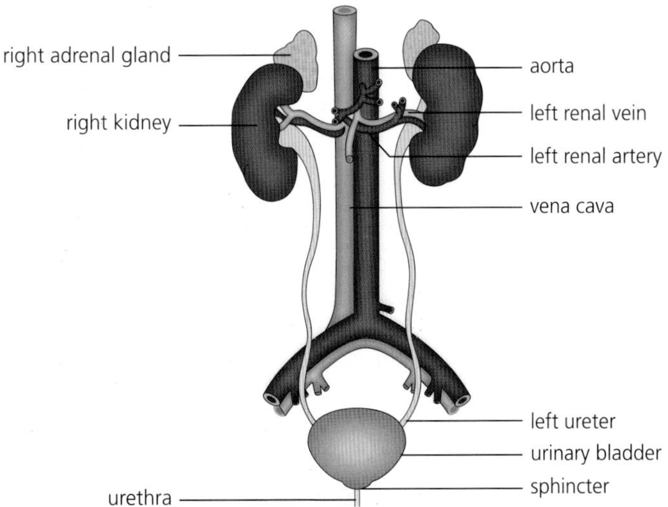

Figure 6.8 The human urinary system.

Ultrafiltration and selective reabsorption by nephrons

Figure 6.9 (overleaf) shows the location of one nephron within a kidney. You can see that it begins and ends in a blood-rich outer layer, called the cortex, passing in the meantime into the central less-blood-rich region, called the medulla. You will soon understand what causes the appearance of the cortex and medulla.

TIP
You do not need to be able to recall details of the gross structure of the kidney.

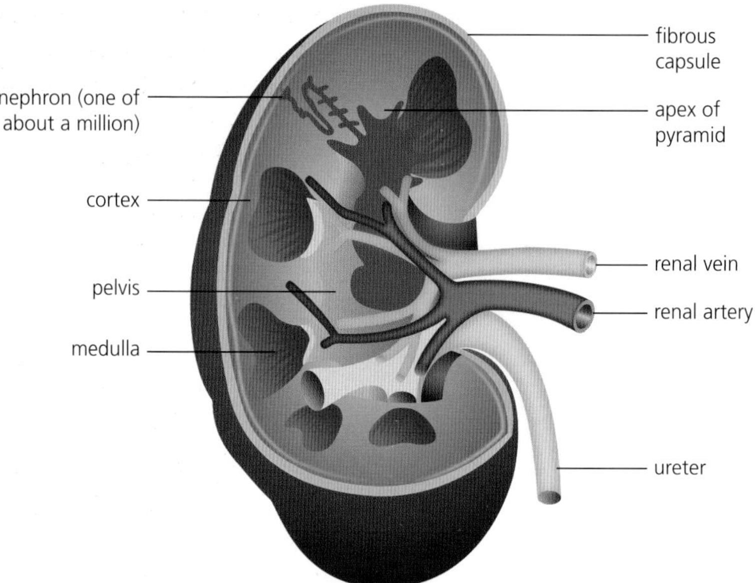

Figure 6.9 A section through a kidney with one nephron shown.

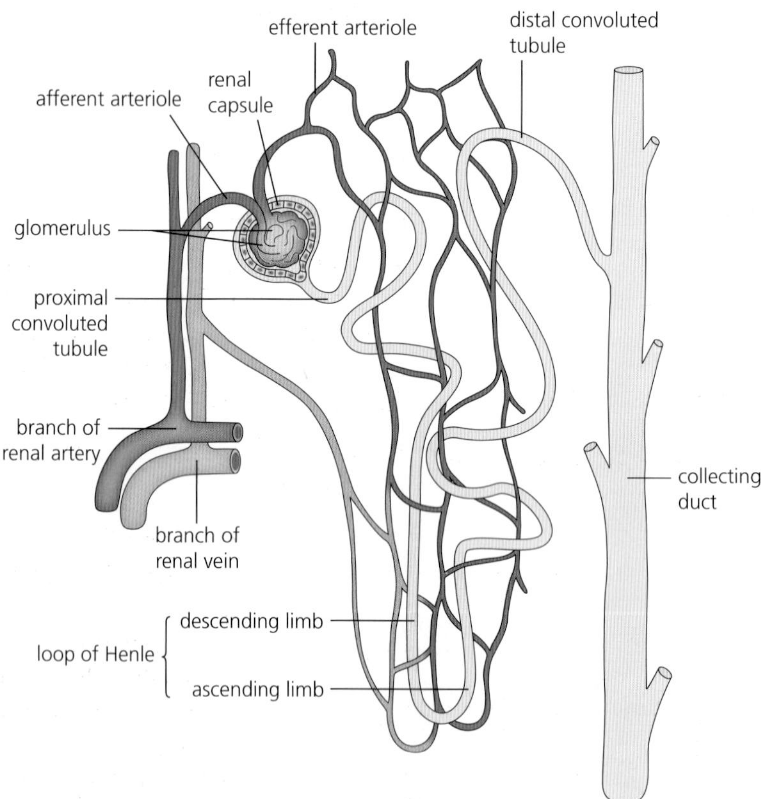

Figure 6.10 A nephron and its blood supply.

A human kidney contains an estimated 1 million nephrons; Figure 6.10 shows one of them. You can see that it has several sections: the renal capsule, proximal convoluted tubule and distal convoluted tubule are all present in the cortex of the kidney; the loop of Henle and the collecting duct are in the medulla. You can also see that within each renal capsule (see Figure 6.11) is a capillary network, called the **glomerulus**. Each glomerulus is supplied with blood by an afferent arteriole that branches from a renal artery. Blood leaves the glomerulus via an efferent arteriole that forms a second capillary network that wraps around the rest of the nephron. This is much narrower than the afferent arteriole, so this creates a high pressure in the capillaries of the glomerulus. As the blood flows through the glomerulus, the pressure forces water from the blood plasma, along with some molecules and ions, across a filtering system into the renal space. This process is called **ultrafiltration**.

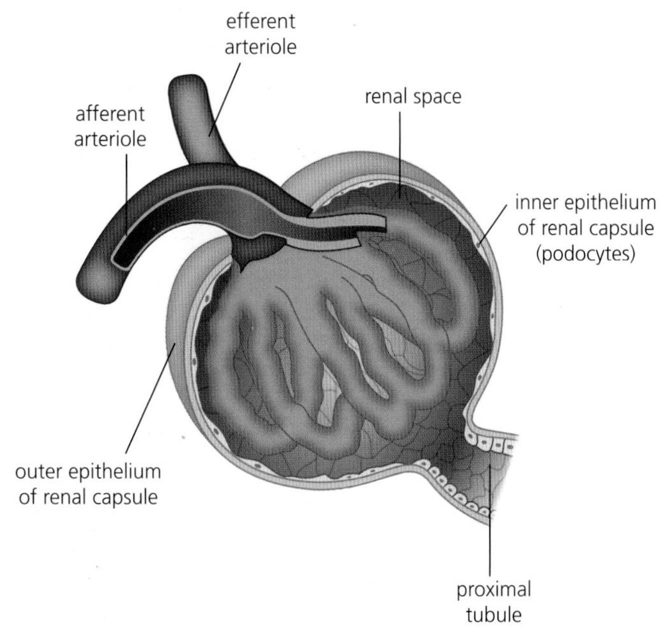

Figure 6.11 The renal capsule.

TIP

Ultrafiltration is similar to the process by which tissue fluid is formed. You learned about this in *AQA A-level Biology 1 Student's Book* Chapter 9.

Ultrafiltration

Blood enters a glomerulus via an afferent arteriole that is wider than the efferent arteriole through which it leaves. As a result, a high pressure is created in the glomerulus, which forces water, ions and other small molecules into the renal space. This is the process of ultrafiltration. The fluid that is forced into the renal space is called the **glomerular filtrate**.

The filtering system of the renal capsule is shown in Figure 6.12. It has three layers.

- The capillary endothelium has large gaps that allow blood plasma through but not blood cells.
- The basement membrane, which is a mesh of protein molecules that supports the capillary endothelium, acts as a fine filter. It allows plasma proteins with a molecular mass of 68 000 daltons or less to pass through, but not larger ones.
- The **podocytes** forming the lining of the renal capsule have large gaps between them that allow the glomerular filtrate through into the renal space.

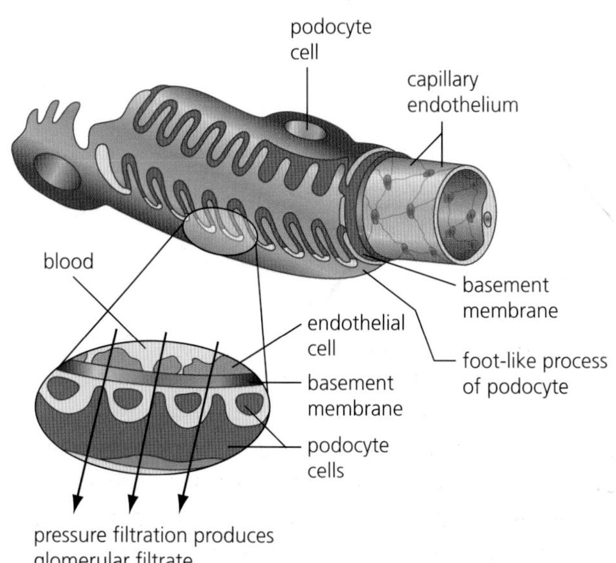

Figure 6.12 The filtering system of kidney renal capsule.

Ultrafiltration in the renal capsule is non-selective except by size. Any substance that is small enough (i.e. below the renal threshold) can be filtered out of the blood whether it is useful or not. This means that useful substances such as glucose and amino acids are filtered out, along with inorganic ions, water and urea. Many of the useful substances are reabsorbed in the proximal convoluted tubule.

Reabsorption in the proximal convoluted tubule

Many of the substances in the glomerular filtrate are reabsorbed by active transport into the cells lining the proximal convoluted tubule. These normally include glucose, amino acids and some inorganic ions. Some of the urea is reabsorbed by diffusion. Figure 6.13 shows how the structure of an epithelial cell lining the proximal convoluted tubule is adapted for reabsorption.

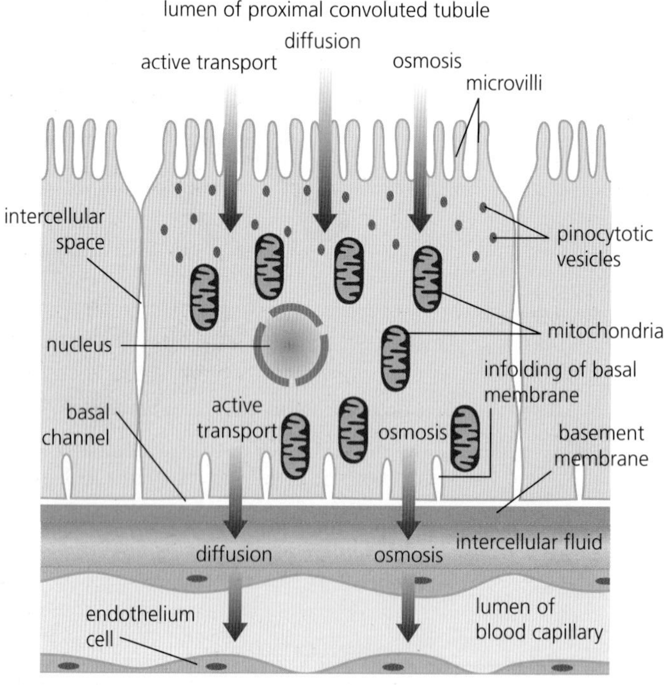

Figure 6.13 Epithelial cell lining the proximal convoluted tubule.

Water is also absorbed by cells lining the proximal convoluted tubule. The active transport of glucose, mineral ions and amino acids into the tubule cells lowers their water potential. As a result, water enters the tubule cells by osmosis. The glucose, amino acids and mineral ions absorbed by the tubule cells then diffuse into the blood capillaries around the proximal convoluted tubule and water moves into the blood capillaries by osmosis. Usually all of the glucose and amino acids in the filtrate are reabsorbed in the proximal convoluted tubule. However, in a person with diabetes, the glucose carrier proteins in the membrane cannot reabsorb all the glucose from the filtrate, so some is lost in the urine.

TEST YOURSELF

10 A person with a damaged basement membrane in the renal capsule has large proteins in their urine. Explain why.

11 Use information from Figure 6.13 to explain two ways in which an epithelial cell of the proximal convoluted tubule is adapted for the reabsorption of glucose and other essential nutrients.

12 Describe in terms of water potential how water is reabsorbed in the proximal convoluted tubule by osmosis.

13 Haemoglobin has a molecular mass of 68 000 daltons but it is not normally found in the urine. Suggest why.

The role of the loop of Henle

The ability of humans to produce urine that is more concentrated than blood plasma is due to the activity of the loop of Henle. The loop of Henle creates a high concentration of inorganic ions deep in the medulla of the kidney. It has two parts, the descending limb and the ascending limb.

- The descending limb of the loop of Henle is permeable to water but not very permeable to mineral ions such as sodium and chloride.
- The ascending limb of the loop of Henle has thick walls, which are impermeable to water. The narrow part of the ascending limb allows mineral ions to move passively into the medulla, but the wide ascending limb actively transports sodium chloride into the medulla.

TIP

You will remember that you learned about countercurrent exchange in Year 1, when you studied gas exchange in fish gills (see *AQA A-level Biology 1 Student's Book* Chapter 7).

The fluid in the descending limb of the loop of Henle flows in the opposite direction to the fluid in the ascending limb. The result is that the concentration gradient between the two limbs is maintained all the way along the loop.

You can see in Figure 6.14 that the loop of Henle creates a water potential gradient that allows water to be reabsorbed from the glomerular filtrate by osmosis. Some water has already been absorbed by osmosis in the proximal convoluted tubule, so the filtrate in the wide part of the descending limb (A) has the same water potential as the fluid in the surrounding tissues. However, because the surrounding tissue fluid in the medulla of the kidney has a high concentration of mineral ions, this creates a water potential gradient so that water is drawn out of the narrow part of the descending limb (B) by osmosis as the filtrate passes down the loop of Henle. The mineral ion concentration in the medulla increases towards the tip of the loop, so water can pass out along the whole length of the descending limb. This water is then carried away in the surrounding capillaries. The filtrate is now reduced in volume and contains a higher concentration of salts. The ascending limb is permeable to sodium chloride but impermeable to water. As the filtrate passes up the thin part of the ascending limb (C), sodium and chloride ions diffuse into the surrounding tissue fluid. Higher in the ascending limb (D), chloride ions are actively transported out of the limb and sodium ions follow into the tissue fluid. These processes maintain the high sodium and chloride ion gradient in the surrounding tissue fluid that is needed for water reabsorption in the collecting ducts.

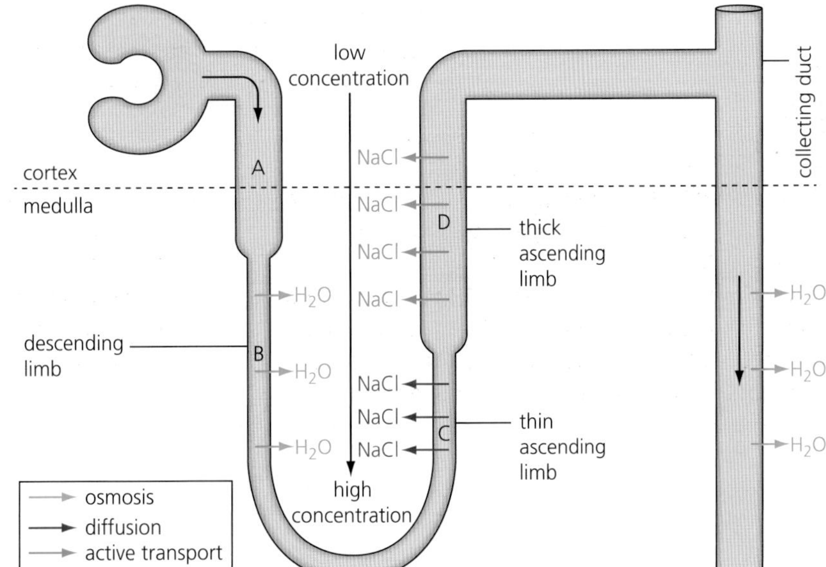

Figure 6.14 A single loop of Henle.

Control of water reabsorption by the distal convoluted tubule

Most of the useful substances, such as glucose and amino acids, have already been absorbed in the proximal convoluted tubule. However, some mineral ions and water remain. In the distal convoluted tubule some of the remaining mineral ions and water are reabsorbed. Almost all of the water that is required has already been reabsorbed, but this is the part of the nephron where water reabsorption is controlled.

Hormones control the amount of reabsorption of water by affecting the permeability of the distal convoluted tubule and collecting duct to water. Figure 6.15 shows how.

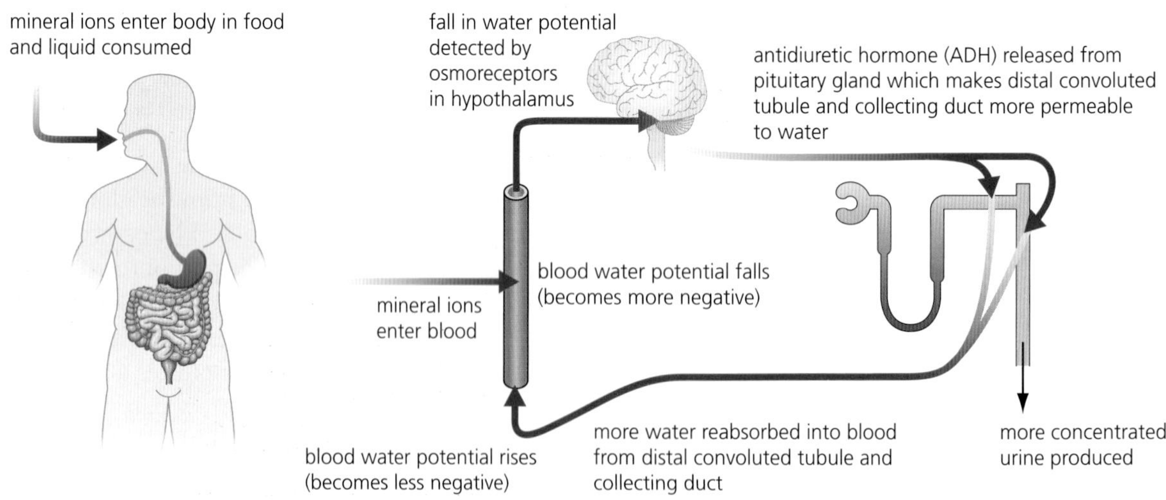

Figure 6.15 The role of ADH in water reabsorption.

Special receptor cells in the hypothalamus of the brain, called **osmoreceptors**, are sensitive to the stimulus of changes in the water potential of the blood. When mineral ions and other solutes in food and drink or loss of water from the body cause the water potential of the blood to fall (i.e. become more negative), these osmoreceptors send impulses to the **posterior pituitary** to release antidiuretic hormone (**ADH**) into the blood. ADH increases the permeability of both the distal convoluted tubule and the collecting duct. This happens because ADH binds to specific receptor molecules on the cells lining the distal convoluted tubule and collecting duct causing protein channels, called **aquaporins**, to move into their cell-surface membranes. The increase in the number of aquaporins allows more water to pass through the membrane by osmosis. As the water potential in the tubule is higher than the water potential of the blood, water is reabsorbed from the distal convoluted tubule and collecting duct. As some water has been reabsorbed from the tubule by osmosis, a smaller volume of more concentrated urine is produced.

When high fluid intake causes the water potential of the blood to rise, the osmoreceptors send impulses to the pituitary gland, which inhibits ADH release by the posterior pituitary. This fall in ADH release reduces the water permeability of both the distal convoluted tubule and the collecting duct, so less water is reabsorbed from the urine. This is another example of negative feedback. Secretion of ADH leads to an increase in water potential in the blood, which leads to a reduction in ADH secretion. You will probably have worked out that the lack of ADH causes the aquaporins to leave the cell-surface membranes of cells lining the ducts and move to the interior of these cells. Therefore a larger volume of more dilute urine is produced. This is summarised in Figure 6.16.

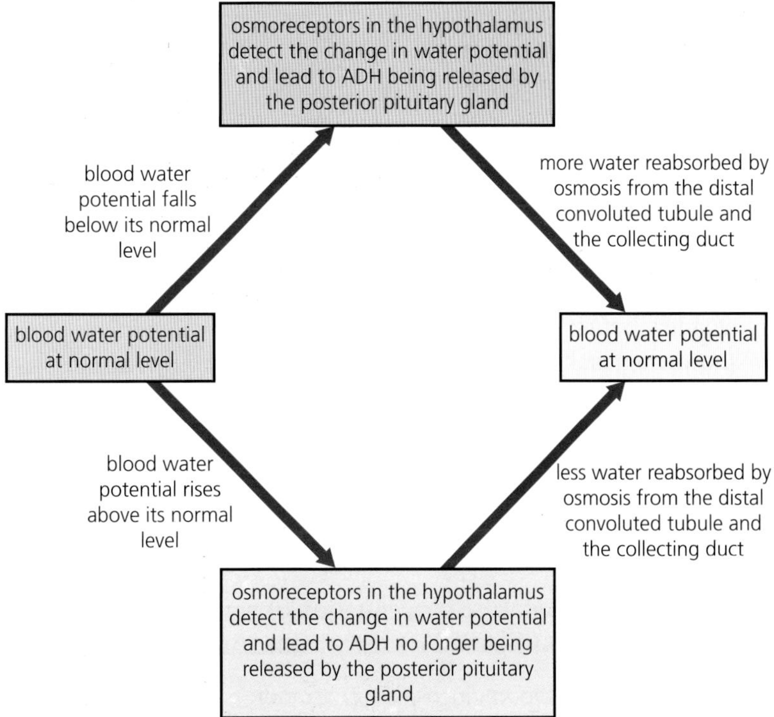

Figure 6.16 The control of blood water potential by the kidney.

TEST YOURSELF

14 Filtrate entering the loop of Henle has the same water potential as the blood. Explain why.

15 Explain how the high concentration of salts in the tissue fluid around the loop of Henle causes water to be drawn out of the thin part of the descending limb.

16 Suggest why the active transport of chloride ions out of the wide ascending limb causes sodium ions to follow.

17 Alcohol inhibits the release of ADH. Explain why a person working outdoors on a hot day should avoid drinking alcohol.

18 Use Figure 6.16 to explain the meaning of negative feedback.

Practice questions

1 One test for diabetes is called the glucose tolerance test. The person taking the test has nothing to eat, and only water to drink, for several hours before the test. Their blood glucose concentration is tested. They are given a sugary drink, and then their blood glucose concentration is measured regularly over the next few hours. The graph shows the results obtained for two patients.

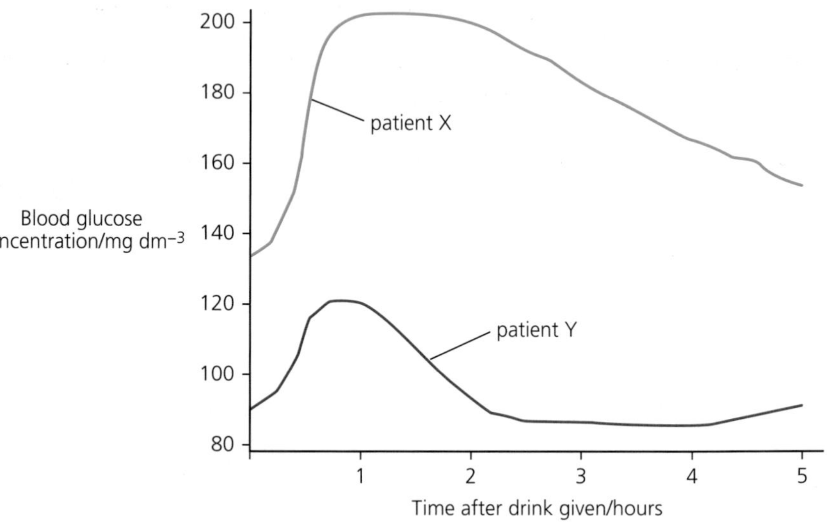

a) The patients had nothing to eat, and only water to drink, for several hours before the test. Explain why. *(1)*

b) Which of patients X and Y has diabetes? Explain your answer. *(2)*

c) i) The blood glucose concentration of patient Y fell between 1 and 2 hours after the sugary drink. Explain why. *(3)*

ii) Patient Y's blood glucose concentration rose slightly between 4 and 5 hours after the sugary drink even though the patient did not have anything to eat and drank only water. Explain how. *(2)*

2 The table shows some substances present in food and their effect on blood glucose concentration.

Substance present in food	Component(s) of substance	Effect on blood glucose concentration
Glucose	Glucose	Large increase
Starch	Glucose	Large increase
Sucrose		Moderate increase
Lactose	Glucose and galactose	Moderate increase
Cellulose	Glucose	No increase

a) Complete the table to show the component(s) of sucrose. *(1)*

b) i) Cellulose and starch have different effects on blood glucose concentration. Explain why. *(3)*

ii) Glucose and lactose have different effects on blood glucose concentration. Explain why. *(3)*

c) Glycogen and glucagon are both compounds that are involved in regulating blood glucose concentration. Explain their different roles. *(5)*

a) i) Describe how glomerular filtrate is formed in the kidney nephron. *(4)*

ii) Normally glucose is not present in the urine. However, people with diabetes may have glucose in their urine. Explain why. *(3)*

b) Furosemide inhibits carrier proteins that reabsorb sodium ions in the ascending limb of the loop of Henle. Explain the effect that this drug will have on water reabsorption in the kidney. (3)

4 The graph shows the effect of the loss of liver and kidney function on the concentration of urea in the blood.

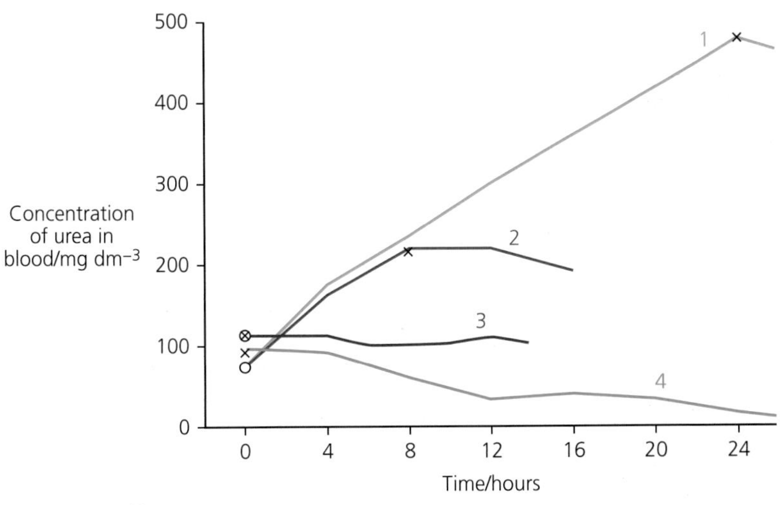

Key:

- kidney function lost at time 0, liver function lost 24 hours later
- kidney function lost at time 0, liver function lost 8 hours later
- kidney and liver function both lost at time 0
- liver function lost at time 0, kidney function not lost
- × time when liver function was lost
- o time when the kidney function was lost

a) What is the evidence from patient 1 and 2 that urea is produced in the liver? (3)

b) Explain why

i) the concentration of urea stays approximately the same for 13 hours in patient 3 (2)

ii) the concentration of urea in the blood falls steadily in patient 4. (2)

Stretch and challenge

5 To what extent does the length of the loop of Henle relate to the environment the animal lives in? Use reference materials to research the relationship between the length of the loop of Henle and water availability in the animal's environment. A good starting point would be desert animals such as the kangaroo rat.

6 Examine the role of the distal convoluted tubule in regulating blood pH. Contrast its role in regulating blood pH with the lungs and circulatory system.

7 Contrast negative feedback and positive feedback. Research hypothermia and hyperthermia, and explain how these are examples of positive feedback. Use your knowledge of physiology to justify the first aid treatments for these conditions.

7 Genes, alleles and inheritance

PRIOR KNOWLEDGE

- Genes are passed from one generation to the next in both plants and animals. Simple genetic diagrams can be used to show this.
- Genetic diagrams can be used to predict the outcomes of monohybrid crosses.
- Using these diagrams it is possible to predict and/or explain the outcome of crosses between individuals for each possible combination of dominant and recessive alleles of the same gene.
- Homozygous describes a cell or organism where both variants of a gene are the same, whereas heterozygous describes a cell or organism where both variants of a gene are different.
- The term phenotype refers to the appearance of an organism and the term genotype refers to its genetic makeup.
- In human body cells, one of the 23 pairs of chromosomes carries the genes that determine sex. In human females the sex chromosomes are the same (XX); in human males the sex chromosomes are different (XY).
- Some characteristics are controlled by a single gene. Each gene may have different forms called alleles.
- An allele that controls the development of a characteristic when it is present on only one of the chromosomes is a dominant allele.
- An allele that controls the development of characteristics only if the dominant allele is not present is a recessive allele.
- A gene occupies a fixed position, called a locus, on a particular chromosome.
- A gene is a base sequence of DNA that codes for the amino acid sequence of a polypeptide.
- Random mutation can result in new alleles of a gene.
- Meiosis produces daughter cells that are genetically different from each other.
- The process of meiosis consists of two nuclear divisions, resulting usually in the formation of four haploid daughter cells from a single diploid parent cell.
- Genetically different daughter cells result from the independent segregation of homologous chromosomes.
- Crossing over between homologous chromosomes results in further genetic variation among daughter cells.

TEST YOURSELF ON PRIOR KNOWLEDGE

1 Match each word with its definition.

Table 7.1

Word	Definition
1 Heterozygous	**A** An allele that controls the development of a characteristic when it is present on only one of the chromosomes
2 Mutation	**B** A cell in which the two alleles of the same gene are different
3 Genotype	**C** A different form of a gene
4 Phenotype	**D** A sequence of DNA bases that codes for one polypeptide
5 Homozygous	**E** A change in the base sequence of DNA resulting in a new allele
6 Haploid	**F** An allele that controls the development of a characteristic only when the dominant allele is not present
7 Diploid	**G** A cell that contains one set of chromosomes
8 Dominant	**H** The genetic makeup of an individual
9 Recessive	**I** The characteristics of an organism resulting from its alleles plus environment
10 Allele	**J** A cell in which both alleles of a gene are the same
11 Gene	**K** A cell that contains two sets of chromosomes

2 Draw a diagram to show how sex is determined at fertilisation in a human.

3 Describe what happens during crossing over and explain how it results in variation.

4 What is independent segregation?

5 A tall pea plant with genotype Tt is crossed with a dwarf pea plant with genotype tt. What are the possible genotypes and phenotypes in their offspring?

Introduction

Each gene occupies a position on a particular chromosome, called its **locus**. Ever since this was confirmed, less than 100 years ago, geneticists have tried to locate the position of genes on chromosomes.

The first gene maps were based on the results of genetic crosses, like the ones you will read about in this chapter. The reasons that the fruit fly has been used so often in genetics investigations are that it has a short life cycle, is easy to culture in large numbers (which gives statistically reliable results) and has only four pairs of homologous chromosomes.

Very early on, geneticists noticed that fruit flies sometimes inherited the same combination of alleles of different genes from generation to generation. For example, they inherited the alleles for black body colour, vestigial wings and cinnabar eyes as if they were in a single block of genes. The geneticists reasoned that this was because the genes were located very close together on the same chromosome. As a result, the combination of alleles in one parent would be inherited as a single unit, unless they were separated by crossing over in the first prophase of meiosis during gamete formation.

Extension

Cross-over values

By performing large numbers of crosses, geneticists estimated the frequency of crossing over between genes on sister chromatids of homologous pairs during meiosis. They called this the cross-over value. They reasoned that genes with a small cross-over value were closer together and thus less likely to be separated by crossing-over events than those with a larger cross-over value. Thus, cross-over values allowed geneticists to judge the relative positions of genes on chromosomes. Figure 7.1 shows the type of chromosome map they produced for the black body, cinnabar eyes and vestigial wing alleles on chromosome 3 of the fruit fly, *Drosophila melanogaster*.

Figure 7.1 A map showing the position of three alleles (b), (cn) and (vg) on chromosome 3 of *Drosophila*, based on cross-over values.

A second type of chromosome map resulted from studies using optical microscopes. When chromosomes are stained with chemical dyes, bright-coloured and dark-coloured bands appear along their length. A gene's cytogenetic location can be related to these bands on the chromosome. Figure 7.2 shows human chromosome 19. The bands, caused by Giemsa stain, are very obvious.

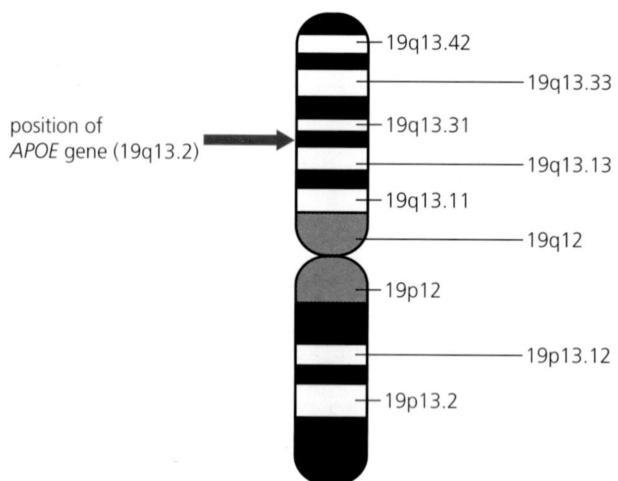

Figure 7.2 A drawing of human chromosome 19 prior to DNA replication. The restriction near the middle of the chromosome is called the centromere and it divides the chromosome into two 'arms'. The position of one gene – the *APOE* gene – is shown.

Extension

The Human Genome Project

The Human Genome Project (HGP) enabled biologists to produce a molecular map of DNA that gives more detailed information than the cytogenetic method above. The HGP was an international project completed in 2003. During the project a number of research teams around the world used different techniques to find the base sequence of each human chromosome. As a result the position of a human gene can be described in terms of the base pairs it occupies along a DNA molecule.

Phenotype and genotype

Phenotype The features of an organism that result from an interaction between the expression of its genes and their interaction with the environment.

Genotype The alleles of a gene (genetic constitution), or all of the alleles of all of the genes, that an individual inherits.

When we study an organism, we notice a number of distinctive features that it possesses. These might be observable features, such as flower colour or length of beak, or they might be chemical differences, such as the inability to produce lactase in lactose-intolerant people. These observable or measurable features make up the phenotype of an organism.

Figure 7.3 summarises how an organism's phenotype results from an interaction between its genes and the environment in which it lives. An organism's genetic constitution is called its genotype. We can refer to the genotype meaning all of an organism's genes or we can refer to the genotype controlling a single characteristic.

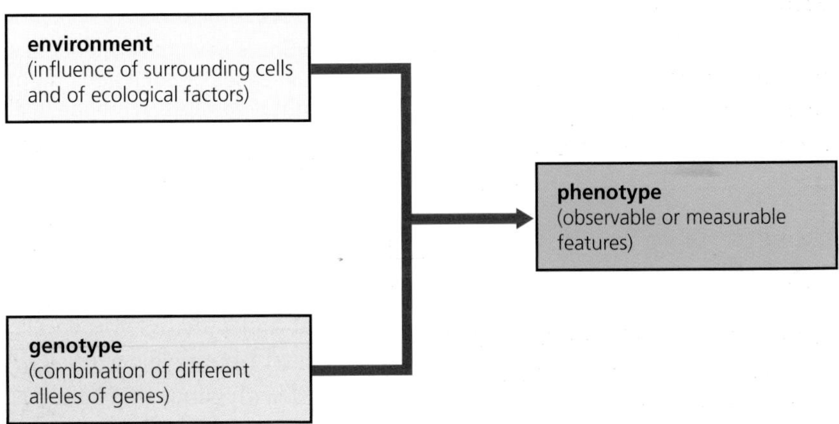

Figure 7.3 An organism's phenotype results from an interaction between its genotype and its environment.

Figure 7.4 This Himalayan rabbit has inherited genes that result in the production of black fur. Only the fur on its nose, ears, tail and paws is black. Why is the rest of its fur white? The answer lies in the interaction of genotype and environment, in this case temperature, in producing the phenotype.

The Himalayan rabbit is an intriguing example of the interaction of genotype and the environment (see Figure 7.4). This rabbit has a gene that encodes an enzyme that results in black fur. As you can see only a few parts of the rabbit's body have black fur; the rest of its fur is white. The enzyme encoded by the black-fur gene has a low optimum temperature. At temperatures above 34 °C this enzyme is denatured. Over most of the rabbit's body the enzyme is inactivated by body heat, resulting in white fur. Only its extremities – the nose, ears, tail and paws – are cold enough for the enzyme to be active.

Many non-scientists seem to believe that an organism's phenotype is controlled entirely by its genotype. For example, certain newspapers tell their readers that scientists have found a gene that 'causes' coronary heart disease (CHD), giving people hope that a genetic cure for CHD can be found. Readers of these newspapers seem to find it convenient to blame their genes (their genotype) for a high risk of CHD and to ignore the environmental risk factors over which they have control, such as a diet rich in saturated fats, smoking tobacco or failure to take exercise.

Scientists locate gene that causes some breast cancers

OBESITY GENE FOUND

Figure 7.5 Some newspaper headlines suggest that there are simple genetic causes of diseases or conditions. Although possession of a particular allele of a gene might *predispose* someone to a disease such as cancer, environmental factors, such as diet or smoking, are also involved.

Genes and alleles

A gene is a sequence of bases in a DNA molecule that encodes another functional molecule. As you will see in Chapter 10, some genes encode molecules of ribonucleic acid (RNA). However, in this chapter we are only concerned with genes that encode functional polypeptides. Thus, in this chapter, we are using the term **gene** to mean a sequence of bases in a DNA molecule that encodes a functional polypeptide. How many bases there are in the DNA sequence determines how many amino acids are in the encoded polypeptide.

Allele An alternative form of a gene.

Often a gene can have more than one form, each with a slightly different base sequence. An **allele** is one of two or more different forms of a gene. Look back to page 125; base pairs 50 100 901 to 50 104 488 on chromosome 19 carry the genetic code for apolipoprotein E: it is the *APOE* gene. However, this gene can have at least three different base sequences, called **e2**, **e3** and **e4**. These different sequences are alleles of the *APOE* gene. Figure 7.6 shows how different alleles of a gene result in the formation of polypeptide chains with slightly different amino acid sequences.

In Western Europe, the most common allele of the *APOE* gene is e3. The base sequence of this allele results in the production of a functional lipoprotein that carries excess cholesterol from the blood to the liver; that is, it is fully functional apolipoprotein E.

People with a copy of the e4 allele of the *APOE* gene produce a different polypeptide that is not as efficient at carrying cholesterol to the liver and which results in an increased risk of atherosclerosis (an accumulation of fatty deposits and scar-like tissue in the lining of the arteries). The progressive narrowing of the arteries that can result from atherosclerosis increases the risk of heart attacks and strokes. People with a copy of the e4 allele are also at increased risk of developing clumps of proteins, called amyloid plaques, in their brains. As a result, these people have an increased

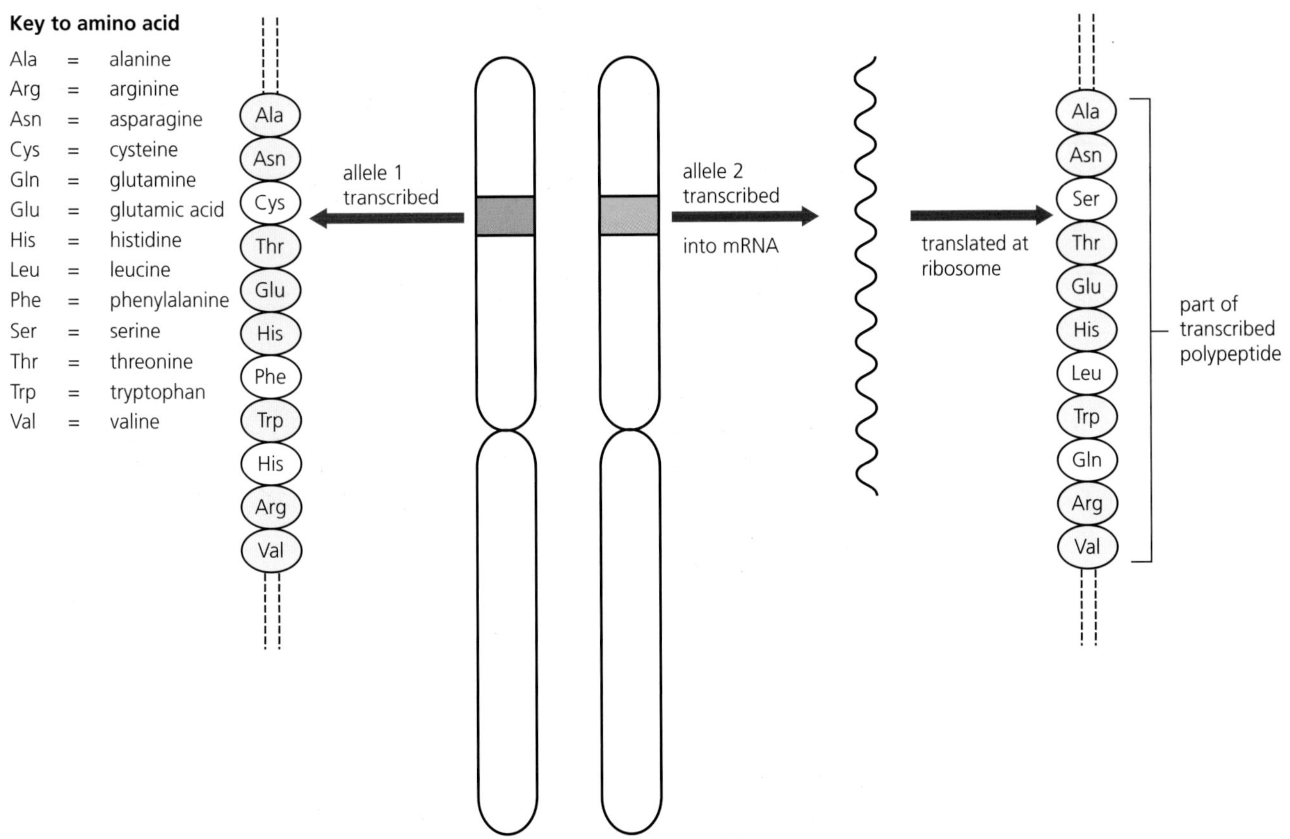

Figure 7.6 A gene may have one or more different alleles. Each allele has a slightly different sequence of DNA bases, resulting in the formation of a polypeptide chain with a different sequence of amino acids.

risk of developing a form of Alzheimer's disease. This risk is even greater if they possess two copies of the e4 allele.

People with a copy of the e2 allele of the *APOE* gene produce yet another polypeptide that increases their risk of developing hyperlipoproteinaemia type III. This is a condition associated with a very high concentration of cholesterol and triglycerides in the blood.

TIPS

- Make sure you know the difference between a gene and an allele, and that you use both terms correctly.
- There are no specific examples of inheritance that you need to learn, but make sure you understand the principles involved.

TEST YOURSELF

1. What is the source of the heat that denatures the enzyme that causes the Himalayan rabbit in Figure 7.4 to produce black fur?
2. Explain the difference between a gene and an allele.
3. Explain why it is wrong to say that the e4 allele causes heart attacks and strokes.

Homozygous and heterozygous genotypes

Look at Figure 7.7 to remind yourself of something you learned in the first year of your A-level Biology course. In sexual reproduction in humans, diploid cells divide by meiosis to produce haploid gametes: egg cells and sperm cells. Each haploid cell contains only one of the homologous chromosomes from each pair. Since these chromosomes carry the same genes in the same order, a haploid cell carries only one copy of each gene. At fertilisation a zygote is formed, which is diploid. This zygote, and every cell formed from it by mitosis, carries two copies of each gene.

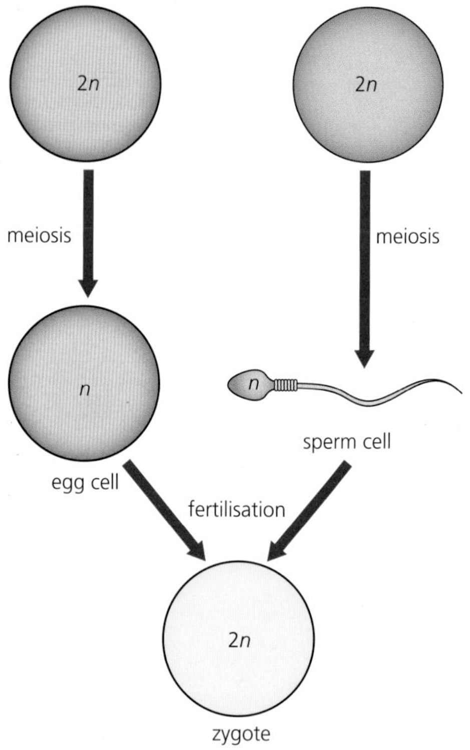

Figure 7.7 Egg cells and sperm cells have only one copy of each chromosome and thus only one copy of each gene carried on a chromosome. At fertilisation, the zygote cell becomes diploid: it has two copies of each gene.

In diploid organisms, such as humans, every **somatic** (body) cell has the same genotype. We can refer to the genotype of a somatic cell and the genotype of an individual interchangeably, since they are the same.

- If an individual possesses two identical alleles of the same gene, we say the genotype is **homozygous** for that gene. We represent a homozygous genotype using only one type of symbol, such as AA, aa, C^AC^A or C^aC^a.
- If an individual possesses two different alleles of the same gene, we say the genotype is **heterozygous**. We represent a heterozygous genotype using two different symbols, such as Aa or C^AC^a.

Homozygous Possessing the same alleles of genes at the same locus on homologous chromosomes.

Heterozygous Possessing different alleles of genes at the same locus on homologous chromosomes.

Dominant, recessive and codominant alleles

We have seen that a gene is a sequence of DNA bases that encodes a functional polypeptide. However, if a gene has two alleles, often one of them will be **dominant** and one will be **recessive**. This means that one (the dominant allele) will code for the functional polypeptide, whereas one (the recessive allele) will code for a different polypeptide, which may be non-functional. As a result, the second allele has an effect on the phenotype that is different from the normal allele of this gene.

Imagine an organism that is heterozygous for one gene. The dominant allele will contribute to the phenotype, but the recessive allele will contribute to the phenotype only if the dominant allele is not present. Look at

Table 7.2. It shows the effect of the three alleles of the human gene encoding the ABO blood group.

Table 7.2 The gene controlling the ABO blood group in humans has three alleles: I^A, I^B and I^O. The symbol 'I' stands for immunoglobulin, a type of globular protein.

Allele of gene controlling ABO blood group	Polypeptide in the surface membranes of red blood cells encoded by allele
I^A	antigen A
I^B	antigen B
I^O	neither antigen A nor antigen B

Notice from Table 7.2 that the I^A and I^B alleles encode functional polypeptides that are located in the cell-surface membrane of red blood cells. These polypeptides act as antigens, meaning that they will cause an immune response in the blood of another mammal that lacks the same polypeptide. In contrast, the I^O allele does not encode a functional polypeptide in the cell-surface membrane.

Table 7.3 shows the genotypes that are possible for the human blood group gene. Remember that, although there are three alleles of this gene, only two of them can be present in any one cell.

Table 7.3 The possible genotypes for the human ABO blood group and the phenotypes associated with them. Here, blood group is the phenotype.

Genotype	Antigen present in surface membrane of all red blood cells	Name of blood group
I^AI^A	antigen A	group A
I^AI^O	antigen A	group A
I^BI^B	antigen B	group B
I^BI^O	antigen B	group B
I^AI^B	antigen A and antigen B	group AB
I^OI^O	neither antigen A nor antigen B	group O

What can we learn from Table 7.3? First, notice that the name of the blood group derives from the polypeptides (antigens) in the surface membranes of the red blood cells. Now look at the first row of the table. This tells us that someone who is homozygous for the I^A allele has only antigen A in the surface membranes of their red blood cells. You might have expected this, but now look at the second row. This tells us that, like the homozygous I^AI^A person, a heterozygous I^AI^O person has only antigen A in the surface membranes of their red blood cells and is also blood group A. From this we conclude that the I^A allele is dominant and the I^O allele is recessive.

The third and fourth rows of the table tell a similar story about the I^B and I^O alleles. We conclude that the I^B allele is dominant and the I^O allele is recessive. Not surprisingly, only when the genotype is homozygous I^OI^O does the effect of the I^O allele show in the phenotype.

Now look at the fifth row of the table. In the heterozygous genotype I^AI^B, both antigen A and antigen B are present in the cell-surface membranes of red blood cells. When both the alleles of a gene in a heterozygote show their effect in the phenotype we call them **codominant** alleles.

TEST YOURSELF

4 The genotypes for individuals show two genes, such as AA, Aa or aa. Explain why there are two genes in the genotype.
5 Explain why only two alleles of a gene can be present in a diploid cell.
6 What is an antigen?
7 Which of the alleles in Table 7.2, I^A, I^B or I^O, is recessive?
8 a) What is an allele?
b) Alleles I^A and I^B are codominant. Explain the meaning of this statement.

Monohybrid inheritance involving only dominant and recessive alleles

In studying **monohybrid inheritance**, we follow the inheritance of a single character that is controlled by one gene. As we learn more about gene action, we find that few characteristics are actually controlled by only a single gene. Figure 7.8 shows examples of Fast Plants®, a form of *Brassica rapa* that was bred for research activities and is often used in school and college investigations because it has a very short life cycle. The plants on the right produce a purple pigment, called anthocyanin, in all parts of the plant. The plants on the left do not produce anthocyanin.

Figure 7.8 Fast Plants® normally produce a purple pigment like those on the right of the photograph. These plants have a dominant allele that enables them to produce anthocyanin. The plants on the left of the photograph are homozygous for a recessive allele of this gene. They do not produce anthocyanin.

Constructing a genetic diagram to explain a monohybrid cross

Figure 7.9 shows what happens when a homozygous Fast Plant® that produces anthocyanin is crossed with a plant that does not produce anthocyanin. The dominant allele for anthocyanin production is given the symbol A; the recessive allele is given the symbol a. The **genetic diagram** has been laid out in a particular way.

The first row in the diagram shows the phenotypes of the parents; in this case one can produce anthocyanin and one cannot. The next line shows the genotypes of the parents.

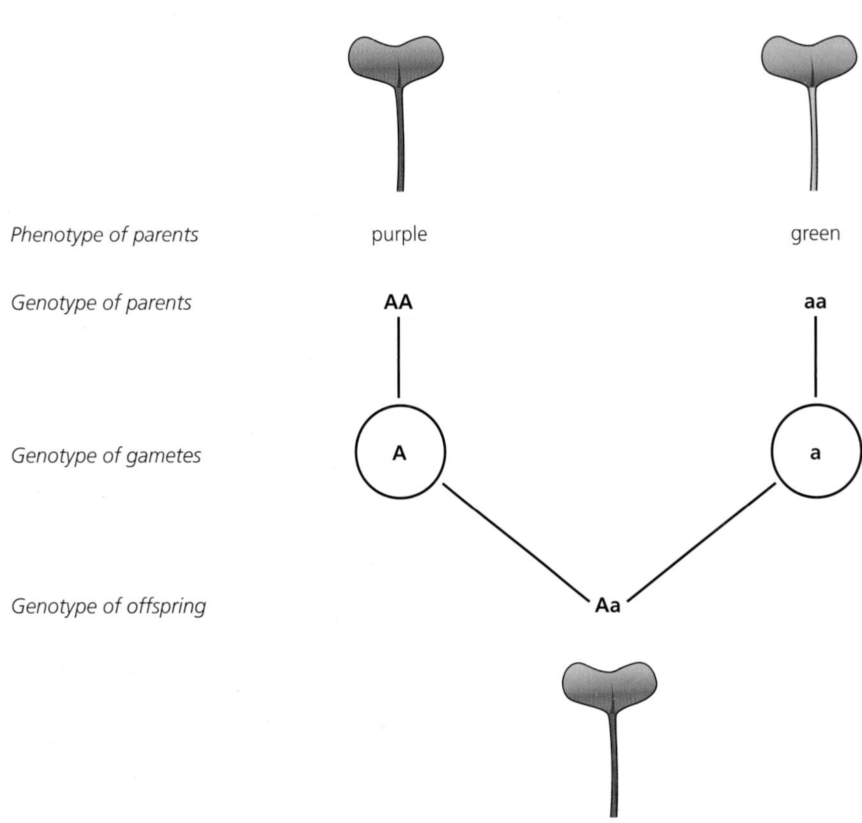

Figure 7.9 A genetic diagram showing monohybrid inheritance of anthocyanin production when a purple Fast Plant® is crossed with a green Fast Plant®. You should use the layout of this genetic diagram when answering questions involving monohybrid inheritance.

We know that the parent that cannot produce anthocyanin must be homozygous, since the allele that results in no anthocyanin being produced is recessive (a). If the dominant allele (A) was present in the genotype, the plant would be able to produce anthocyanin. The next row of the genetic diagram shows the genotype of the gametes that each parent can produce. It helps to make the diagram clear if we put these genotypes in a circle.

The genotypes of the gametes contain only one allele, A or a, because gametes are always haploid. Therefore, the gametes produced during monohybrid inheritance will always contain only one of the alleles controlling the character we are investigating. The next row in Figure 7.9 shows the genotype of the offspring. Since all the gametes of one parent contain the A allele and all the gametes of the other parent contain the a allele, all the offspring must have the genotype Aa: they are heterozygotes. Their phenotype shows the effect of the dominant A allele, so they are all able to produce anthocyanin.

Figure 7.10 shows what happens if these heterozygotes are allowed to interbreed. It uses the same layout as Figure 7.9.

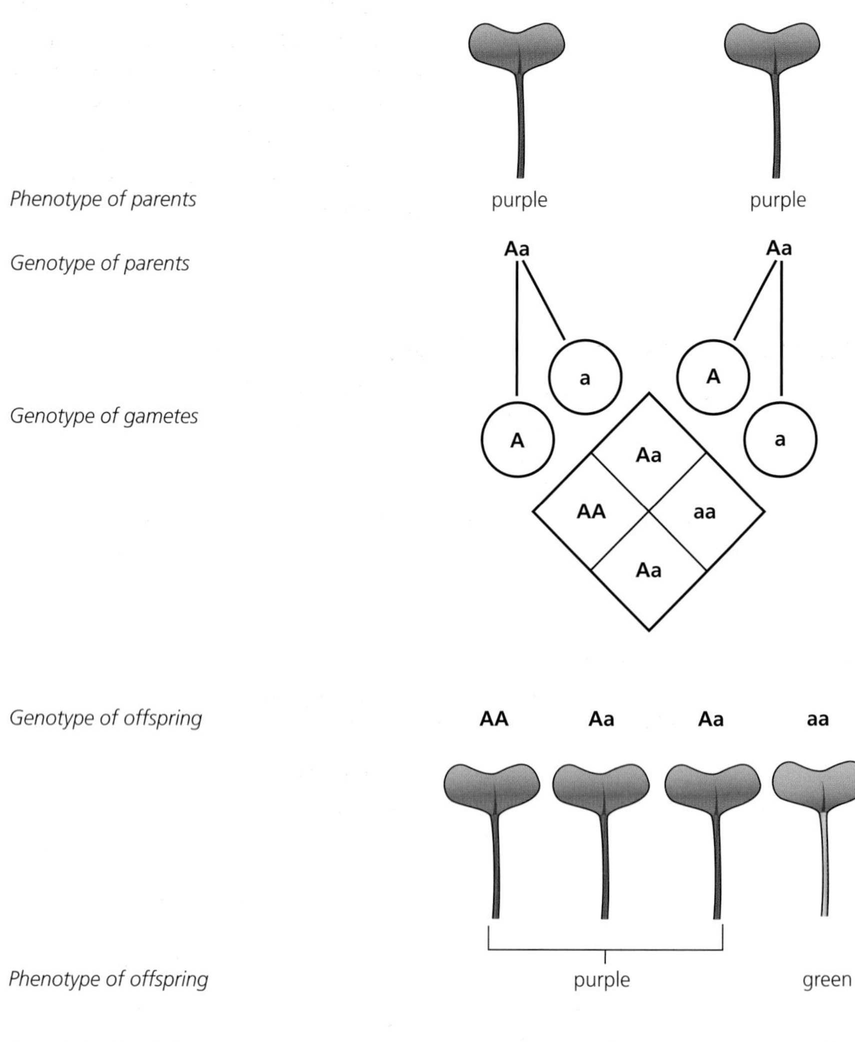

Figure 7.10 A genetic diagram showing the result of a cross between two heterozygous Fast Plants®. In a monohybrid cross involving a dominant and a recessive allele, a ratio of 3 : 1 is always expected in the offspring of heterozygous parents.

During the first division of meiosis, homologous chromosomes pair together. They are then pulled apart, one to each end of the spindle, when spindle fibres contract. One chromosome carries the A allele and the other carries the a allele, so the gametes with A and a will be produced in equal numbers.

Look at the fourth row in Figure 7.10. Instead of using the notation we used in Figure 7.9 to show fertilisation, we have used a Punnett square. If we represented fertilisation by drawing lines between the two gametes from one parent and the two gametes from the other, we would end up with a network of crossing lines. Not only does this look untidy, it can confuse us as we complete the genetic diagram. We are much less likely to make a mistake, and so get our answer wrong, by using a Punnett square.

In a monohybrid cross between two heterozygous parents, there are three possible genotypes among their offspring. In this case, they are AA, Aa and aa. Since the A allele is dominant, plants with genotypes AA and Aa can produce anthocyanin, so there are only two phenotypes, purple and green.

The expected ratio of phenotypes in Figure 7.10 is shown as 3 : 1. To make this prediction we have assumed that fertilisation between the gametes of the parents occurs at random. If there are a large number of fertilisations, we expect gametes containing an A allele from one parent to fuse with equal frequency with gametes containing the A allele or a allele from the other parent. This gives us the expected ratio of phenotypes in the offspring.

EXAMPLE

Monohybrid inheritance in *Drosophila*

The first photo at the beginning of this chapter (page 123) shows several adult fruit flies (*Drosophila melanogaster*). These particular flies have long wings, but other adults have small, stunted wings, called vestigial wings. Wing length is controlled by a single gene that has two alleles, one resulting in long wings and one resulting in vestigial wings.

Two heterozygous, long-winged *Drosophila* were mated together. There were 100 flies in the offspring generation. About three-quarters of the offspring had long wings and about one-quarter had vestigial wings.

1 Which of the two alleles for wing length is dominant and which is recessive? Explain your answer.
Long wings (wild type) is dominant because some of the offspring have vestigial wings but neither parent does. Therefore vestigial wings is recessive. We know this because when you breed long wings and vestigial together you get long-winged offspring. However, few vestigial-winged flies would survive in the wild; therefore, if the gene was dominant the species might become extinct.

2 How many of the offspring generation would you expect to have vestigial wings? Use a genetic diagram to explain your answer.
The fact that some of the offspring are vestigial winged when neither parent is tells you that the allele for vestigial wings is recessive. However, vestigial winged flies have two recessive alleles, one from each parent. Therefore each parent must be heterozygous for wing length.
Let L be the symbol for long wings and l the symbol for vestigial wings.
The parent flies must both be heterozygous.
So of 100 flies, 75 would be expected to have long wings and 25 would be expected to have vestigial wings.

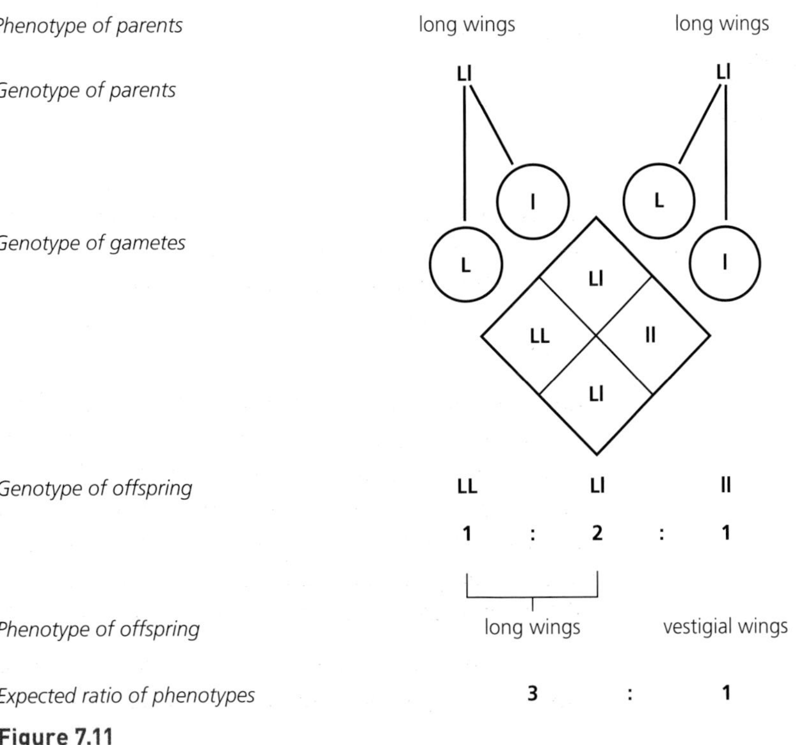

Figure 7.11

TEST YOURSELF

9 Two short-haired rabbits were interbred and they had eight offspring. Two were long-haired and six were short-haired.
  a) Which allele is dominant, long or short hair? Explain your answer.
  b) Give the genotypes of the parents.
  c) Use a Punnett square to show how these parents produced these offspring.

10 In humans, cystic fibrosis is a genetic condition determined by a recessive allele. Two parents who do not have cystic fibrosis have a baby with cystic fibrosis.
  a) Explain how this happened. Use a Punnett square in your answer.
  b) What is the chance that a second baby born to this couple will also have cystic fibrosis?

Monohybrid inheritance involving codominant alleles

Snapdragons are commonly grown in gardens. Some snapdragon plants produce red flowers and some produce white flowers. Flower colour is controlled by a single gene that has two alleles, one for red flowers and one for white flowers. In this example, the two alleles are codominant. Figure 7.12 shows what happens when a snapdragon with red flowers is crossed with a snapdragon with white flowers.

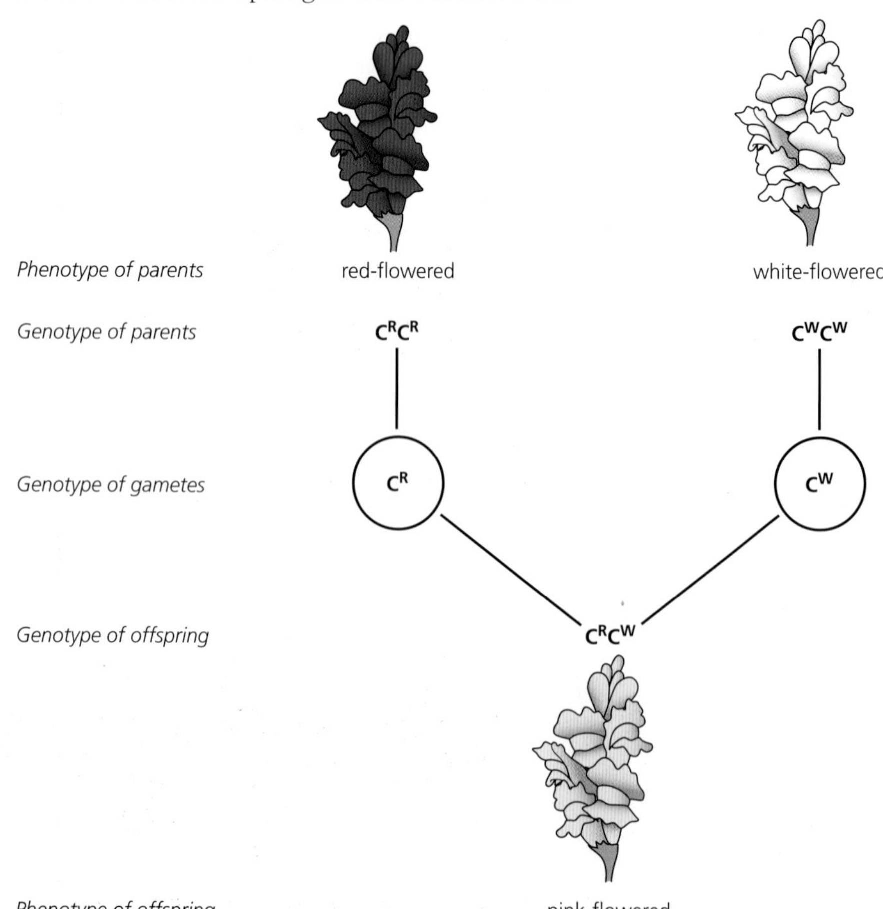

Figure 7.12 Flower colour in snapdragons is controlled by one gene that has two codominant alleles. The offspring of a red-flowered plant and a white-flowered plant all have pink flowers.

TIP

Note that when alleles are codominant we use a different kind of symbol for the alleles. In this example we have used the letter C (to represent the gene for colour) with 'W' for white or 'R' for red in superscript. The W and R are capital letters because neither of them is recessive or dominant.

When the red-flowered plant is crossed with the white-flowered plant, all the offspring are heterozygous. If the allele for red flowers was dominant, all these heterozygotes would have red flowers; if the allele for white flowers was dominant, all these heterozygotes would have white flowers. As you can see in Figure 7.12, the heterozygotes produce flowers that are neither red nor white; they are pink. Why are they pink? Because the allele for red flowers and the allele for white flowers are codominant: they both show their effect in the phenotype of a heterozygote. Some red and some white pigment is produced, making the flowers pink.

You can see the effect of codominant alleles in Figure 7.13. This diagram shows two different crosses, one between two plants with pink flowers and another between a plant with pink flowers and a plant with white flowers.

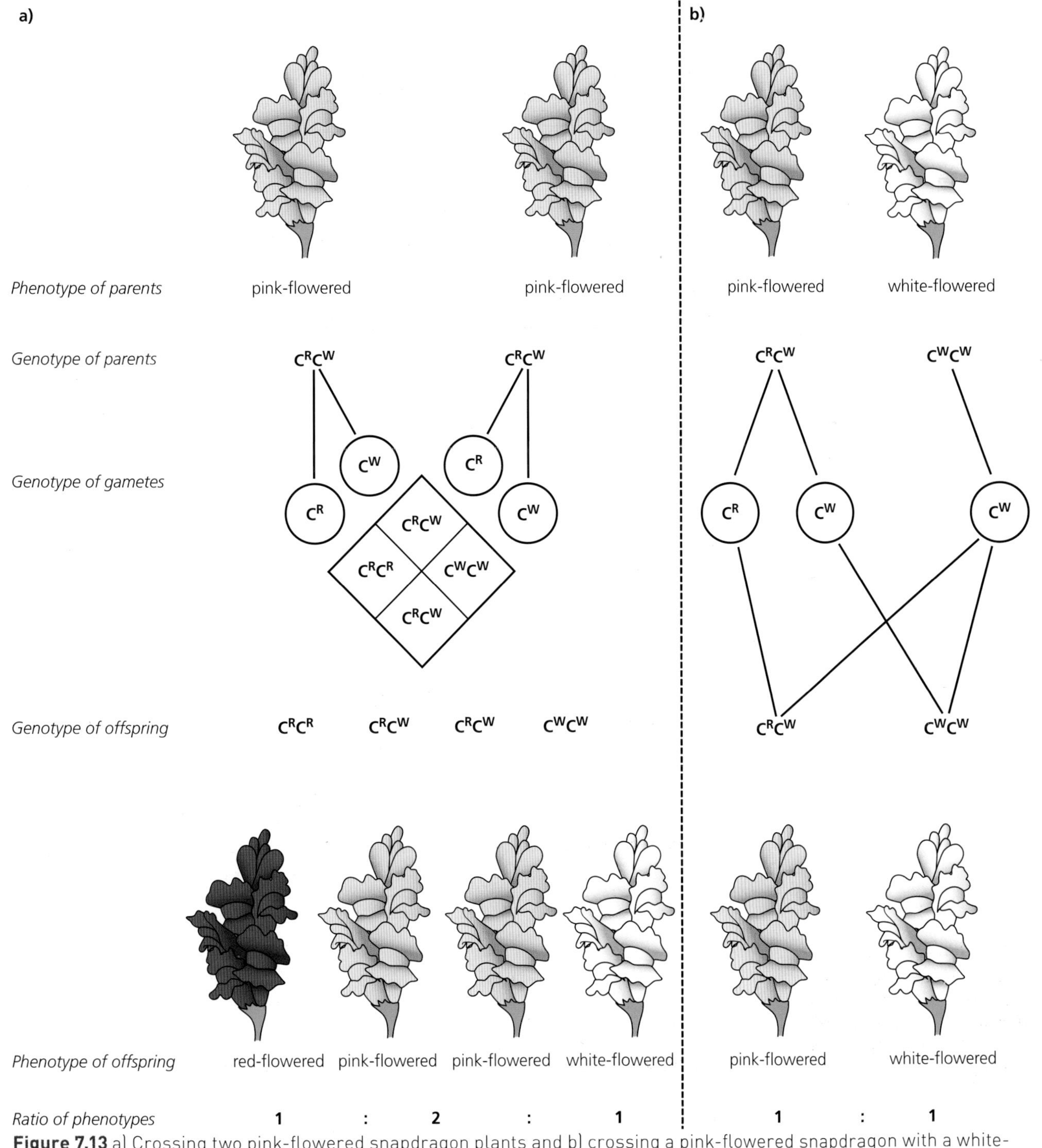

Figure 7.13 a) Crossing two pink-flowered snapdragon plants and b) crossing a pink-flowered snapdragon with a white-flowered snapdragon.

You can see that, instead of the 3 : 1 ratio we expected in monohybrid inheritance with a dominant and recessive allele (Figure 7.10), we get a ratio of 1 : 2 : 1 (Figure 7.13a).

EXAMPLE

Inheritance of coat colour in horses

Like all mammals, horses have hair. We refer to a horse's hair as its coat. Some horses have a reddish-brown-coloured coat, called bay, and some have a white-coloured coat. One gene for coat colour is involved, but it has two alleles, C^R and C^W. If a horse is homozygous C^RC^R all the hairs in its coat will be reddish-brown and the horse will have a bay coat. If a horse is homozygous C^WC^W all the hairs in its coat will be white and the horse will have a white-coloured coat. However, if a horse is heterozygous, C^RC^W, half the hairs in its coat will be reddish-brown and half will be white. The result is a horse with a pinkish-colour coat. This coat colour is referred to as red-roan. Coat colour in horses is an example of monohybrid inheritance with codominant alleles of a single gene. Figure 7.14 shows horses with these three coat colours.

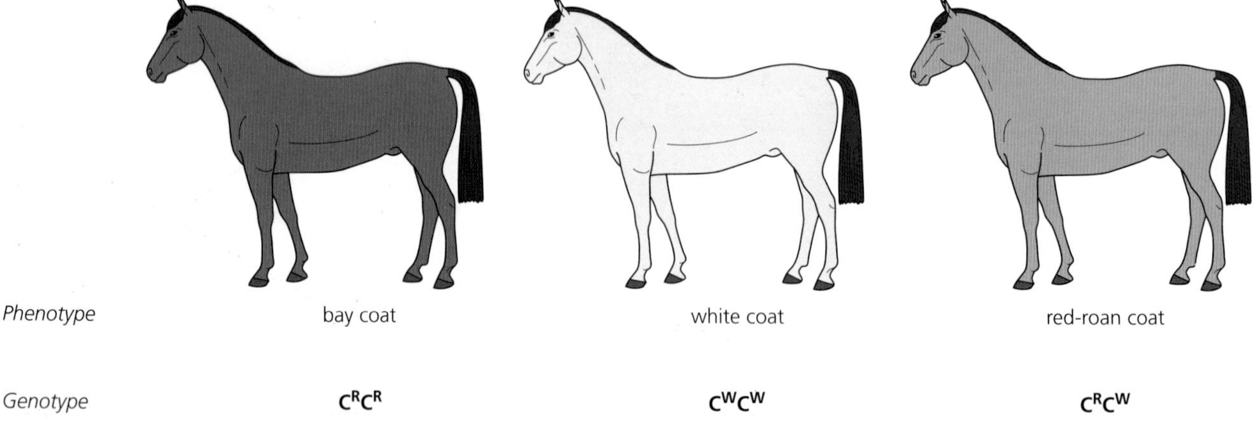

Figure 7.14 Horses with a reddish-brown coloured coat (bay) are homozygous C^RC^R; horses with a white-coloured coat are homozygous C^WC^W. Half the hairs in the coat of heterozygotes, C^RC^W, are reddish-brown and half are white, producing a coat colour called red-roan.

1 In the coat of a red-roan horse, half the hairs are reddish-brown and half are white. What does this suggest about the alleles of the coat-colour gene in the hair-producing cells?

It suggests the alleles are codominant.

When a stallion with a bay coat was mated with a mare with a white coat, all the offspring were red-roan.

2 Why, in the above statement about Figure 7.14, was it unnecessary to explain whether the parent horses were homozygous or heterozygous?

A horse with bay coat or white coat must be homozygous, since the heterozygotes are red-roan.

3 What ratio of coat colour would you expect in the offspring of red-roan parents? Use a genetic diagram to explain your answer.

You would expect offspring that were red-roan, bay and white in a ratio of 2 : 1 : 1. Figure 7.15 shows the genetic diagram.

Parent phenotype red-roan × red-roan

Parent genotype C^RC^W × C^RC^W

Gametes: C^W, C^R and C^R, C^W

Offspring: C^RC^R, C^RC^W, C^RC^W, C^WC^W

Offspring genotypes	C^RC^W		C^RC^R		C^WC^W
	2	**:**	**1**	**:**	**1**
Offspring phenotypes	red-roan	:	bay	:	white
Expected ratio of phenotypes	**2**	**:**	**1**	**:**	**1**

Figure 7.15 Genetic diagram for coat colour in horses.

Monohybrid inheritance involving multiple alleles and codominance

We have looked at monohybrid inheritance involving genes with only two alleles. Many genes have more than two alleles; we refer to them as **multiple alleles**. We saw in Table 7.3 (page 133) the genotypes and phenotypes involved in the inheritance of the human ABO blood group. This feature involves multiple alleles (three alleles of one gene) and it involves codominance. Figure 7.16 shows a cross involving the ABO blood group. One parent is blood group AB and the other is blood group O.

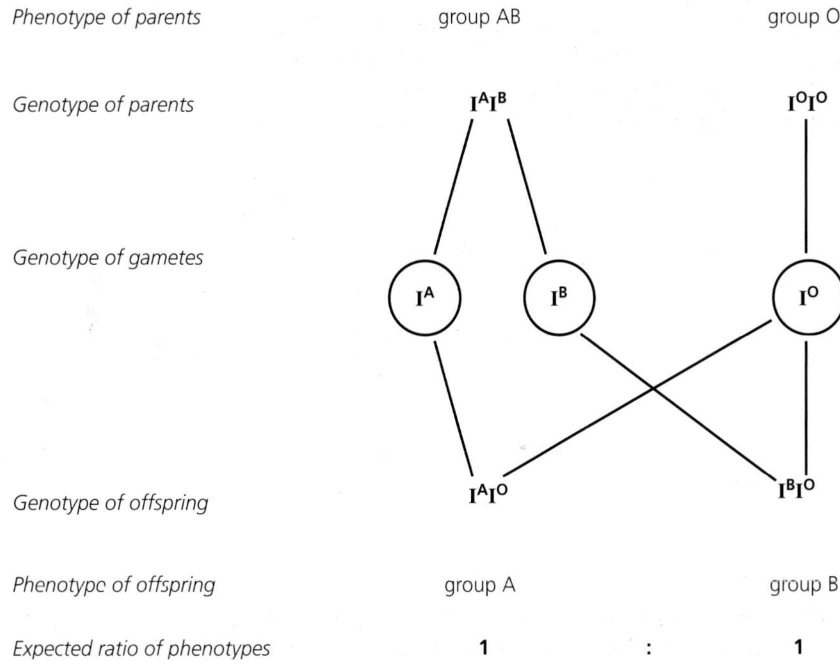

Figure 7.16 A parent of blood group AB and a parent of blood group O will have children of blood group A or blood group B, but none with the same blood group as the parents.

TIP

Please look at Chapter 13 on maths skills, to find out more about statistical tests and to see a worked example of this test.

Because data from genetic crosses is in discrete categories, the best way to present the totals for each phenotype would be on a bar graph. You would use a chi-squared (χ^2) test to find out whether the number of each phenotype matched the expected ratio.

TEST YOURSELF

11 What ratio of blood groups would you expect among the offspring of a mother with the genotype I^AI^O and a father with the genotype I^BI^O? Explain your answer.

12 In Andalusian fowl, F^B is the allele for black plumage and F^W is the allele for white plumage. These alleles show codominance. The heterozygous condition results in blue plumage. List the genotypic and phenotypic ratios expected from the crosses:

a) black × blue

b) blue × blue

c) blue × white

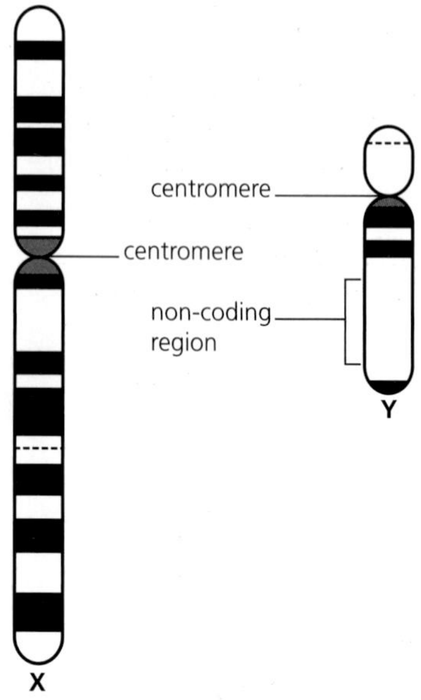

Inheritance of sex

In humans, the members of a homologous pair of chromosomes are usually the same size and carry the same genes in the same order. This is not, however, true of chromosome pair 23. Figure 7.17 shows chromosome pair 23 from a human male. One of the chromosomes (the Y chromosome) is very short and carries few genes; the other (the X chromosome) is long and carries many more genes. Although the short Y chromosome carries very few genes, one of them is crucial in determining sex. This gene, called the testis-determining gene (or *SRY*), turns the developing sex organ of a 7-week-old embryo into a testis, and the embryo develops into a male. Because the X chromosome carries some genes which the Y chromosome does not, a male is effectively haploid for these genes.

A human male has one X chromosome and one Y chromosome; his genotype is XY and half his sperm cells will carry an X chromosome and half will carry a Y chromosome. A human female has two X chromosomes; her genotype is XX and all the eggs she ever produces will carry an X chromosome.

Figure 7.17 Chromosome pair 23 in a human male. The two homologous chromosomes are different sizes and only share a small number of genes. The shorter Y chromosome does not have a copy of many of the genes present in the X chromosome.

Monohybrid inheritance involving a sex-linked character

A sex-linked character is one that is controlled by a gene located on one of the sex chromosomes. A sex-linked gene is on the X chromosome. Examples of sex-linked characters in humans include haemophilia (a disorder in which the blood of sufferers clots only slowly), red-green colour blindness and Duchenne muscular dystrophy.

Figure 7.18 represents what might happen when a woman with blood that clots normally has children with her husband who has haemophilia. The symbols we are using in this diagram represent two things: the sex chromosome and the alleles of the gene for blood clotting. The allele for normal blood clotting is dominant, so we represent it as H. The allele for slow blood clotting is recessive, so we represent it as h. However, this gene is located on the X chromosome.

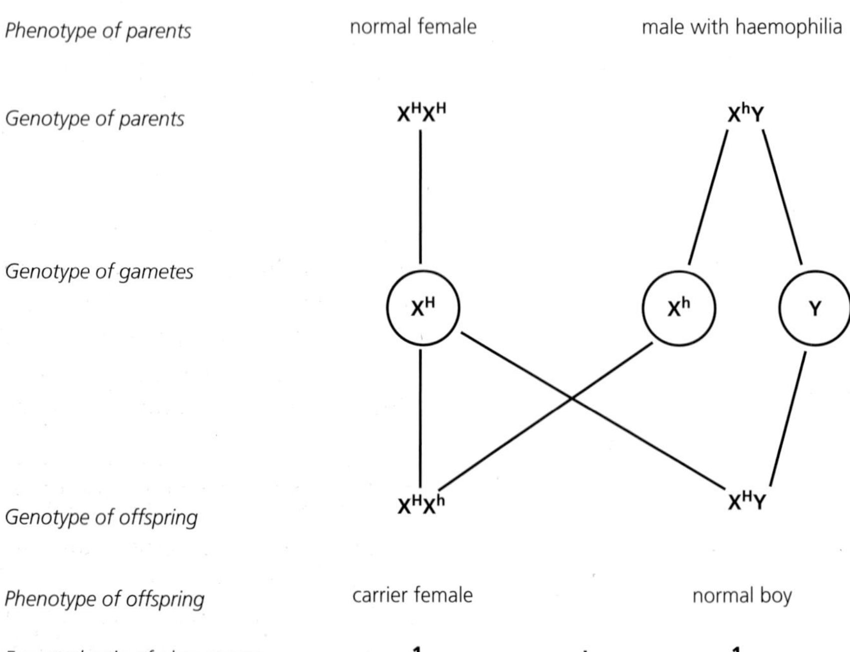

Figure 7.18 Possible children produced by a normal woman whose husband has haemophilia.

We show this using the symbol X^H for an X chromosome with the H allele and X^h for an X chromosome with the h allele. Since the Y chromosome has no copy of the gene for blood clotting, it is shown simply as Y.

Notice in Figure 7.18 that any boy born to this couple will have blood that clots normally; he will not have haemophilia like his father. Any girl born to this couple will be heterozygous for the blood-clotting gene. We call her a **carrier** for haemophilia but, since the H allele is dominant, her blood clots normally. Figure 7.19 shows what might happen if this girl who is a carrier grows up, marries a man whose blood clots normally and has children.

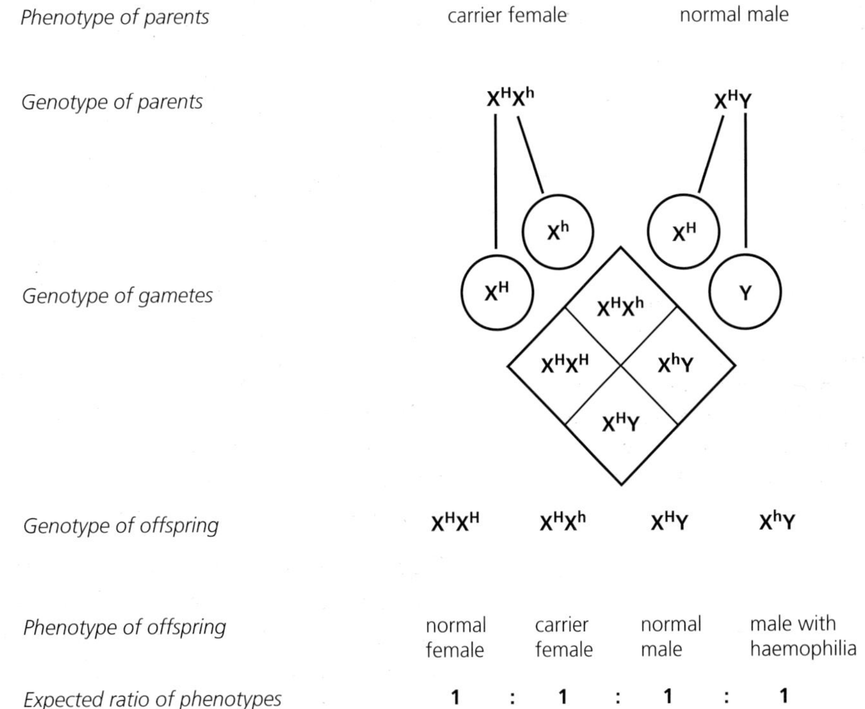

Figure 7.19 Possible children produced by a woman who is a carrier for haemophilia and her normal husband.

TEST YOURSELF

13 Use a genetic diagram to explain why there is a 1 in 2 chance of a child being a girl. Use the symbols X and Y to denote X and Y chromosomes.

14 Suggest why sex-linked genes are usually on the X chromosome rather than on the Y chromosome.

15 Would it be possible for a girl to have haemophilia? Explain your answer.

Notice that the woman in Figure 7.19 can produce a son whose blood clots normally and a son who has haemophilia. The son with haemophilia inherits the recessive gene for haemophilia on the X chromosome from his mother, whose own father had the disorder (see Figure 7.18). This is typical of characteristics that are controlled by genes that are located on the X chromosome: the condition appears in males in alternate generations.

Interpreting pedigrees

So far, we have used genetic diagrams to explain a pattern of inheritance. An alternative method for presenting information about inheritance is shown in Figure 7.20 overleaf. This diagram shows a **pedigree**. In pedigrees such as this, females are represented by circles and males by squares. Couples who have children are linked by a horizontal line and the children are represented in a hierarchical fashion below them. In the case of humans, we cannot breed them, so a pedigree is a record of the family tree of crosses in the past.

You can see in the figure that individual 1 is represented by a square, so he is male. The square contains the symbol A, so he has blood group A. Individual 2 is shown by a circle containing the symbol B, so she is a female with blood group B.

Figure 7.20 This pedigree shows the inheritance of ABO blood groups in three generations of a family.

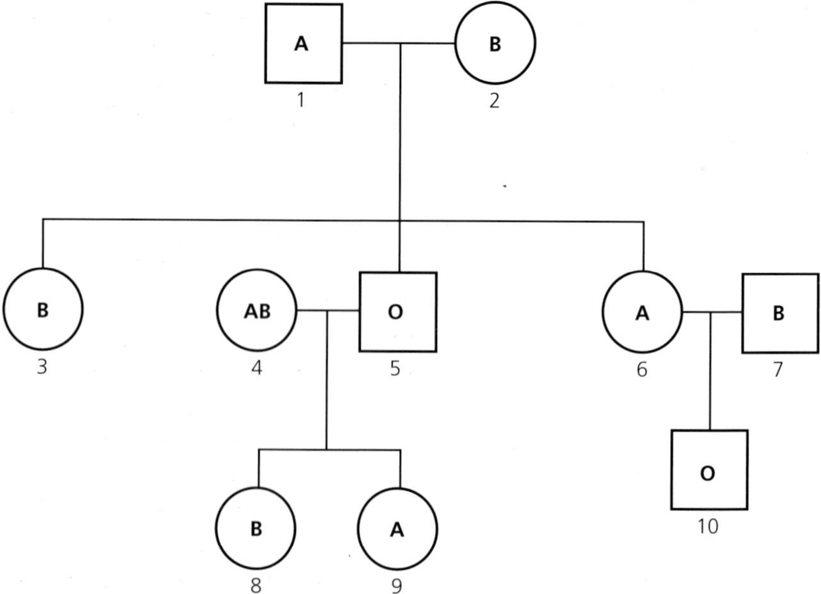

We can see who, out of individuals 3, 4, 5, 6 and 7, are the children of couple 1 and 2. The diagram shows a horizontal line between individuals 1 and 2. This indicates that they had children together. The vertical line from individuals 1 and 2 goes directly downwards to connect to each of their children, so individuals 3, 5 and 6 are the children of the parents 1 and 2. Individuals 4 and 7 are not connected by vertical lines to the parents 1 and 2, so are not their children. You can see from the horizontal lines that individual 4 is the partner of 5, and individual 7 is the partner of 6.

A pedigree only tells us phenotypes, it does not tell us genotypes. When we interpret a pedigree, we have to use clues provided by the phenotypes of some individuals to work out the genotypes of as many people in the pedigree as we can. The skill you need to develop is to identify which individuals provide these clues.

Individual 5 is blood group O, so we know his genotype is I^OI^O. Having found this, we can work backwards and forwards through the pedigree to work out other genotypes. Since individual 5 has the genotype I^OI^O, he must have inherited one I^O allele from each parent. However, his father (individual 1) has blood group A, so his father's genotype must be I^AI^O. The mother of individual 5 (individual 2) has blood group B, so her genotype must be I^BI^O.

Individual 3 is blood group B. She must have inherited the I^B allele from her mother (individual 2). If she had inherited the I^A allele from her

TIP

If you are asked for the probability of something, you should give a decimal fraction, as in the example above. A probability has a maximum value of 1, a minimum value of 0 and intermediate values that are decimal fractions, for example 0.125. An answer such as '1 in 8', although representing the chance of something happening, is not a correct way of expressing probability.

father, she would be blood group AB. Since she is blood group B, she must have the genotype I^BI^O. Individual 6 is blood group A. She must have inherited the I^A allele from her father. She cannot have inherited an I^B allele from her mother or she would be blood group AB, so she must have the genotype I^AI^O.

Since individual 10 is blood group O, we know he must have the genotype I^OI^O. This means that his mother (6) and father (7) must carry the I^O allele, so they are I^AI^O and I^BI^O, respectively. Consequently, they are equally likely to produce offspring with the genotype I^AI^O, I^AI^B, I^BI^O or I^OI^O. If you are not sure why, draw a genetic diagram of $I^AI^O \times I^BI^O$ to see what offspring can result. This gives a probability of a child with blood group O of 0.25. The probability of a child being a boy is 0.5. So, the probability of a boy with blood group O is $0.25 \times 0.5 = 0.125$.

Dihybrid inheritance

Dihybrid inheritance involves a phenotype that is inherited as the result of two different genes. This involves the same principles you learned carrying out monohybrid crosses. However, the number of possible phenotypes increases because there are more different ways in which the alleles of two different genes can combine.

Dihybrid inheritance with no linkage

Let's look at the inheritance of two characters in pea plants: stem height and flower colour. Each character is controlled by a different gene. The two genes are located on different chromosomes; that is, they are unlinked.

- The gene for height has two alleles, T (tall) and t (dwarf).
- The gene for flower colour has two alleles, R (red) and r (white)

Figure 7.21 shows the results of crossing a homozygous tall, red-flowered pea plant with a dwarf, white-flowered plant.

Notice that the offspring of these two plants are all tall and red-flowered plants. This is because the parents can only produce one kind of gamete, so all the offspring have a dominant allele for height and a dominant allele for flower colour. However, when these mature offspring are interbred, each parent can produce four different kinds of gamete. This is because, during meiosis, either allele of the gene for height can end up with either allele of the flower colour gene. At fertilisation, these gametes give 16 different combinations in the offspring.

If you look carefully at the Punnett square in the diagram, you will see that the phenotypic ratio for each character is what you would expect from your previous work:
3 tall : 1 dwarf
3 red-flowered : 1 white-flowered.

However, due to independent assortment of non-homologous pairs of chromosomes, this dihybrid cross gives a phenotypic ratio of:
9 tall, red-flowered plants
3 tall, white-flowered plants
3 dwarf, red-flowered plants
1 dwarf, white-flowered plant.

Figure 7.21 Dihybrid inheritance in peas.

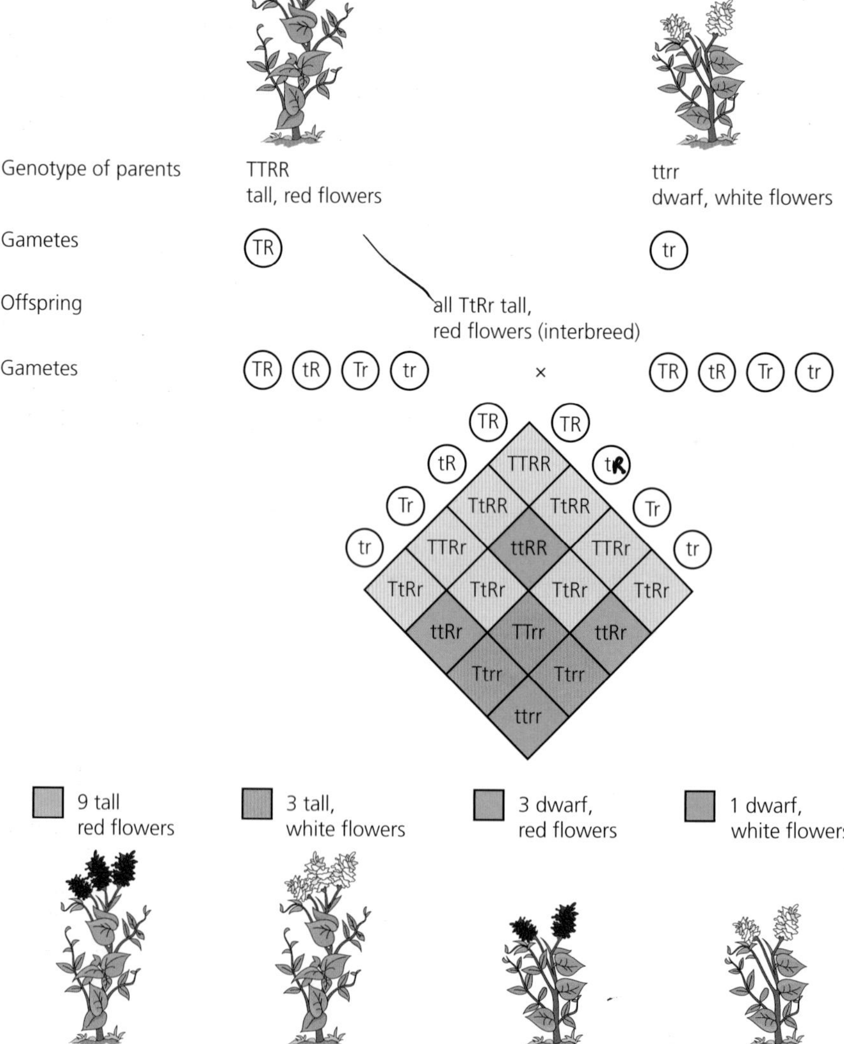

TEST YOURSELF

16 In squash, the gene for fruit colour has two alleles. The allele for white fruit colour (W) is dominant over that yellow fruit colour (w). The gene for fruit shape also has two alleles. The allele for disc-shaped fruit (D) is dominant over that for sphere-shaped fruit (d).
A plant that produced white, disc-shaped fruit was crossed with a plant that produced yellow, sphere-shaped fruit. All the offspring produced white, disc-shaped fruit.

a) Give the genotypes of the two parent plants and of the offspring.

b) The offspring plants were interbred. Use a Punnett square to show the possible genotypes and phenotypes of their offspring, and the ratio in which you would expect them to be produced.

17 In shorthorn cattle, the gene for possession of horns has two alleles. The allele for the polled (hornless) condition, H, is dominant over that for horned (h). The gene controlling coat colour has two codominant alleles, C^R and C^W. The genotype C^RC^R results in a red coat and C^WC^W results in a white coat but the heterozygote, C^RC^W, is roan.

a) A homozygous polled white male is crossed with a horned, red female. What will be the appearance of the offspring?

b) If you bred two of the offspring together, what would be the possible genotypes and phenotypes of their offspring? Use a Punnett square.

18 In humans, red-green colour blindness is caused by a sex-linked recessive allele. A woman with blood group O who has normal vision but had a father with red-green colour blindness, marries a man with normal colour vision and blood group AB. Use a Punnett square to show the possible genotypes and phenotypes of their children.

Epistasis

In cells, enzymes control reactions. Many metabolic pathways, such as respiration and photosynthesis, consist of several steps, with each step controlled by a different enzyme. This means that each enzyme in the pathway depends on the previous enzyme to provide its substrate. If any of the enzymes in a pathway is non-functional, the pathway comes to a halt.

Figure 7.22 shows a metabolic pathway involving two enzymes. You can see that a different gene codes for each enzyme. The effect of enzyme B depends upon the action of enzyme A. Homozygous recessive individuals for gene A (aa) produce an inactive enzyme A, so they are unable to catalyse the reaction that results in a pale blue pigment. Even if this individual produces an active form of enzyme B, the individual will be colourless.

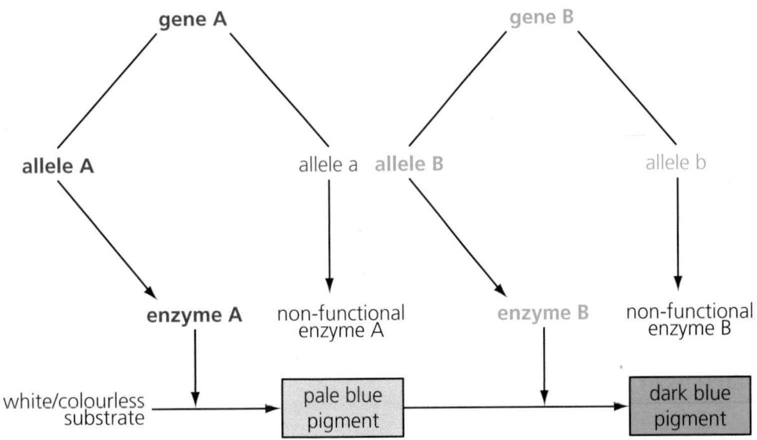

Figure 7.22 A metabolic pathway with two enzymes.

This is called **epistasis**, which means that the expression of one gene affects the expression of another. This occurs in metabolic pathways controlled by enzymes coded for by different genes. Epistasis reduces the number of possible phenotypes.

Figure 7.23, overleaf, shows inheritance of comb shape in chickens, which is another example of epistasis. Comb shape is controlled by two genes, each with two alleles, that interact to produce four different phenotypes:

the rose gene has two alleles, R and r
the pea gene has two alleles, P or p.

Table 7.4

Shape of comb	Genotype
Rose	RRpp or Rrpp
Pea	rrPP or rrPp
Walnut	RrPp or RRPp or RrPP
Single	Rrpp

Figure 7.23 Inheritance of comb shape in chickens.

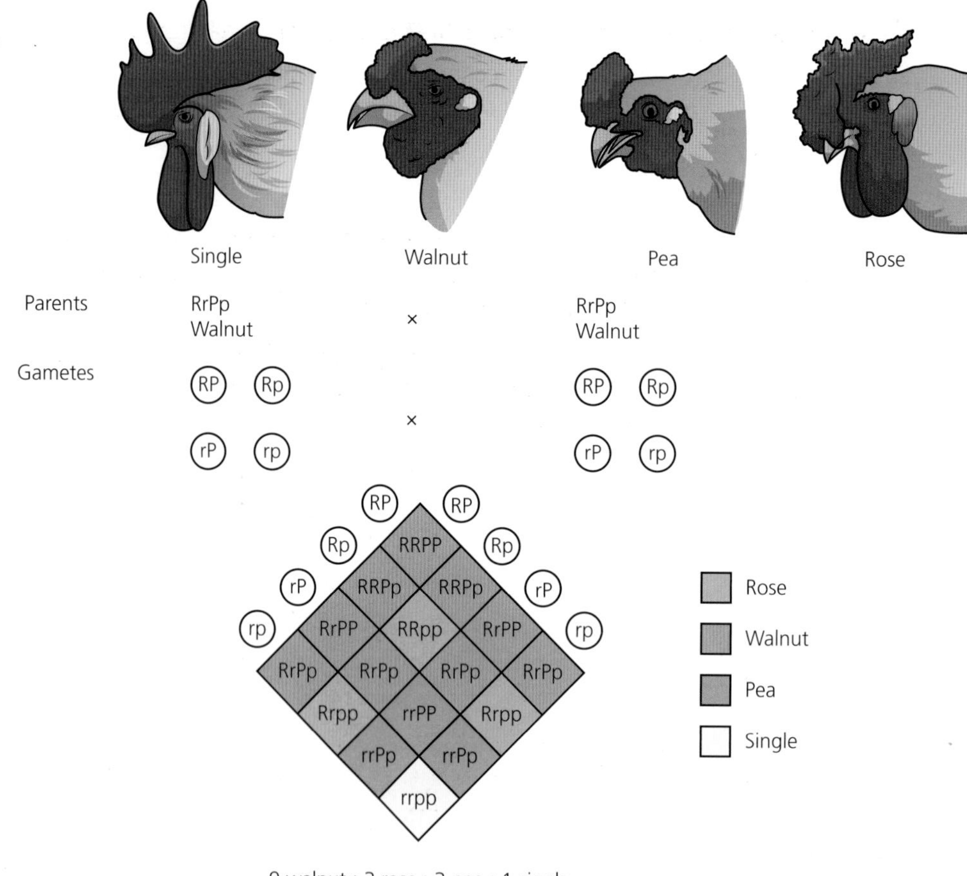

TEST YOURSELF

19 One gene for fur colour in mice has three alleles, C^A, C^B and C^C. The results of some crosses between mice are shown in Figure 7.24.

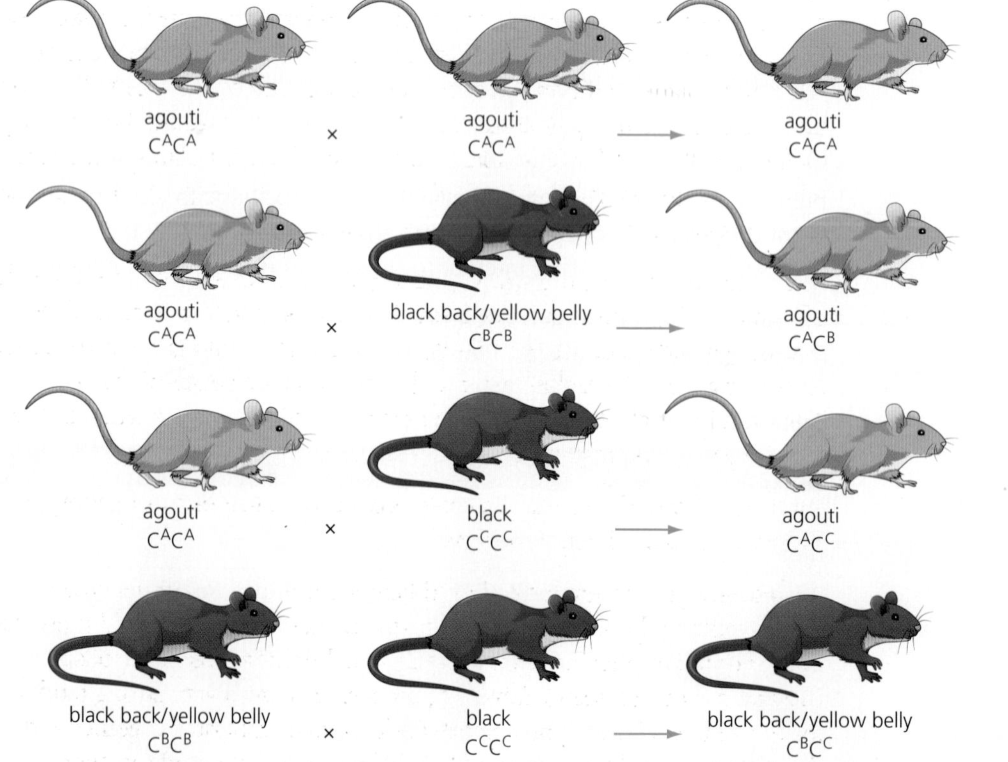

Figure 7.24

a) Complete the table to show the possible genotype(s) for each phenotype

Phenotype	Possible genotypes
Agouti	
Black back yellow belly	
Black	

b) Use a Punnett square to show the possible offspring that would be produced from a cross between an agouti mouse of genotype C^AC^C and a mouse with a black back and yellow belly of genotype C^BC^B.

20 The diagram shows a metabolic pathway in sweet peas. Two plants with the genotype CcPp were interbred. Use a Punnett square to show the ratios of phenotypes and genotypes in their offspring.

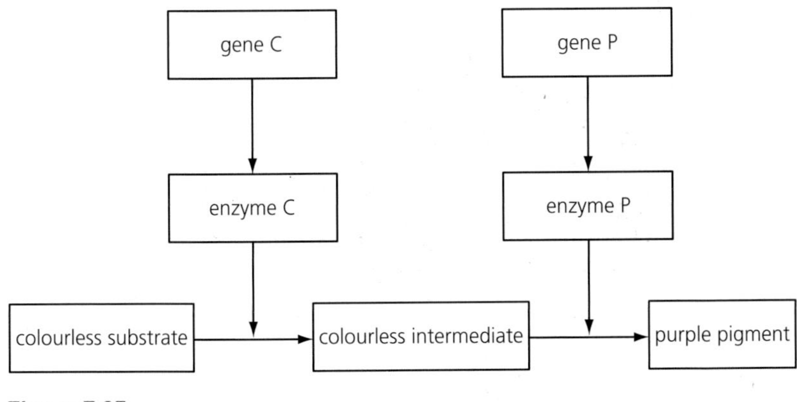

Figure 7.25

Dihybrid cross with autosomal linkage

Recombinant offspring An organism that contains a different combination of alleles from either of its parents.

In Figure 7.21 you saw that, when two tall, red-flowered (TtRr) pea plants were crossed, some of the offspring showed the same traits as their parents, i.e. tall and red-flowered, but others were dwarf and/or had white flowers. This is only possible because the gene for height in pea and the gene for flower colour are on different chromosomes. This allows independent assortment to occur so the parent plants could produce four different gametes. The T or t allele passed into the gametes independently of the R or r allele. As a result, the offspring included tall, white-flowered plants, dwarf, red-flowered plants and dwarf, white-flowered plants. These are called recombinant offspring because they are displaying new combinations of these traits.

In one kind of maize, there is a gene that determines seed colour. One allele of this gene (Y) codes for an enzyme that results in yellow seeds, while the recessive allele (y) codes for a faulty enzyme that results in colourless seeds. Another gene codes for an enzyme that controls the shape of the seeds. The dominant allele of this gene (S) results in smooth seeds whereas the recessive allele (s) results in wrinkled seeds.

Table 7.5 shows the result of a cross between a maize plant of genotype YySs and a plant of genotype yyss.

If these two genes were located on different chromosomes, independent assortment would occur. As we have seen earlier, this cross would then result in four different phenotypes in roughly equal proportions in the offspring. Instead, as you can see in Table 7.5, the results show a very strong tendency for the yellow allele, Y, and smooth allele, S, to be inherited together and for the colourless allele, y, and wrinkled allele, s, to be inherited together.

Table 7.5 Inheritance of seed colour and shape in one variety of maize.

Gametes			Gametes produced by YySs parent			
			YS	Ys	yS	ys
	Gametes produced by yyss parent	ys	YySs	Yyss	yySs	yyss
Phenotype			Yellow smooth	Yellow wrinkled	Colourless smooth	Colourless wrinkled
Proportion expected with independent assortment (%)			25	25	25	25
Actual results (%)			48.2	48.2	1.8	1.8

The explanation is that the gene loci for the seed colour gene is on the **same chromosome** as the seed shape gene. As they are on the same chromosome, alleles of the two genes cannot be separated by independent assortment and are inherited together,

Nevertheless, you can see in Table 7.5 that there are a few recombinants produced in this cross. Recombinant offspring have a different combination of alleles from both parents. This can be explained by crossing over during meiosis (*see AQA A-level Biology 1 Student's Book*, Chapter 11, page 195). If a chiasma is formed between the seed colour gene and the seed shape gene, the chromatids of homologous chromosomes exchange pieces and a new combination of alleles is produced.

Therefore the genes for seed colour and seed shape in this variety of maize are said to be **linked**; that is, they are located on the same chromosome. As this chromosome is not a sex chromosome – an **autosome** – we refer to this type of linkage as **autosomal linkage**.

Linkage group Sets of genes on the same chromosome which tend to be inherited together.

Geneticists have studied many characteristics in organisms such as maize. They have found many genes that are linked. Some of these give a higher proportion of recombinants than the example we have just studied, while others give a lower proportion. Geneticists assume that the more recombinants produced, the further apart two gene loci are. From this information they have discovered linkage groups of genes that tend to be inherited together, and they have worked out the relative positions of the genes on the chromosomes. This is what you learned about at the start of this chapter.

TEST YOURSELF

21 When tall tomatoes with red fruit are crossed with dwarf tomatoes with yellow fruit, all the offspring are tall with red fruit. When these offspring are interbred, they also produce offspring that are tall with red fruit, except for a very few dwarf plants with yellow fruit. What does this tell you about the inheritance of height and fruit colour in tomatoes?

22 In sweet peas, the allele of the gene for flower colour that results in purple flowers (R) is dominant over that for red flowers (r) and the allele of the gene for pollen shape that results in long pollen (L) is dominant over that for round pollen (l).
Two plants heterozygous for flower colour and pollen shape were interbred.

a) Use a Punnett square to show the ratios of genotypes and phenotypes that you would expect in the offspring.

b) The actual results were 296 plants with purple flowers and long pollen, 19 with purple flowers and round pollen, 27 with red flowers and long pollen and 85 with red flowers and round pollen. Explain these results.

Practice questions

1 Tay–Sachs disease is a rare and usually fatal genetic disorder that causes progressive damage to the nervous system. The diagram shows a family in which some individuals have Tay–Sachs disease.

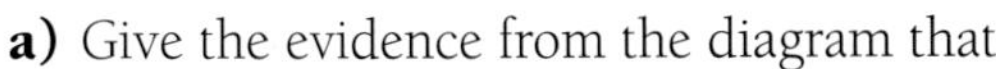

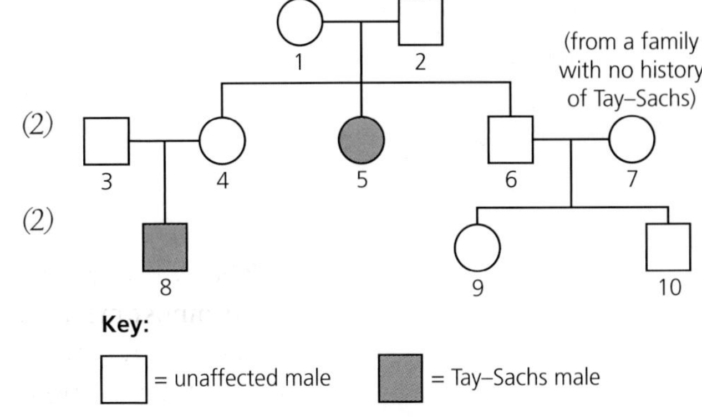

a) Give the evidence from the diagram that

i) Tay–Sachs disease is caused by a recessive allele (2)

ii) Tay–Sachs disease is not sex-linked. (2)

b) Give the possible genotype(s) for

i) individual 4

ii) individual 6

iii) individual 10. (3)

2 a) What is a sex-linked gene? (2)

b) In cats, the X chromosome carries a gene for coat colour. The gene has two alleles that show codominance: X^G results in ginger fur, X^B results in black fur. Cats with the genotype X^GX^B are covered in patches of ginger fur and patches of black fur, referred to as tortoiseshell. A tortoiseshell female was mated with a black male. Use a genetic diagram to show the possible genotypes and phenotypes among the offspring resulting from this cross. (4)

c) Tortoiseshell male cats are very rare. Explain why. (2)

3 Some forms of clover are cyanogenic. This means that they produce cyanide gas when their leaves are damaged. The production of cyanide gas occurs using a metabolic pathway controlled by two enzymes. A different gene encodes each of these enzymes.

a) Explain why the following genotypes of clover cannot produce cyanide gas (i.e. they are acyanogenic): aaBB, aaBb, AAbb, Aabb. (1)

b) Two clover plants, each with the genotype AaBb, were crossed.

i) Give all the possible genotypes among the gametes that each parent could produce. (1)

ii) Use a genetic diagram to show the genotypes of the offspring of this cross. (2)

iii) What proportion of the offspring are likely to be acyanogenic? (1)

4 The ABO blood group in humans is determined by a single gene with three alleles: I^A, I^B and I^O.

a) i) Distinguish between the terms gene and allele. *(1)*

ii) What is meant by codominance? *(1)*

b) The diagram shows the inheritance of ABO blood groups in one family. The blood group of some individuals is given in the pedigree.

Using information in the diagram and the symbols I^A, I^B and I^O, give the phenotype and genotype of each of the following individuals in the pedigree. Give reasons for your answers. *(4)*

Individual	Phenotype	Genotype	Explanation
2			
5			
4			
9			

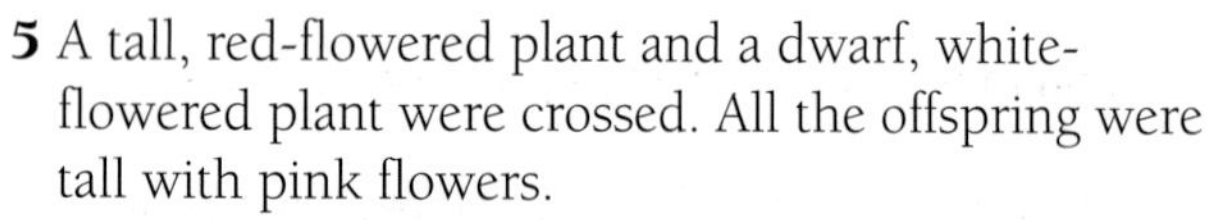

5 A tall, red-flowered plant and a dwarf, white-flowered plant were crossed. All the offspring were tall with pink flowers.

a) What was the genotype of the tall, pink-flowered plants? *(1)*

b) These tall, pink-flowered plants were interbred to produce a large number of offspring. Use a genetic diagram to show the ratio of phenotypes and genotypes in their offspring. *(3)*

c) What results would you expect if you crossed a tall, red-flowered plant with a dwarf, white-flowered plant, and the genes for height and flower colour were on the same chromosome? Explain your answer. *(2)*

d) A plant breeder wanted to find out whether a particular tall plant was homozygous or heterozygous for height. Explain how the breeder could carry out a genetic cross to solve this problem. *(3)*

Stretch and challenge

6 Geneticists can produce a chromosome map from the results of crosses involving autosomal linkage. Explain how they would do this.

7 Explain how sex is determined in humans. Turner's syndrome, Klinefelter's syndrome and Swyer syndrome are all abnormalities of sex chromosomes in humans, for example. Compare and contrast how these abnormalities occur.

8 Explain how sex is determined in birds, insects, reptiles and in some species of animal that are able to change sex.

8 Gene pools, selection and speciation

PRIOR KNOWLEDGE

- Two organisms belong to the same species if they are able to produce fertile offspring.
- Genetic diversity can be defined as the number of different alleles of genes in a population and is a factor enabling natural selection to occur.
- Natural selection in the evolution of populations involves four principles.
- First, random mutation can result in new alleles of a gene.
- Second, many mutations are harmful but in certain environments the new allele of a gene might benefit its possessor, leading to increased reproductive success.
- Third, advantageous alleles are inherited by members of the next generation.
- Finally, as a result advantageous alleles usually increase in frequency in the population over many generations.

TEST YOURSELF ON PRIOR KNOWLEDGE

1. What is a mutation?
2. Scientists carrying out captive breeding of zoo populations of endangered species aim to interbreed animals that are as unrelated to each other as possible. Explain why.
3. Orangutans from the island of Borneo, *Pongo pygmaeus*, and orangutans from the island of Sumatra, *Pongo abelii*, are classified as separate species. However, in zoos it is possible for them to interbreed and produce fertile offspring.
   a) Suggest why they were originally classified as separate species.
   b) Responsible zoos do not allow these two species of orangutan to interbreed. Suggest why.

Introduction

In the 20th century, the rabbit population in the UK grew so large that rabbits became a serious agricultural pest. They also became a pest across mainland Europe and in Australia, where they had been introduced to provide a food source for early settlers.

In the late 19th century, laboratories in Uruguay had discovered a virus that caused skin lesions in the local rabbits but which killed European rabbits. This virus was the myxomatosis virus and the disease that killed European rabbits was myxomatosis. During the 1940s, a number of research organisations studied the use of myxomatosis as a potential biological control method to reduce rabbit populations.

Figure 8.1 You might have seen our common rabbit species, the European rabbit, feeding in fields or parks early in the morning or during the late evening. It might surprise you to learn that rabbits were introduced by the Normans from mainland Europe about 800 years ago and were originally bred for food and fur. With changes in agricultural practices, rabbits became serious agricultural pests in the 20th century.

Initially, governments were unwilling to sanction this method of biological control. One reason for the UK government's reluctance was public opinion: although farmers wanted to be rid of rabbits, most people in the UK had been brought up on cultural traditions that were sympathetic to rabbits. Perhaps you had a fluffy toy rabbit, or read stories involving rabbit characters, as a child. Another reason was that scientists at the time feared the virus might 'jump the species barrier' and infect animals other than rabbits. If this idea seems familiar, exactly the same concern has been expressed more recently about the avian (bird) influenza virus.

Before governments decided to sanction the use of myxomatosis to control rabbits, events took their course. A retired physician, who owned a rabbit-infested farm, released the virus in France in mid-1952. From France, the virus reached England, at Edenbridge in Kent, in 1953 and spread rapidly throughout the UK, killing almost 90% of the rabbit population. In Australia, the virus 'escaped' from a government laboratory in the Murray Valley sometime during the Christmas–New Year holiday in 1950–51. It spread quickly throughout the Murray-Darwin basin, killing millions of rabbits.

Although myxomatosis initially devastated rabbit populations, they have since recovered. Rabbit populations are nothing like as large as they were, but they are now in balance with the myxomatosis virus. How has this balance happened? One reason is that rabbits have become more resistant to the myxomatosis virus. This has occurred through a process called natural selection, in which rabbits with alleles giving resistance to myxomatosis had greater reproductive success than rabbits that were susceptible to myxomatosis.

It might surprise you to learn that natural selection has also changed the virus population. In England, the myxomatosis virus is transmitted from rabbit to rabbit by fleas, when they bite rabbits to feed on their blood. However, in Australia, the virus is transmitted by mosquitoes. A mosquito will only bite a living rabbit, so any virus that kills its host too quickly is unlikely to be passed on to another host by a mosquito. In Australia, natural selection has favoured myxomatosis viruses that are less virulent, that have hosts that live longer and provide greater opportunity for the virus to be passed to another host via a mosquito bite.

Gene pools

Members of a species do not live alone; they live in populations. A **population** is a group of organisms of the same species occupying a particular space at a particular time and potentially able to interbreed. In Chapter 7 we looked at the inheritance of genes by individuals. Now we will look at the alleles of a gene controlling a single characteristic in a population.

Figure 8.2 shows two adults of a single species of the banded snail, *Cepaea nemoralis*. One of the snails has a yellow shell and the other has a pink shell. This shell colour is controlled by two alleles of a single gene. The pink allele (C^P) is dominant over the yellow allele (C^y). Imagine there are 1000 snails in a population living in a beech woodland. Since every snail has two alleles in its genotype for shell colour (C^PC^P, C^PC^y or C^yC^y), we can say that there are 2000 alleles in the shell-colour genotypes of snails in this population. This represents the gene pool for this characteristic in this population. In general, we can define the **gene pool** as all of the alleles of all of the genes that are present in a population at any given time.

Figure 8.2 The banded snail is common in woodlands and grasslands in the UK. The difference in shell colour is controlled by two alleles of a single gene.

Allele frequencies in the gene pool of a population

We saw above that a population of 1000 individuals of banded snail has a gene pool of 2000 alleles of the gene for shell colour. If all the snails in this population had the genotype C^PC^P, all the alleles in the gene pool would be C^P. In other words, the **frequency** of the C^P allele in this gene pool would be 1. If all the snails had the genotype C^PC^y, the frequency of the C^P allele in the gene pool of this population would be 0.5 and the frequency of the

C^y allele would also be 0.5. Therefore, an allele frequency of 1 is equivalent to 100% of the alleles of that gene in the population, and an allele frequency of 0.5 is equivalent to 50% of the alleles in that population.

TIP

Note that C^PC^y, where colour pink is dominant over colour yellow, is the standard notation used by the vast number of researchers working on populations of *Cepaea nemoralis*. However, at A-level you would use the notation given in the question, which might be different.

Hardy–Weinberg: the principle and the equation

Hardy and Weinberg were two scientists who looked at **allele frequencies** within populations. The principle that carries their names, the **Hardy–Weinberg principle**, predicts that the frequency of the alleles of one gene in a particular population will stay the same from generation to generation. In other words, there will be no genetic change in the population over time. The principle is based on mathematical modelling and makes the following five important assumptions.

1 **The population is large**. In small populations, chance events can cause large swings in allele frequencies.

2 **There is no movement of organisms into the population (immigration) or out of the population (emigration).** Any such movement of organisms would result in new alleles entering the gene pool or existing alleles leaving the gene pool.

3 **There is random mating between individuals in the population.** This ensures that there is an equal probability of any allele of a gene being passed on to the next generation.

4 **All genotypes must have the same reproductive success.** This also ensures that there is an equal probability of any allele of a gene being passed on to the next generation.

5 **There is no gene mutation.** Any mutation of genes would cause some alleles in the gene pool to change to different alleles of the same gene.

Using the Hardy–Weinberg equation

We cannot always tell the genotype of an individual from its phenotype. For example, you learned in Chapter 7 that an organism showing the effect of a dominant allele could be homozygous for that allele or it could be heterozygous. This is where the Hardy–Weinberg equation is helpful. It allows us to calculate the frequencies of alleles in the gene pool and of genotypes and phenotypes in the population. Therefore we can use the Hardy–Weinberg equation to find the probability of certain phenotypes in a future generation of a population.

We can always tell which phenotype has the homozygous recessive genotype because the recessive allele only shows its effect in a homozygote. In a heterozygote, the effect of the recessive allele is masked by that of the dominant allele.

TIP

Make sure that you are able to find the frequency of the alleles in a population as a number between 0 and 1 when presented with numbers or percentages.

The Hardy–Weinberg equation is applied to two alleles of a gene. The symbol A represents the dominant allele, which will always show its effect in the phenotype, and the symbol a represents the recessive allele. Since we are dealing with allele frequencies, we need symbols to represent the frequencies of the two alleles in the gene pool. We use the following symbols

p = frequency of the A allele in the gene pool
q = frequency of the a allele in the gene pool.

The gene pool for this gene consists of only two alleles, A and a. The frequency of the gene itself is 1.0, since all organisms have the gene. Consequently, the frequencies of the two alleles added together must also be 1, i.e. $p + q = 1$.

To summarise so far, we have a population with a gene pool for a particular gene that has two alleles. The frequency of the dominant allele, A, is p and the frequency of the recessive allele, a, is q.

So, what is the frequency of genotype AA in the population of individuals? The frequency of the A allele is p, so the frequency of the AA genotype must be $p \times p$, or p^2. We can work out the frequency in the population of each of the three possible genotypes, AA, Aa and aa.

The Hardy–Weinberg equation is:

$$p^2 + 2pq + q^2 = 1$$

In a single population

the frequency of the AA genotype is p^2
the frequency of the aa genotype is q^2
the frequency of the Aa genotype is $2pq$.

If you do not feel confident about the mathematics, do not worry. So long as you remember the above frequencies, you will find the calculations involving the Hardy–Weinberg equation are very easy sums to do.

Let's go back to our beechwood population of 1000 banded snails. When a group of students investigated this population, they found that 160 of the snails had yellow shells and 840 had pink shells. We always start with the phenotype which gives away its genotype, i.e. the homozygous recessive. In this case, yellow-shelled snails are homozygous recessive (C^yC^y).

So, there were 160 snails with yellow shells in the population of 1000 snails. To work out the frequency of yellow-shelled snails in this population you must remember that frequencies are always given as a decimal value. There were 160 yellow-shelled snails in the population of 1000 snails. The frequency of these yellow-shelled snails is 160/1000 = 0.16.

Now we have worked out that the frequency of homozygous recessive snails is 0.16, we need to calculate the frequency of the recessive allele in this population. The Hardy–Weinberg equation tells us that the frequency of the homozygous recessive individuals is q^2. So, if we know the frequency, we simply need to use a suitable scientific calculator to find its square root. In this case the frequency of the yellow-shelled snails $q^2 = 0.16$. Therefore, the frequency of the allele for yellow shells $q = \sqrt{0.16} = 0.4$.

We can now calculate the frequency of the A allele in the gene pool. The Hardy–Weinberg equation tells us that $p + q = 1$. We have just calculated the value of q, so we find the value of p as $1 - q = 1 - 0.4 = 0.6$.

If you did these calculations correctly, you can pat yourself on the back. We are now in a position to work out the frequency of the three genotypes controlling snail shell colour in the beechwood population.

- The frequency of the genotype $C^PC^P = p^2 = 0.6 \times 0.6 = 0.36$.
- The frequency of the genotype $C^PC^y = 2pq = 2 \times 0.6 \times 0.4 = 0.48$.
- The frequency of the genotype $C^yC^y = q^2 = 0.4 \times 0.4 = 0.16$.

We know that our calculation must be correct because $p^2 + 2pq + q^2 = 0.36 + 0.48 + 0.16 = 1$.

TIP

Always check you have the right answers by adding your values for p^2, q^2 and $2pq$. If your arithmetic is right they will add up to 1.

We know from the data above that the students found 160 yellow-shelled snails in the woodland population. From this we can work out how many snails in the population were homozygous with pink shells and how many were heterozygous with pink shells. Hopefully, you will now find the last part of the calculation very easy; all we need to do is to multiply the frequency by the number of snails in the population.

- The frequency of homozygous pink-shelled snails is 0.36. There were 1000 snails in the population, so the number of homozygous pink-shelled snails is $0.36 \times 1000 = 360$.
- The frequency of the heterozygous snails is 0.48, so the number of heterozygous pink-shelled snails is $0.48 \times 1000 = 480$.

We can always check that we are correct, because if we add up the number of different snails we have calculated, we should find they give us the total in the population. In this case, $160 + 360 + 480 = 1000$.

TEST YOURSELF

1 Would you expect the Hardy–Weinberg principle to have held true over the past 20 years in the human population of the town in which you live? Explain your answer.

2 In humans, the ability to roll the tongue is determined by the dominant allele T. Non-rollers are homozygous recessive (tt).

In a school, the following data were obtained:

Tongue-rollers	Non-rollers	Total
490	210	700

How many of the tongue-rollers were heterozygous for tongue-rolling?

3 For one of the genes controlling coat colour in mice, the allele for agouti coat (A) is dominant over that for non-agouti (a). In a sample, 16% were found to have a non-agouti coat.

a) What are the frequencies of the agouti and non-agouti alleles in the population?

b) What percentage of the population would you expect to be homozygous for A, and what percentage would you expect to be heterozygous?

4 'Woolly hair' is common among Norwegian families. People with this condition have hair that is tightly kinked and easily broken. The gene controlling this condition has two alleles. The allele for woolly hair (H) is dominant over that for normal hair (h). In a population of 1200 people, 1092 individuals had woolly hair. Use the Hardy–Weinberg formula to calculate the frequency of each of the genotypes HH, Hh and hh.

5 In a population of 20 000, 16 800 people were found to have the Rhesus positive blood group. The allele for Rhesus positive is dominant over Rhesus negative. How many people in this population are heterozygous for the Rhesus blood group?

Natural selection resulting in differential reproductive success

The Hardy–Weinberg principle predicts that allele frequencies in a large population remain stable from generation to generation. One of the assumptions on which the Hardy–Weinberg principle is based is that all genotypes in the population have equal reproductive success. This will not be true if organisms with one particular genotype have a phenotype that makes them:

- more likely to die before reproducing
- unable to grow sufficiently well to reproduce successfully
- unable to attract a mate.

In all these cases, organisms with one particular phenotype are less likely to reproduce successfully than others in the population with different phenotypes and will leave fewer, or no, offspring. We say there is **differential reproductive success** between the phenotypes in the population.

Differential reproductive success is common in populations. In beech woodlands, the yellow-banded snails shown in Figure 8.2 are very conspicuous against the pink leaf litter lying on the ground. The snails with pink shells are better camouflaged and so are more difficult to find. Song thrushes are birds that eat banded snails. Like us, song thrushes have colour vision. A large number of investigations have shown that song thrushes find more of the conspicuous yellow-shelled snails than they do pink-shelled snails in beech woodlands. As a result, fewer yellow-shelled snails survive to reproduce. This means that fewer of the C^y alleles are passed on to the next generation. The process by which the frequency of alleles in a population changes, because different genotypes have differential reproductive success, is called **natural selection**. Table 8.1 summarises the process of natural selection.

Table 8.1 An explanation of natural selection in a beech woodland population of banded snails. The events in the left-hand column can be applied to any example of natural selection.

Sequence of events leading to natural selection	Application of these events to selection of yellow-banded snails in beech woodlands
Within a population there is variation in phenotypes. These phenotypes result from genetic variation in the genotypes controlling this characteristic and environmental factors.	The C^P and C^y alleles of the shell colour gene result in snails with pink shells and snails with yellow shells.
There is differential reproductive success between the different phenotypes. In other words, some phenotypes are better adapted to the environment, so they survive longer and therefore are more likely to reproduce more.	Yellow-shelled snails are more conspicuous than pink-shelled snails among beech litter. Song thrushes find yellow-shelled snails more easily than they find pink-shelled snails. Fewer yellow-shelled snails survive to reproduce.
Organisms with greater reproductive success leave more offspring than those with less reproductive success.	In a beech woodland, pink-shelled snails have more offspring than yellow-shelled snails.
Organisms with greater reproductive success will be more likely to pass their combination of alleles on to their offspring. As a result, the frequency of particular alleles will increase in the population; i.e. natural selection has occurred. **Evolution is a change in the allele frequencies in a population**.	In a beech woodland, the frequency of the pink allele will increase and the frequency of the yellow allele will decrease. Pink-shelled snails are at a selective advantage in beech woodlands.

Directional, stabilising and disruptive selection

Natural selection operates only on phenotypes that organisms possess and results in populations being better adapted to their environment. It can result in a change in a population from one phenotype to another or it can result in a reduction of variation of a particular phenotypic character about an optimum modal value.

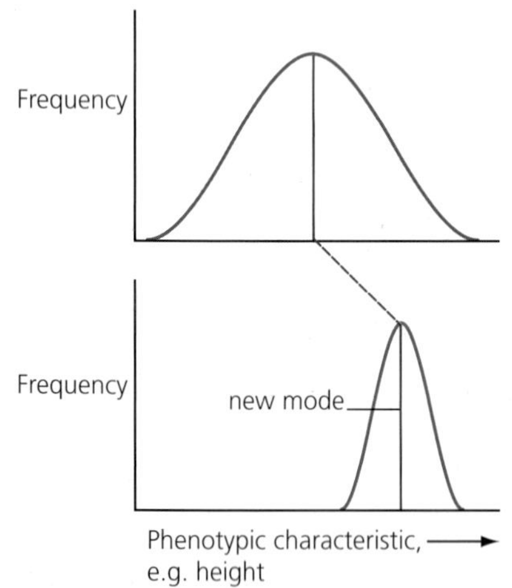

Figure 8.3 The graphs show variation in one characteristic of a population. The upper graph shows the range of phenotypes before natural selection; the lower graph shows the range of phenotypes after directional selection has occurred.

Directional selection

Directional selection is usually associated with a change in the environment. It acts against one of the extremes and/or in favour of the other extreme in a range of phenotypes, e.g. height. As a result, one or more phenotypes becomes rarer and alternative phenotypes become more common (Figure 8.3).

Look first at the upper of the two graphs. It represents a frequency distribution of phenotypes in a population. The graph is a normal distribution with a fairly large standard deviation. The mode (most frequent value) is marked in red. The graph represents the frequency distribution of this population before natural selection has occurred. The lower graph is a frequency distribution of the population after directional selection has occurred. The standard deviation of this curve is less than the upper curve and its mode has shifted to the right on the *x*-axis. Natural selection has caused a change in the allele frequencies in the population, favouring organisms with a characteristic towards the upper end of the range of the distribution.

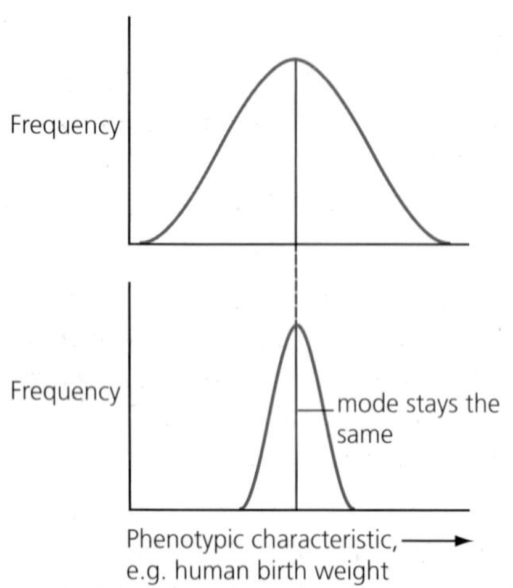

Figure 8.4 Stabilising selection reduces the variation of one phenotype in a population. The upper graph shows the range of phenotypes before natural selection; the lower graph shows the range of phenotypes after stabilising selection has occurred.

Stabilising selection

Stabilising selection is normally associated with a stable environment in which populations have already become well adapted to the environment. It acts against both extremes in a range of phenotypes. As a result, variation about the mode is reduced (Figure 8.4). Again, the upper graph shows the frequency distribution of this population before natural selection. The lower graph shows the frequency distribution of the same population after natural selection has occurred. This time, the mode is in the same position: this is the most advantageous phenotype. Stabilising selection has reduced the variation about this modal value.

Stabilising selection occurs on birth mass in humans. Babies with very low or very high birth masses have a higher infant mortality rate than those with a birth mass near the mode of the range.

TEST YOURSELF

6 Populations of banded snails also live in grassland, where they are preyed on by song thrushes. Which shell colour – yellow or pink – would you expect to be more common in grasslands? Explain your answer.

7 Look back to Figure 8.3 on the previous page. How can you tell that the lower curve has a smaller standard deviation than the upper curve?

8 The National Health Service (NHS) was introduced to the UK in July 1948. Suggest why natural selection on human birth mass has had less of an effect in this country since 1948 than before this date.

Disruptive selection

This is where both extreme phenotypes are selected for. You can see this in Figure 8.5.

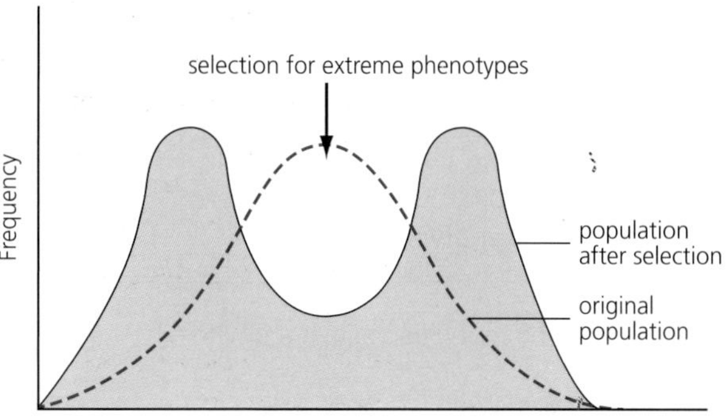

Figure 8.5 Disruptive selection. In disruptive selection, both extremes of the phenotype are selected for, and the intermediate phenotype is selected against.

An example of this would be where pale-coloured snails and dark-coloured snails are both camouflaged in an environment, but intermediate-coloured snails are not. Therefore intermediate-coloured snails are more susceptible to predators, meaning that snails that are either pale or dark are more likely to survive than the intermediate-coloured snails.

EXAMPLE

Natural selection and sickle-cell anaemia

Many large towns and cities in the UK have sickle-cell anaemia clinics. Sickle-cell anaemia is caused by a mutant allele of the gene controlling the production of β-globulin, one of the polypeptides in a haemoglobin molecule (Figure 8.6). Although the mutant allele changes only one amino acid in the β-globulin chain, its effect is striking.

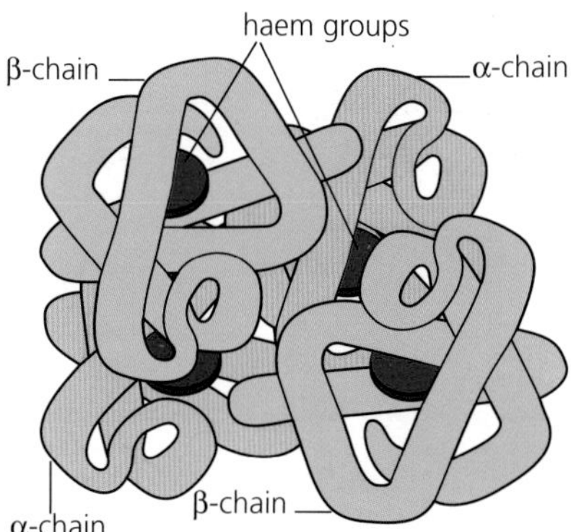

Figure 8.6 A haemoglobin molecule consists of four polypeptides, two α-globulin chains and two β-globulin chains, and four haem groups that bind to oxygen.

When the concentration of oxygen is low, haemoglobin molecules with the mutant β-globulin chains have abnormally low solubility and form fibres within red blood cells. This causes the red blood cells to change from disc-shaped cells to sickle-shaped cells. You can see these differences in Figure 8.7. The sickle shape reduces the surface area of red blood cells.

Sickle cells are also targeted for destruction by the immune system, so have a much shorter life span than normal red cells (about 15 days compared with the normal 120 days). Both these differences result in anaemia. Sickle cells are also liable to get stuck in capillaries, so that nearby tissues become starved of oxygen.

1 Sickle cells have a smaller surface area and shorter life span than normal red blood cells. Anaemia occurs when the blood cannot carry enough oxygen. Explain why both these differences result in anaemia.

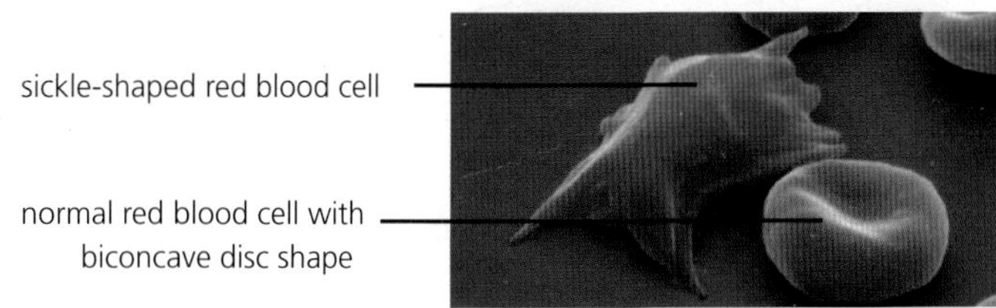

Figure 8.7 Red blood cells from a person suffering sickle-cell anaemia. Sickle cells have haemoglobin molecules with β-globulin chains that differ by one amino acid from those in normal disc-shaped cells.

Having a smaller surface area reduces the volume of oxygen that can diffuse into the cells, so the cells carry less oxygen than normal cells, leading to anaemia. Also, if the cells have shorter life span than normal cells, this means that, at any time, there will be fewer red blood cells in the blood. Therefore less oxygen is transported in the blood, causing anaemia.

The production of β-globulin is controlled by a single gene that has two alleles. We will call the β-globulin gene Hb and its alleles Hb^A and Hb^S. Table 8.2 shows the possible genotypes and phenotypes for this characteristic.

Table 8.2 The genotypes and phenotypes associated with sickle-cell anaemia.

Genotype	Phenotype
Hb^AHb^A	All red blood cells have normal haemoglobin and are disc-shaped.
Hb^AHb^S	The red blood cells contain a mixture of normal and abnormal haemoglobin. People with this genotype have the sickle-cell trait, but are healthy and usually show no symptoms of sickle-cell anaemia.
Hb^SHb^S	All red blood cells contain abnormal haemoglobin, which causes all red blood cells to become sickled in low oxygen concentrations. People with this genotype have sickle-cell anaemia, the complications of which can shorten their lives.

2 What does Table 8.2 tell you about the Hb^A and Hb^S alleles?
They are codominant. You can tell this because the heterozygote has a different phenotype from both the homozygotes. Both alleles produce a functional protein.

3 Sickle-cell anaemia can shorten people's lives. Explain why you might expect natural selection to reduce the frequency of this allele in human populations.
People with sickle-cell anaemia, without medical treatment, are likely to die in childhood. Therefore they are unlikely to pass on their alleles, which will reduce the frequency of the sickle-cell allele in the population.

Despite its apparent disadvantages, sickle-cell anaemia is common among people of African, Middle Eastern or southern Mediterranean descent. To explain this, we need to look at how malaria is spread. Malaria is common in Africa and in parts of the Middle East and the southern Mediterranean. Malaria is caused by a single-celled organism, called *Plasmodium falciparum*. This parasite is transmitted when a female *Anopheles* mosquito bites and sucks blood from an infected person and then injects saliva, containing several parasites, as she bites another person. Once in the blood system of a human host, the parasites enter the host's red blood cells where they use oxygen in the red blood cells for their own respiration.

4 Suggest how the behaviour of a red blood cell that is infected by *P. falciparum* will differ in someone who suffers from sickle-cell anaemia and someone who does not.
The P. falciparum *parasite will respire inside the cell, using up oxygen. Therefore infected red blood cells will carry less oxygen. The reduction in oxygen concentration caused by respiration of* P. falciparum *causes the red blood cells of a sickle-cell anaemia sufferer to become sickled. This causes them to be destroyed by the body's immune system within about 15 days. As they are destroyed, the* P. falciparum *within them is also destroyed. This makes people with the sickle-cell trait less susceptible to malaria than people with no sickle-cell condition.*

5 In an area where malaria is endemic, which genotype, Hb^AHb^A, Hb^AHb^S or Hb^SHb^S, will be at a selective advantage? Explain your answer.
Hb^AHb^S will be at an advantage. This is because Hb^SHb^S are likely to die prematurely from sickle-cell anaemia, whereas people with Hb^AHb^A are more likely to die of malaria. The heterozygotes have enough normal red blood cells to avoid the worst symptoms of sickle-cell anaemia, and they have resistance to malaria.

6 What is the probability that two parents, both of whom have the sickle-cell trait, will have a child that suffers from sickle-cell anaemia? Use a genetic diagram to justify your answer.

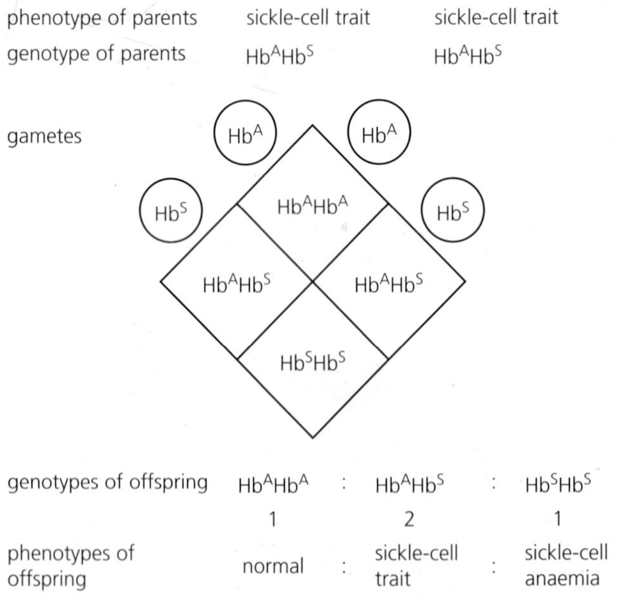

phenotype of parents: sickle-cell trait, sickle-cell trait
genotype of parents: Hb^AHb^S, Hb^AHb^S
gametes: Hb^A, Hb^S; Hb^A, Hb^S
Hb^AHb^A | Hb^AHb^S | Hb^AHb^S | Hb^SHb^S

genotypes of offspring	Hb^AHb^A	:	Hb^AHb^S	:	Hb^SHb^S
	1		2		1
phenotypes of offspring	normal	:	sickle-cell trait	:	sickle-cell anaemia

Figure 8.8 Genetic diagram to show the probability that two parents, both of whom have the sickle-cell trait, will have a child that suffers from sickle-cell anaemia.

Figure 8.8 shows the genetic diagram for working out the probability that two parents with the sickle-cell trait will have a child who has the condition. The probability of two parents who have sickle-cell trait having a child with sickle-cell anaemia is 1 in 4, or 25%.

Speciation

Natural selection brings about changes in the allele frequencies in a population (in other words, evolution). This can lead to the appearance of new species.

Figure 8.9 shows four closely related species in the taxonomic family Canidae. The two dogs belong to different breeds but are both the same species called *Canis familiaris*. Different breeds of dog originated when humans bred ancestral dogs that had useful characteristics. Jack Russell terriers (Figure 8.9c) were originally bred for hunting foxes, and bullmastiffs (Figure 8.9d) were bred to find and immobilise poachers. When humans breed animals or plants for their useful characteristics, we call this **artificial selection, or selective breeding**. The other animals in Figure 8.9, the wolf (*Canis lupus;* Figure 8.9a) and the jackal (*Canis aureus;* Figure 8.9b), are naturally occurring species. They arose by evolution, of which natural selection is a part. The formation of new species by natural selection is called **speciation**.

Figure 8.9 These animals belong to the same taxonomic family, Canidae. a) A wolf; b) a jackal; c) a Jack Russell terrier; d) a bullmastiff. The different species resulted from natural selection and the different breeds of dogs resulted from artificial selection by humans.

Figure 8.10 This male fruit fly has very short (vestigial) wings, which make him less successful in courtship than males that have long wings.

You learned in the first year of your Biology A-level course that organisms belong to the same **species** if they breed together in their natural habitat and produce fertile young. So how can one species give rise to another during speciation?

The accumulation of genetic differences is vital for speciation to occur by natural selection. Figure 8.10 shows a fruit fly, belonging to the genus *Drosophila*. You might have seen fruit flies like this on over-ripe fruit in your home or around your dustbin in summer. There are many different species of fruit fly. The differences between them are so small that you would probably not be able to tell one from another. However, the flies can tell. Before mating, male and female fruit flies undergo a courtship ritual. The male performs a dance, during which he vibrates his wings, changes his body position and licks the female. The whole sequence, including the response of the female, is controlled by many genes and is species-specific. If a male does not have the correct courtship dance, the female does not allow him to mate. The male in Figure 8.10 is homozygous for a mutant allele of the gene for wing length. As a result, he has very short (vestigial) wings. This will not kill him, but it means that his courtship dance is less likely to attract a female to mate with him. He is less likely to pass on his alleles of the gene for wing length.

As a result of gene mutation, the male fruit fly in Figure 8.10 will probably die without leaving any offspring. His mutated genes will not be passed on. However, suppose these genetic changes occurred not in an individual but in an entire population that was isolated from another population of the same species. We now have conditions in which speciation can occur.

For two populations to become two different species, they have to become reproductively isolated. The accumulation of differences in their separate gene pools might eventually result in these two populations not being able to reproduce and produce fertile offspring. Reproductive isolation may occur if a mutation causes one population to reproduce at a different time of year from the other population. This would prevent the two populations from being able to interbreed, so they would then become different species.

Allopatric speciation

Allopatric speciation is the formation of new species from populations of the same species that are geographically isolated. To understand this, let's imagine a population that has become isolated. This might happen, for example, if a pregnant rabbit was washed on a piece of driftwood from the mainland to an offshore island where there were no other rabbits. Here, our female would give birth to a group of offspring. Rabbits do not have the same taboos as humans about mating with their brothers and sisters, so these rabbits would mature and breed with each other. In a short space of time, there would be a population of rabbits on the island. Because rabbits are not strong swimmers, there would be no interbreeding between the rabbits in the island population and those in the mainland population. We can say that there is unlikely to be **gene flow** between the island population's gene pool and the mainland population's gene pool. Each population of a species lives in a different, though often similar, environment. Therefore we would expect different selection pressures to act on local populations.

Gene mutations occur at a constant, low rate in any population. Some gene mutations are beneficial, resulting in organisms with phenotypes due to the beneficial allele of a gene reproducing more successfully than others. As a result, the frequency of these beneficial alleles in the gene pool of this population will increase. In our imaginary example, this process will happen separately in the mainland and island populations of rabbits. Over a period of time, differences in their gene pools may accumulate. Look back to the definition of species that we used earlier. If any of the accumulated differences in the two gene pools prevent rabbits from the mainland and island populations mating, then, by our definition, we now have two species of rabbit. If any of the accumulated differences in the two gene pools result in hybrids between the mainland and island populations being infertile then, again according to our definition, we have two species of rabbit, because they are reproductively isolated.

TIP

You will find out about how genes mutate in Chapter 10.

TIP

Remember that mutations occur at random.

TEST YOURSELF

9 Canidae is the name of a taxonomic group, called a family. Name the taxonomic group represented by a) *Canis* and b) *familiaris*.

10 Islands in an archipelago often have species of a single genus that are unique to each island and all different from the species on the mainland. Explain why.

11 Explain the difference between the terms haploid, diploid and polyploidy.

12 What are the genetic causes of variation between the different members of a population?

13 Give examples of factors that might lead to some individuals surviving and others not surviving.

TIP

Remember that individuals cannot adapt to a change in the environment. Individuals that by chance have phenotypes that are, well adapted to the environment are more likely to survive and pass on their alleles. You should never write things like 'The species needed to adapt, so...'.

Sympatric speciation

Sometimes a new species can arise without two populations being geographically isolated. An example of how this could happen is the evolution of copper tolerance in certain plants.

EXAMPLE

The evolution of copper tolerance

The mine waste that formed the hill shown in Figure 8.11 contains high concentrations of copper ions. The fields around the copper waste contained populations of a species of grass called bent grass but for many years none grew on the hill.

1 Seeds from bent grass plants are dispersed great distances by the wind. Suggest why no bent grass plants grew on the hill in Figure 8.11.
There were copper ions present in the soil on the hill. These are toxic to bent grass, so it is unable to grow there.

After several years, bent grass plants began to grow on the hill. Experiments found that these plants were tolerant of copper in the soil.

2 Suggest how copper tolerance had originated in the copper-tolerant plants.
A mutation occurred in a plant that made it able to grow in the presence of copper. This plant survived, and passed on its allele for copper tolerance to its offspring, so this population of bent grass plants were able to grow on the hill.

3 The genes that enabled bent grass to become copper-tolerant spread through the population as a result of natural selection. However, the populations growing in the fields around the hill are not copper-tolerant. Suggest why.
Where copper is present in the soil, it is an advantage to be copper-tolerant. However, the allele for copper tolerance does not confer an advantage on plants growing where copper is not present, and in these areas copper tolerance may actually be a disadvantage.

Figure 8.11 This hill was formed when waste was tipped from a nearby copper mine.

4 Suggest how you could test whether these two populations have become separate species.
You could interbreed plants from the two different populations to see whether they produce fertile offspring. If they do, then they are the same species. In this example, the copper-tolerant grass is not yet a separate species from the copper-intolerant grass. However, over time it is possible that a mutation might occur in one of the populations that affects their reproduction. For example, one population might develop a mutation that causes them to flower at a different time of the year from the other population. If this occurs, the two populations will no longer be able to interbreed. At this point we can say that they are separate species.

Genetic drift

Genetic drift is an effect that can occur in small populations. It occurs when a few individuals, just by chance, either fail to reproduce or have more offspring than others. This then changes the allele frequency in the next generation. A good way to explain this is to imagine that you have a bag of jelly babies, with 100 red and 100 green sweets in it. Without looking, you remove 10 sweets from the bag. Next you pass the bag to a friend, who does the same. If you repeat this a few times, and then look at the sweets each person has, it is likely that the samples will be very different. It might be that you have five red and five green sweets, but it is even possible that you have 10 sweets of one colour and none of the other colour. If the jelly babies you have taken out were real organisms, this process would have separated them from the rest of the population and prevented them from breeding. You can imagine that the allele frequencies in the small populations could be very different from the allele frequencies in the original population, just by chance.

Genetic drift may occur because of genetic bottlenecks or the founder effect, both resulting from chance events in small populations.

TIP
You do not need to recall the terms 'genetic bottleneck' or 'founder effect'.

Northern elephant seals provide an example of a **genetic bottleneck**. These seals were hunted by humans so much that, by the end of the 19th century, only about 20 seals were left. Their population is much bigger now – hundreds of thousands – but their genetic diversity (in terms of the variety of alleles) is still very limited because all the present-day seals have arisen from a small number of individuals.

The occurrence of polydactyly among the Old Amish population of Pennsylvania in the USA provides an example of the **founder effect**. Polydactyly is a genetic condition in which people have additional fingers and sometimes toes. In most human populations this condition is very rare, but in this Amish population it is much more common than in other groups of humans. The reason for this is that the Amish population arose from about 200 German immigrants in the 18th century, one or two of whom possessed the allele for polydactyly. This is a high frequency of the allele in a very small population. Since then, the Amish people have tended to marry people within their own population, so a relatively small variety of alleles has remained in the gene pool.

TEST YOURSELF

14 Explain why genetic drift is important in small populations but not in large populations.

15 In a small region of north-west Venezuela there is an unusually high frequency of a severe inherited neuromuscular disorder known as **Huntington's disease**. Approximately 150 people in the area during the 1990s had this rare fatal condition and many others were at high risk of developing it. This disease usually does not strike until early middle age, after most people have had their children. All of the victims of Huntington's disease in this region trace their ancestry to a single woman who moved into the area in the 19th century. She had a very large number of descendants and there is now a population of about 20 000 people with a high risk of having this genetically inherited condition.

a) Explain why the disease is unusually common in this area.

b) Huntington's disease does not normally show any symptoms until after people have had their children. What effect will this have on the frequency of the Huntington's allele in this population?

Practice questions

1 In mice, the allele for black fur colour (B) is dominant to the allele for white fur colour (b). Two black and two white mice were kept in a large cage in the laboratory under ideal breeding conditions. After a few months, there were 149 black mice and 84 white mice in the cage.

a) Use the Hardy–Weinberg equation to calculate the expected percentage of heterozygous black mice in this population. *(3)*

b) Give two assumptions that are made when using the Hardy–Weinberg equation. *(2)*

2 **a)** Explain the meaning of the term gene pool. *(1)*

In a type of stag beetle, the allele for gold body colour, G, is dominant over that for black body colour, g.

b) In a stag beetle with black body colour, how many copies of the g allele would you expect to find in a nucleus taken from a muscle cell in the beetle's leg? Explain your answer. *(1)*

c) In a population of stag beetles, 182 had gold body colour and 18 had black body colour. Showing your working in each case, calculate the frequency of:

i) the g allele *(1)*

ii) the G allele. *(1)*

3 **a)** *Howea forsteriana* and *Howea belmoreana* are both kinds of palm found on Lord Howe Island in Australia. It is believed that these two species evolved sympatrically. Explain how sympatric speciation occurs. *(4)*

b) Name the taxonomic group represented by:

i) *Howea*

ii) *forsteriana.* *(1)*

4 Scientists have found a new species of river dolphin in the Araguaia and Tocantins rivers. The scientists have studied its DNA. They say that its genes suggest that the species formed 2.08 million years ago when the Araguaia-Tocantins basin was cut off from the rest of the Amazon river system by huge rapids and waterfalls, isolating the dolphins from the rest of the population.

a) What is a species? *(1)*

b) Suggest how this new species of river dolphin has developed from ancestral dolphins in the Amazon. *(4)*

c) Suggest how studying DNA provided the scientists with information about how long ago this new species formed. *(3)*

5 Scientists carried out an investigation on a dozen tiny islands in the Bahamas. They measured the leg length of a large number of tiny *Anolis sagrei* lizards. In half of the islands, they introduced a larger lizard species, which preys on *A. sagrei*, but cannot climb trees. When they returned to the islands, they found that on the islands where the larger lizard had been introduced the leg length of the *A. sagrei* lizards had changed. Some had longer legs, enabling them to run faster, while others had shorter legs, enabling them to climb trees faster.

a) i) Why did the scientists introduce the larger lizard to only half of the islands? *(2)*

ii) What kind of selection was shown in this example? Explain your answer. *(1)*

b) The scientists released the larger species on to the islands because they knew that, every few years, the islands are subjected to flooding which wipes out all ground-living species. Explain why this was important. *(3)*

c) In the few months after flooding, would you expect the leg length variation in *A. sagrei* to change? Explain your answer. *(2)*

Stretch and challenge

6 Suggest how you could adapt the Hardy–Weinberg equation to account for selection pressure against one phenotype.

7 Suggest why the cystic fibrosis allele is relatively common among white Europeans when homozygous recessive infants would have died in childhood until modern medicine was available.

9 Populations in ecosystems

PRIOR KNOWLEDGE

- Physical factors that may affect organisms are temperature, availability of nutrients, amount of light, availability of water and availability of oxygen and carbon dioxide.
- Quantitative data on the distribution of organisms can be obtained by random sampling with quadrats and sampling along a transect.
- A sample size should be large to ensure validity and reproducibility.
- The mean, median and mode are all figures that can be calculated to help in the evaluation of data.
- Biodiversity can relate to a range of habitats, from a small local habitat to the Earth.
- Species richness is a measure of the number of different species in a community.
- An index of diversity describes the relationship between the number of species in a community and the number of individuals in each species.
- Farming techniques reduce biodiversity.

TEST YOURSELF ON PRIOR KNOWLEDGE

1 How can you sample organisms using a line transect?

2 Insert the right word in the table to match its definition.

Term	Definition
	The middle value in a series of numbers.
	A calculated 'central' value of a set of numbers. To calculate it, you add up all the numbers, then divide by how many numbers there are.
	The number which appears most often in a set of numbers.

3 Give two ways in which farming reduces biodiversity.

Introduction

The World Wildlife Fund (WWF) published its latest Living Planet Index in October 2014. It publishes this index every 2 years. The report immediately hit the headlines because it showed that the world's populations of wildlife, including mammals, birds, reptiles, amphibians and fish, have dropped by more than a half over the past 40 years.

The WWF arrived at this index by measuring over 10 000 representative animal populations and found that these populations had fallen by 52% between 1970 and 2010. The previous report showed a drop of 28%

between 1970 and 2009. The populations that had fallen the most were those of freshwater fish, down by 76% over this period.

The Director General of the WWF International explained that this issue needs to be dealt with urgently over the next few decades. He said that humans must stop using the natural resources of the planet and destroying natural habitats as if they can be replenished, as these actions are jeopardising our future.

Sustainability The ability to use resources to meet the needs of the present without compromising the ability of future generations to meet their own needs.

In the report, Kuwait was named as the country with the worst record over the past four decades, consuming most resources per head of any country in the world. They were closely followed by Qatar and the United Arab Emirates. The USA was also named as having a bad track record, alongside other countries including Denmark, Belgium, Trinidad and Tobago, Singapore, Bahrain and Sweden. On the other hand, some of the poorer countries had better records for sustainability, including India, Indonesia and the Democratic Republic of Congo. The UK came 28th of all the countries mentioned in the report, but it still uses far more resources per head of population than most countries.

Figure 9.1 Green turtles are found in Kuwait. Hatchlings like these find it increasingly difficult to reach maturity and breed successfully. Global sea turtle numbers in general have fallen by 80% since 1970.

Some people may think that worrying about wildlife populations is a luxury while there are human populations suffering poverty and disease. However, as biologists we need to explain that human populations are dependent on the stability of populations of other organisms. We need to find ways of coexisting with the wildlife on our planet and using resources more sustainably.

Organisms and their environment

Nettles are common British plants. You can find them growing in woods, on grazing land, by the sides of ponds and streams, and in areas of wasteland in towns and cities. Figure 9.2 shows a nettle plant and some of the animals that can be found on it.

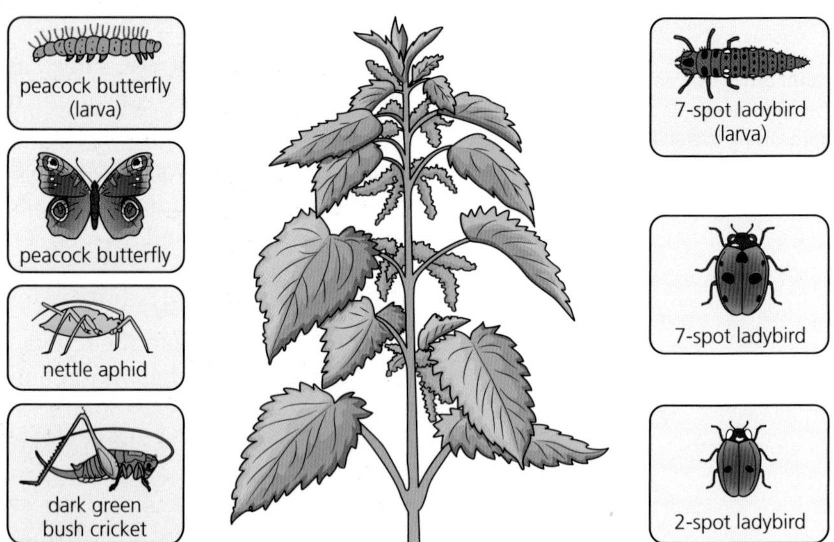

Figure 9.2 Some common animals associated with nettles.

A **population** is a group of organisms of one species, occupying a particular space at a particular time, that can potentially interbreed. All the two-spot ladybirds in a nettle patch are a population. All the nettle plants are another population. A **community** is all the populations of different species living in the same place at the same time. When we talk about the community in a nettle patch, we mean the population of nettles, the population of aphids, the populations of ladybirds and all the populations of other organisms that we have not mentioned, such as the bacteria and mycorrhizal fungi that live in the soil surrounding the roots of the nettles.

The community, together with the non-living components of its environment such as soil, water mud and rock, form an **ecosystem.** Ecosystems can vary in scale, from a patch of nettles or a pond to a mangrove forest or an entire coral reef.

The **habitat** of an organism is the place where it lives. The two-spot ladybird lives on nettle leaves.

The set of conditions that surrounds all of the insects in a patch of nettles consists of abiotic factors and biotic factors.

- **Abiotic (physicochemical) factors** make up the non-living part of the environment. Nettles grow well, for example, where there is a high concentration of phosphate ions in the soil. Warm, humid conditions result in large numbers of aphids on the leaves. The concentration of phosphate ions in the soil and the air temperature and humidity are all abiotic factors. Abiotic factors affect population size by affecting intraspecific competition. For example, if water availability is limited, the population will compete for water, and some of the organisms might die.

- **Biotic factors** are those relating to the other populations in the environment. The numbers of aphids on the nettles will be affected by the numbers of ladybirds because ladybirds feed on aphids. Predation by ladybirds is a biotic factor.

Within a habitat, a species occupies a **niche** that is defined by a combination of abiotic and biotic factors. Its niche describes not only where it is found but what it does there. We can describe the niche of the two-spot ladybird, for example, in terms of the abiotic features of its habitat, such as the temperature range it can tolerate, and the position on the nettle leaves where it is found. We can also include in our definition of its niche the size and species of aphids that it eats. The total combination of tolerances and requirements describes the multiple dimensions of the niche of any species. Each species occupies its own niche which is different from that of any other species in that habitat.

Populations within an ecosystem can only reach a certain size. The number of individuals in a population will be limited by abiotic factors or by biotic factors such as competition or predation. The maximum size that a population can remain sustainable in an ecosystem is called the **carrying capacity**. If the population of a species exceeds the carrying capacity, its number will be reduced until it is at or below the carrying capacity. This means that the number of organisms in a population fluctuates around the carrying capacity.

TEST YOURSELF

1 Suggest reasons why estimates of seabird numbers based on counting breeding birds may not be accurate.
2 Match the words to their definition.

Word		Definition	
1	Population	A	Factors relating to the living part of the environment, e.g. predation
2	Niche	B	All the populations of different species that live and interact together in the same area at the same time
3	Community	C	The set of conditions that surrounds an organism, consisting of abiotic factors and biotic factors
4	Environment	D	Factors relating to the non-living or physical and chemical parts of the environment
5	Abiotic factors	E	The habitat of an organism and its role within it
6	Biotic factors	F	A group of organisms belonging to the same species found in the same area at the same time

Competition

Farmers grow crops for profit. One of the factors they take into consideration is the number of seeds they sow. The number of seeds sown per unit area is called the sowing density. Agricultural scientists have carried out research on how sowing density affects crop yield and weed growth. In one investigation they sowed wheat at different densities and measured the mass of grain produced by each plant and the mass of weeds. Some of their results are shown in the graphs in Figure 9.3.

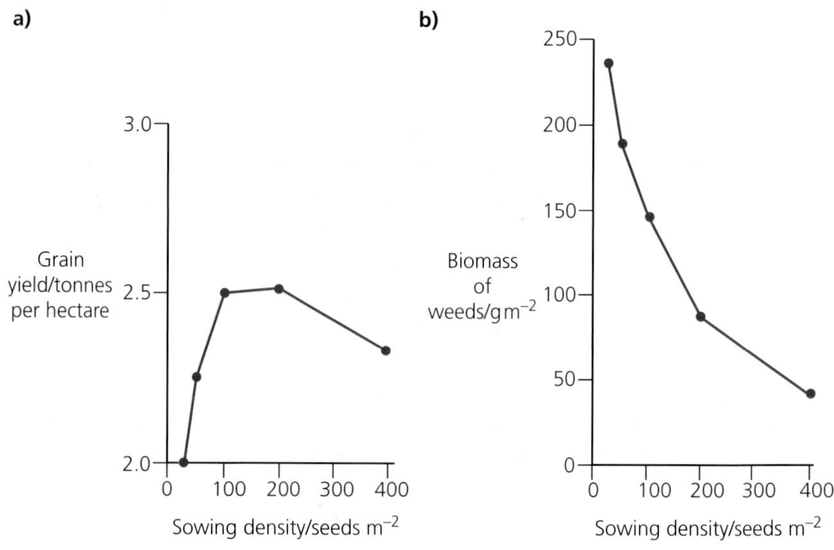

Figure 9.3 Graphs showing the effects of sowing density of wheat on a) grain yield and b) the mass of weeds per square metre.

Intraspecific competition Competition between organisms that belong to the *same* species.

Look first at Figure 9.3a. You can see that at sowing densities greater than 100 seeds m^{-2} the grain yield decreased with increasing sowing density. This can be explained in terms of intraspecific competition, or competition between organisms that belong to the same species. The more wheat plants there are per square metre, the more competition there will be for resources such as light, water and mineral ions. The resources available to each plant will be fewer. This will result in smaller plants that produce less grain. Since they are the same species, with the same niche, they compete for the same things.

Interspecific competition Competition between *different* species.

Now look at Figure 9.3b. This is an example of interspecific competition, which means competition between different species. In this case, the wheat plants are competing with the weeds. The higher the density of wheat plants, the fewer the resources there are available for the weeds.

ACTIVITY

Intraspecific competition

We will look at some more data involving **sowing density**. Table 9.1 shows the results of a trial to find the effect of sowing density of cotton on the mean height of the plants.

Table 9.1 Effect of sowing density on the mean height of cotton plants.

Mean number of seeds sown per hectare	Mean height of cotton plants/m	Standard deviation
67000	1.15	±0.18
33000	1.27	±0.11
17000	1.31	±0.11

1 The cotton seeds in the first row in the table were sown at a density of 67000 seeds per hectare. What was the mean number of seeds sown per square metre? (1 hectare = 10000 square metres)

2 What do the figures in the first two columns of the table suggest about the effect of sowing density on the height of cotton plants? Suggest an explanation for your answer.

Plot the information in this table as a graph.

3 Use your graph to explain whether sowing density has a significant effect on the height of cotton plants.

4 What does the standard deviation tell you?

5 The scientists who carried out this investigation planned to repeat it the following year on a larger scale. Use your answer to question 4 to suggest why.

EXAMPLE

Squirrels and interspecific competition

There are two species of squirrel in the UK: the native red squirrel and the introduced grey squirrel. Red squirrels are smaller than grey squirrels. They store less body fat and they spend more time in the tree canopy. Scientists have suggested several hypotheses to explain why grey squirrels have replaced red squirrels in much of Britain. One hypothesis is that grey squirrels are better adapted to living in oak woods. Acorns are the fruit of oak trees. They mature in the autumn and fall to the ground where squirrels can collect them. Earlier research showed that grey squirrels are much better than red squirrels at digesting acorns.

The graph in Figure 9.4.shows the relationships between the numbers of the two species of squirrel and acorn numbers.

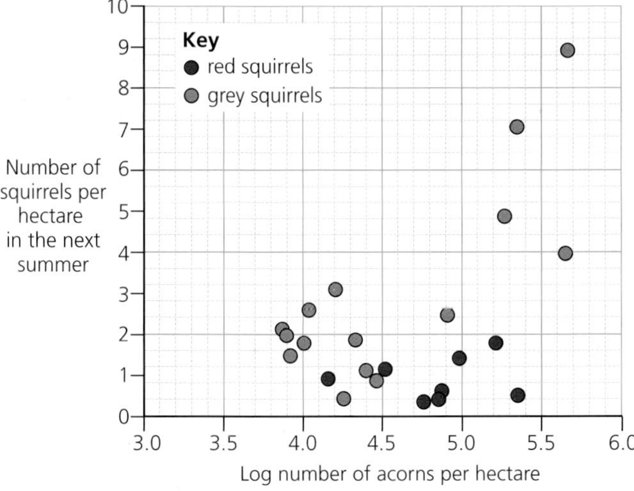

Figure 9.4 Graph showing the relationship between number of acorns produced per hectare and number of squirrels the following summer.

1 Why are the figures for the number of acorns given on a logarithmic scale?
A logarithmic, or log, scale increases in multiples of 10. On the scale shown here a log value of 3 represents 1000; a value of 4 represents 10 000; a value of 5 represents 100 000 and a value of 6 represents 1 000 000. Using a log scale lets us plot a much larger range of values on a single axis.

2 Suggest how the data relating to the number of acorns would have been collected.
The technique used would have to be based on a sampling method. Quadrats were probably used. This would allow the number of acorns in a given area to be found. This figure could then be converted to the number of acorns per hectare.

3 What is the relationship between the log of the number of acorns per hectare and the number of grey squirrels the next summer?
If you were to plot a curve of best fit you would get a line sloping upwards. This shows that there is a positive correlation: the greater the log of the number of acorns per hectare, the greater the number of grey squirrels the next summer.

4 What is the relationship between the log of the number of acorns per hectare and the number of red squirrels the next summer?
Plotting the curve of best fit this time gives a line parallel to the x-axis. This shows that there is no correlation.

5 Use the information in this section to suggest how interspecific competition could account for the absence of red squirrels in most of southern England.
Much of the native woodland in southern England is oak. Grey squirrels are more successful in oak woods because they spend more time on the ground where acorns fall. They can also digest the acorns better and put on more body fat as a result. A greater amount of body fat probably enables grey squirrels to survive over winter in oak woods better than red squirrels.

Predators and their prey

Predators are animals that kill and eat their prey. Figure 9.5 shows a spider that is feeding on a grasshopper. The spider is the predator and the grasshopper is its prey. Obviously, the number of predators influences the number of prey: as the number of predators increases, the number of prey decreases. Equally obviously, the number of prey influences the number of predators. Without enough food, fewer predators will survive.

> **TIP**
>
> When discussing competition, be sure to say whether interspecific or intraspecific competition is taking place, and suggest a factor for which the organisms might be competing.

Figure 9.5 A spider with a grasshopper in its web, an example of a predator and its prey.

EXAMPLE

Scientists use models to help them understand the relationship between the population of a predator and the population of its prey. We will look at one of these models, called the Lotka–Volterra model after the two scientists who originated it (Figure 9.6).

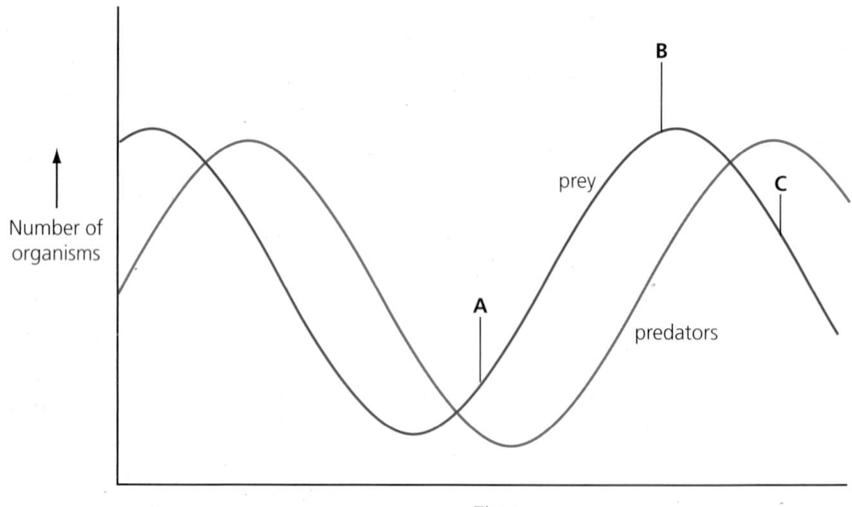

Figure 9.6 The Lotka–Volterra model predicts that the populations of a predator and its prey change over time.

- Look at the part of the graph in Figure 9.6 between points A and B. It shows the prey population increasing. Therefore there will be more food available for the predator population. The predators will breed successfully and their numbers will increase.
- With more predators, however, more prey will be killed so we see, between points B and C, a fall in prey numbers. There is now less food for the predators and their population starts to fall.

- Numbers of prey rise again and the pattern will be repeated.

We will look at a piece of evidence that appears to support this model. Figure 9.7 shows a graph of changes in the numbers of skins of a predator, the lynx, and the number of skins of its prey, the snowshoe hare, bought from trappers by the Hudson Bay Company, Canada, for the years 1890 to 1920.

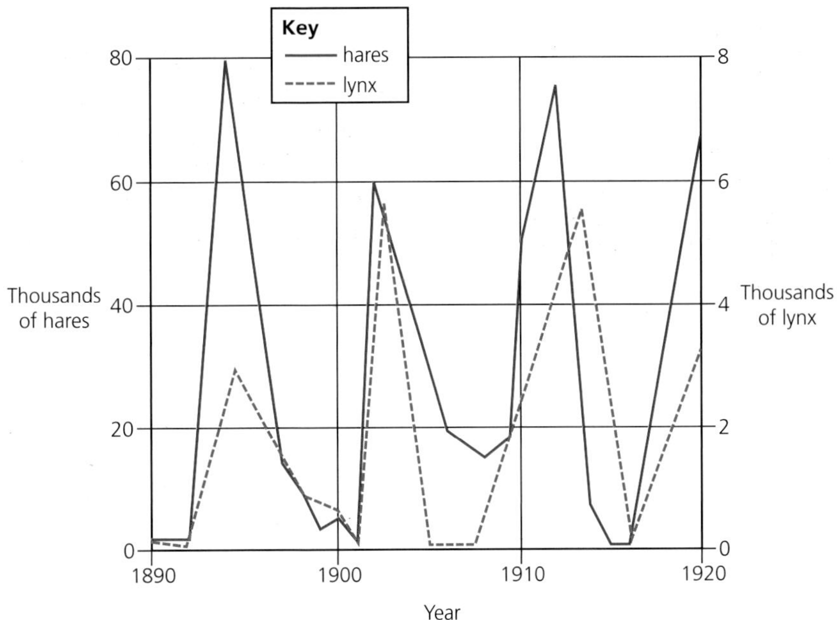

Figure 9.7 Graph showing the numbers of lynx and snowshoe hare skins bought by the Hudson Bay Company between 1890 and 1920.

The curve for the number of lynx skins is different from the curve for the number of snowshoe hare skins in Figure 9.7. There are two features to note. The peak for the lynx skins always comes a little later than the peak for the snowshoe hare skins and the lynx numbers are always lower than the hare numbers. This is usually the case for predators and their prey, although there are some exceptions.

The number of snowshoe hare skins purchased rises between 1908 and 1912. During this period lynx numbers are low, as shown by the number of skins bought by the Company. With few predators to kill them, snowshoe hare numbers therefore rise.

Note that, in one population cycle, lynx numbers rise after snowshoe hare numbers have risen. The rise in lynx numbers is the result of more prey being available. Therefore their numbers will only rise after snowshoe hare numbers have increased.

Between 1910 and 1914 the rise in the lynx numbers causes a fall in the numbers of snowshoe hares. There is therefore less food for the lynx and subsequently their numbers fall.

The changes in the numbers of snowshoe hare skins and lynx skins shown in Figure 9.7 seem to support the Lotka–Volterra model but more recent experimental work has shown that the situation is really much more complicated than this simple model suggests. Here is some information resulting from recent ecological investigations involving snowshoe hares and lynx.

- High populations of snowshoe hares are associated with a shortage of plant food on which the animals feed. When the snowshoe hare population is at its lowest, the plants start to grow again but, for the next 2 or 3 years, toxins in the young shoots are thought to delay the next rise in the population of snowshoe hares.
- When snowshoe hare numbers decrease, lynx eat other prey animals. There is very little evidence of lynx dying of starvation.
- Scientists attached radio collars to lynx. They showed that when the number of snowshoe hares was low, the lynx tended to move away. In one case, they travelled as far as 800 km.
- There are no lynx on Anticosti Island in Eastern Canada. The population of snowshoe hares on the island, however, still shows a regular population cycle.

What we can see from this information is that simple models, such as the Lotka–Volterra model, provide a very useful starting point for analysing population cycles. They cannot provide us with a complete explanation.

TEST YOURSELF

3 Is competition a biotic or an abiotic factor? Explain your answer.

4 Use Figure 9.3a to describe how sowing density affects grain yield in tonnes per hectare.

5 A farmer could use the results of this investigation to find the optimum sowing density for wheat. What conclusion do you think he would draw? Explain your answer.

6 What is the evidence that red squirrels and grey squirrels have similar niches?

Counting and estimating

Every 10 years there has been a census of all the people living in the UK. It is not totally accurate because some people fail to fill in their census forms accurately, or don't fill them in at all, but it does give a good idea of the size of the UK human population.

There are other species where we can get a fairly accurate figure for the population by counting them. For example, seabirds nesting on the Farne Islands off the north-east coast of England can be counted because they nest in large colonies, but the data obtained are much less accurate than for human populations. In the Introduction to this chapter you learned that the WWF estimates world wildlife populations every 2 years, but again there are difficulties in ensuring reasonable accuracy.

We can only count all the animals or plants in a population in a few cases. The organism concerned needs to be large, conspicuous and confined to a relatively small area for a count to be accurate. If it is not possible to count every single organism, we need to take samples. Ecologists usually base their estimates of populations on samples. They must make certain, however, that the samples are representative of the population as a whole. In order to be sure of this, samples must be large enough to be representative and taken at random.

Sample size

The larger the size of a sample, the more reliable the data. Data from a very large sample, however, cannot usually be collected in a short period of time. When we sample an area we have to strike a balance. Look at the graph in Figure 9.8. It shows that a very small sample has very few species in it. As the sample size increases, the number of species that it contains increases. There comes a time when the number of species does not increase much more however much bigger the sample. This sample size is obviously representative of the community as a whole.

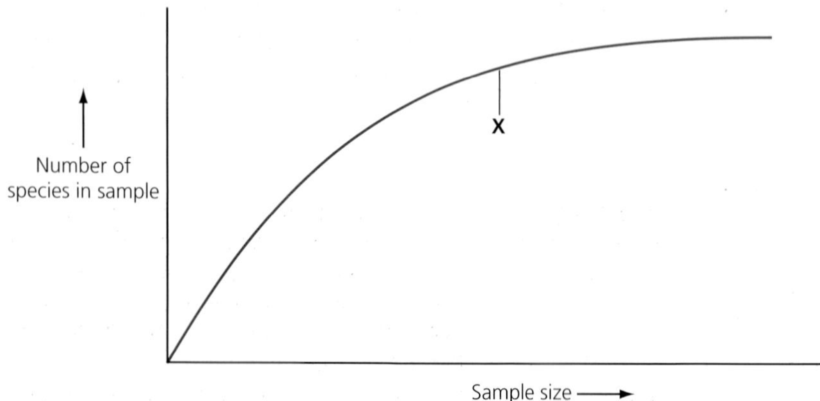

Figure 9.8 The effect of sample size on the number of species recorded. After point X the number of species does not increase much more with increasing sample size. The sample size at X is therefore representative of the community as a whole. A larger sample would simply mean more work.

Extension

Finding the running mean of the number of individuals in a quadrat as each one is sampled allows you to see when the running mean becomes more or less constant. At this point, the number of quadrats sampled is sufficiently large to be giving a representative density for the whole population.

Random sampling

When you studied the topic of variation in *AQA A-level Biology 1 Student's Book* Chapter 11 you learned that, unless samples were taken at random the results may be biased. The same is true of ecological samples. They must be collected at random or, again, the results may be biased. Quadrats are often used to mark out areas to be sampled. When we use a quadrat for sampling, it is important that we use a method that will result in the quadrat being genuinely placed at random in the study area (Figure 9.9).

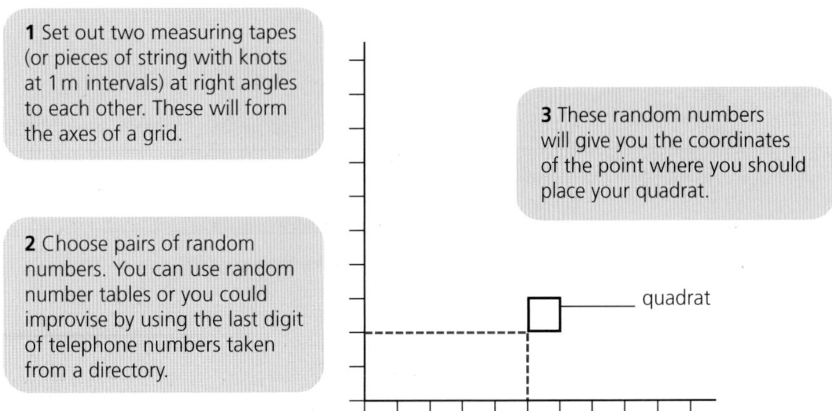

1 Set out two measuring tapes (or pieces of string with knots at 1 m intervals) at right angles to each other. These will form the axes of a grid.
2 Choose pairs of random numbers. You can use random number tables or you could improvise by using the last digit of telephone numbers taken from a directory.
3 These random numbers will give you the coordinates of the point where you should place your quadrat.
quadrat

Figure 9.9 How to place a quadrat randomly.

Quadrats and transects

Two important ways of sampling involve quadrats and transects. Both of these are in general used for plants, but they can also be used for organisms that do not move about very much, such as many of those that live on seashores. Table 9.2 shows when and how these techniques are commonly used. There are three different ways that we use to describe the distribution of organisms once the quadrat or transect is put in position. These are illustrated in Figure 9.10.

Table 9.2 Using quadrats and transects.

Sampling method	Method	What it is used for	How it is used
Quadrat	A frame, usually 0.5 m^2, or a rod 0.5 m long with 10 pins at 5 cm intervals, that can be placed on the ground to sample organisms	Studying the distribution of non-motile plants or animals in a fairly uniform area	Placed at random and used to find population density, frequency or percentage cover. In some investigations, such as those involving the effects of grazing, permanent quadrats may be used. They may remain in place for many years.
Transect	A line through a study area along which samples are taken	Usually used where one or more abiotic factors in the environment gradually vary of if there appears to be a change in communities from one place to another	Placed so that it follows the environmental gradient; for example, up a seashore or from sunlight into shade. This is called a **line transect**. Quadrats may be used continuously along a transect to sample in more detail. This is called a **belt transect**.

a) Population density

b) Frequency

c) Percentage cover

This quadrat measures 0.5 m × 0.5 m. It contains six dandelion plants. The **population density** of dandelions would be 24 plants per m^2. To get a reliable figure you would need to collect the results from a large number of quadrats. If a plant lies partly in and partly out of the quadrat, we normally count it if it overlaps the north or west side of the quadrat, and don't count it if it overlaps the south or east side.

This point quadrat frame is being used to measure **frequency**. The pins of the frame are lowered. Suppose three out of ten pins hit a dandelion plant. The frequency of dandelion plants will be three out of ten, or 30%.

Percentage cover measures the proportion of the ground in a quadrat occupied by a particular species. The percentage cover of the dandelions in this quadrat is approximately 40%.

Figure 9.10 Using a) population density, b) frequency and c) percentage cover to describe the distribution of dandelion plants.

Populations of the same species from two areas with different abiotic conditions can be compared by taking a random sample from both areas. Limpets are molluscs with a cone-shaped shell that attach to rocks with a strong foot. In an investigation, students sampled limpets by placing quadrats randomly on each section of the shore and measuring the height and base diameter of each limpet in the quadrats. After this they found the height : base diameter ratio for each limpet (Table 9.4).

Table 9.4 Height and base diameter of limpets sampled on a rocky shore

Sheltered shore			Exposed shore		
Height/ mm	Base diameter/mm	Height: base ratio	Height/ mm	Base diameter/mm	Height: base ratio
21.7	49.0	0.44	19.2	57.5	0.33
21.8	53.6	0.41	17.4	54.5	0.32
23.3	42.4	0.55	25.5	63.3	0.40
18.2	42.5	0.43	20.5	43.4	0.47
17.9	44.5	0.40	19.6	61.2	0.32
22.5	44.8	0.50	18.4	58.8	0.31
17.4	45.0	0.39	18.6	50.2	0.37
30.4	56.6	0.54	14.6	48.0	0.30
21.5	49.8	0.43	20.6	49.4	0.42
19.5	50.5	0.39	14.7	42.4	0.35
21.5	48.7	0.44	13.5	44.3	0.30
23.0	49.2	0.47	19.0	53.2	0.36
23.3	49.5	0.47	18.3	50.5	0.36
23.0	46.5	0.49	15.2	49.7	0.31
24.6	54.9	0.45	14.2	44.5	0.32

TIP

A ratio would normally be expressed as, for example, 0.44:1, but here we have expressed the ratios using the first figure alone, e.g. 0.44.

TIP

- In an exam you might be asked to recognise the type of data you have been given and to select an appropriate statistical test.
- Look at Chapter 13 on maths skills to find out more about statistical tests and to see a worked example of this test.
- You won't need to do statistical calculations in a written paper.

Because the students were dealing with measured data the best way to present it would be as mean height to base diameter ratios on a bar chart. The bars would be different heights indicating that limpets on the sheltered rocky shore had a higher mean height to base diameter ratio than those on the exposed shore; in other words, those on the exposed shore had grown wider and flatter than those on the sheltered shore. You would use a t-test to find out whether the difference between the means was significant.

REQUIRED PRACTICAL 12

Investigation into the effect of a named environmental factor on the distribution of a given species

This is just one example of how you might tackle this required practical.

Some students decided to investigate the effect of trampling on plant cover by placing a transect across a path. They placed a measuring tape at right angles to the path, including the vegetation at the side of the path.

The students placed a point quadrat at right angles to the measuring tape every 10 cm. The point quadrat had ten pins, and they recorded what each pin was touching. They wanted to convert this to percentage cover, so they multiplied the number of pins touching a specific plant or substrate by 10. Table 9.3 shows their results.

Table 9.3

	Percentage cover				
Distance along quadrat/m	Bare soil	Rock	Grass	Plantain	Dandelion
0	0	0	80	10	10
0.1	0	0	70	10	20
0.2	0	0	90	0	10
0.3	10	0	70	20	0
0.4	20	10	50	20	0
0.5	20	0	60	20	0
0.6	30	10	40	20	0
0.7	40	20	30	10	0
0.8	60	20	10	10	0
0.9	70	30	0	0	0
1.0	60	40	0	0	0
1.1	80	20	0	0	0
1.2	70	10	10	10	0
1.3	70	0	10	20	0
1.4	50	10	20	20	0
1.5	40	10	30	20	0
1.6	30	10	30	20	10
1.7	30	0	40	30	0
1.8	40	10	20	30	0
1.9	30	0	40	20	10
2.0	10	0	70	10	10
2.1	10	0	70	10	10
2.2	0	0	80	10	10
2.3	0	0	90	0	10
2.4	0	0	80	0	20
2.5	0	0	90	0	10

They used these data to plot a kite diagram, which displays the density and distribution of plant or animal species in a particular habitat. It shows the percentage of certain species spread over a certain distance. You can see how to plot a kite diagram in Figure 9.11.

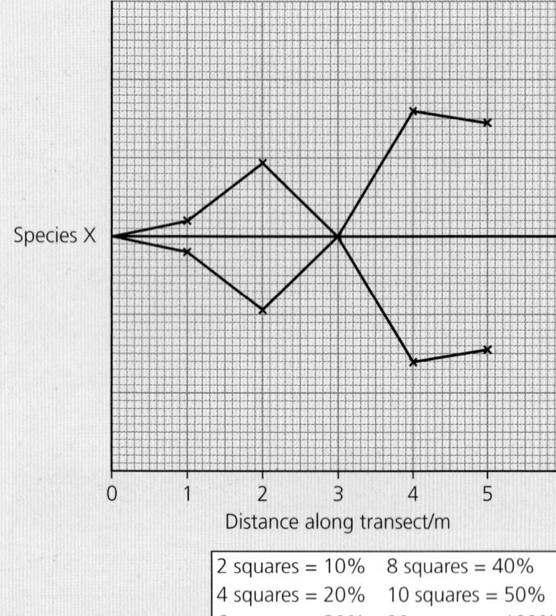

2 squares = 10%	8 squares = 40%
4 squares = 20%	10 squares = 50%
6 squares = 30%	20 squares = 100%

Figure 9.11 How to plot a kite diagram.

TIP

You do not need to recall the details of how to plot a kite diagram.

For each species, you draw a straight line. Then you choose a scale for your kite, so that the width of the kite represents the percentage cover of the species. Crosses are placed each side of the line for each sampling point along the transect. The crosses are joined with a straight line and then the 'kite' is shaded in.

1 Plot a series of kite diagrams to display the data in the table. Set out the graphs as shown in Figure 9.12.

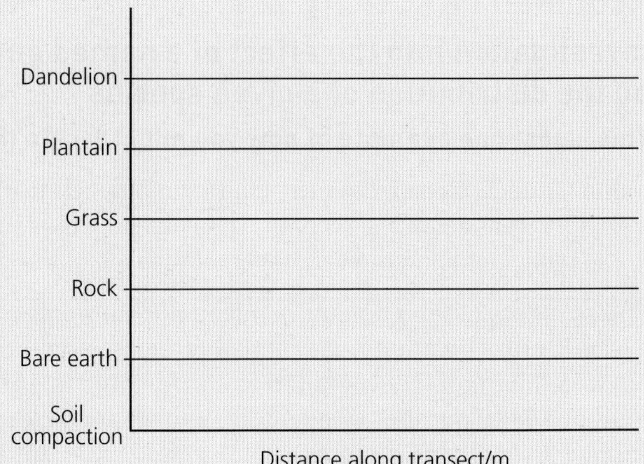

Figure 9.12

2 Suggest how the students could measure soil compaction.

3 What are the advantages and disadvantages of using a point quadrat in this investigation, rather than a quadrat square?

4 What do these data suggest about the effects of trampling on plant cover?

5 Suggest reasons why trampling affects plant cover. Look at Figure 9.13 for some clues.

Figure 9.13 a) *Plantago major* (greater plantain). b) Dandelion

Mark-release-recapture

Most animals move around (are motile) so it is not usually possible to estimate the size of an animal population using a quadrat. Instead we use the mark-release-recapture method. This relies on capturing a number of animals in an area and marking them harmlessly so that they can be recognised again. The marked animals are then released. Some time later a second sample is trapped in the same area and the numbers of marked and unmarked animals in the second sample are recorded. From the data collected the size of the population can be estimated. The calculation relies on the assumption that the proportion of marked animals in a sample (the second sample) is the same as the proportion of marked animals in the whole population.

This equation makes several assumptions:

- the animals all come from the same population
- marking does not harm the animal or make it more likely to be seen by a predator
- there is no migration into or out of the population during the period of the investigation

EXAMPLE

Estimating the size of a population of grasshoppers

Here is how we would estimate the size of a population of grasshoppers:

Number of grasshoppers caught, marked and released = 56

Number of marked grasshoppers in second sample = 16

Total number of grasshoppers in second sample = 48

$$\frac{\text{Number of marked grasshoppers in sample}}{\text{Total number of grasshoppers in sample}} = \frac{\text{Number of marked grasshoppers in population}}{\text{Total number of grasshoppers in population}}$$

$$\frac{16}{48} = \frac{56}{\text{Total number}}$$

Rearranging this equation:

$$\text{Total number} = 56 \times \frac{48}{16}$$
$$= 168$$

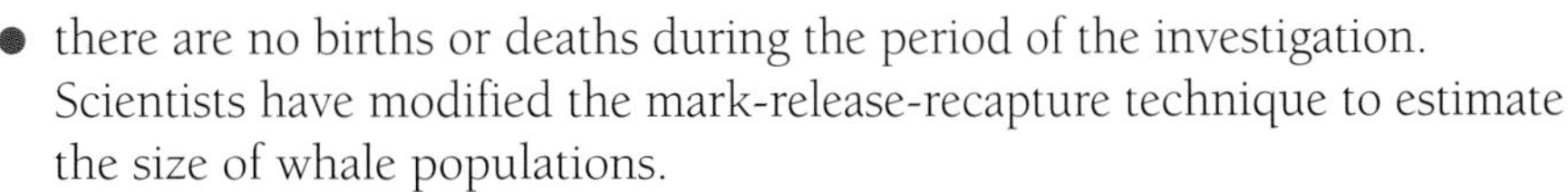

- there are no births or deaths during the period of the investigation. Scientists have modified the mark-release-recapture technique to estimate the size of whale populations.
- Whales cannot be trapped and they cannot easily be marked.
- Scientists can use a small boat to approach a surfaced whale. They can remove a tiny piece of skin from the whale.
- The scientists are able to analyse the DNA from this piece of skin and produce a DNA fingerprint. Each whale has its own unique DNA fingerprint.

Figure 9.14 Because whales spend most of their time below the surface of the sea they cannot be trapped or easily marked, so scientists needed to find a method other than observing and counting the whales in order to estimate the size of whale populations. They developed a DNA fingerprinting method.

Ecosystems are dynamic. This is reflected by fluctuations in populations, among other things. Numbers of a particular species vary around the carrying capacity from place to place and from one time of year to another. These variations may result from differences in abiotic factors. They may also result from changes in biotic factors.

ACTIVITY

Investigating the mark-release-recapture method of analysing population size

Materials:

4 or 5 packets of small chocolate sweets with a coloured sugar coating (e.g. Smarties)
paper bag
egg cup
plastic dish (e.g. margarine tub).

- Tip all the coloured sweets into the plastic dish. Take out the red sweets and count them.
- The coloured sweets represent a population of field mice and the red ones represent the mice trapped on the first evening that are marked and released.
- Tip all the sweets, including the red ones, into the paper bag and shake them gently.
- Use the egg cup to take a sample of sweets from the bag without looking. Tip the sample into the plastic dish. This represents the mice captured the second time.
- Copy Table 9.5 and use the results to calculate the population size.
- Repeat this nine more times so that you can find the mean of 10 population estimates.
- Now that you have estimated the population size, tip all the sweets into the plastic dish and count them.

Table 9.5

Sample number	Total number in sample (a)	Number of marked mice in sample (b)	Number of marked mice in population (c)	Population size = $\frac{ac}{b}$
1				
2				
3				
4				
5				
6				
7				
8				
9				
10				
Mean				

1 How accurate was the first estimate that you calculated?
2 Was the mean estimate after 10 samples more or less accurate?
3 If this was a real population of mice, would it be feasible to sample the population 10 days in a row? Explain your answer.
4 Now think about a variable you might change. For example, does the number of traps set matter? What happens if one or two of the marked animals are eaten by a predator after they are released? What happens if two or three of the unmarked animals die, or if new mice enter the area?

Think about how you can simulate this change, and carry out the calculations again. For example, you can simulate fewer traps by half-filling the egg cup, or setting more traps by using a slightly bigger egg cup. You can remove a couple of red sweets to simulate marked animals being killed, but remember that you wouldn't know this had happened if you were investigating a real population, so your calculations would be based on the original number of marked animals.

5 Based on your investigation, evaluate the use of the mark-release-recapture method of estimating the size of real populations.

TEST YOURSELF

7 Which of the following methods would allow a quadrat to be placed at random?
  A Closing your eyes, turning on the spot and throwing the quadrat over your shoulder.
  B Placing quadrats along a tape at 5 m intervals.
  C Picking random numbers from a hat to give coordinates on a grid.

8 A sample of 40 trout in a fish pond was caught in a net. Each fish was harmlessly marked on one of its fins. The fish were then released back into the pond. One week later a second sample was caught. It contained 17 marked trout and 41 unmarked trout. Estimate the total number of trout in the pond.

9 Use the information on sampling whale populations, page 179, to suggest how scientists could modify the mark-release-recapture technique to estimate the size of a whale population.

10 For the mark-release-recapture method to be reliable, certain assumptions need to be made. Suggest what these are.

Succession

In some places, the populations change over time, creating a new community. This is called succession. It is an ecological process resulting from the activities of the organisms themselves. Over a period of time the organisms modify their environment. These modifications produce conditions better suited to the growth of other species (Figure 9.15).

TIP

Ecosystems are constantly changing because of changes in populations, for example as a result of predation or competition, and changes in abiotic factors. Abiotic factors such as temperature or water availability may vary with the season, or change gradually over a long period of time. This is particularly seen in the case of succession.

Figure 9.15 Succession is an important ecological process associated with slow-flowing streams and rivers.

Some of the plants in the foreground of Figure 9.15a have leaves that float on the surface. Their roots trap particles of silt carried in the water and this forms mud. As the water becomes shallower, these floating plants are replaced by upward growing reeds and other plants.

The first organisms to colonise the bare rock in Figure 9.15b are lichens, but they are only able to grow when water is available. Lichens are involved in breaking the surface of rocks, creating a 'soil' that accumulates under them. This soil holds enough water and mineral ions to allow new species to colonise. As this soil is often washed away, succession is very slow. Ultimately though, the lichens will be out-competed by mosses. These, in turn, will be out-competed by ferns and species of flowering plant.

Succession in sand dunes

Sand dunes occur in many coastal areas and can provide a good example of succession. Succession is a gradual process and the time scale involved prevents us from sitting in one place and observing successional changes. If we walk, however, from the sea shore in to the sand dunes we will pass through different areas which represent different stages in succession (Figure 9.16).

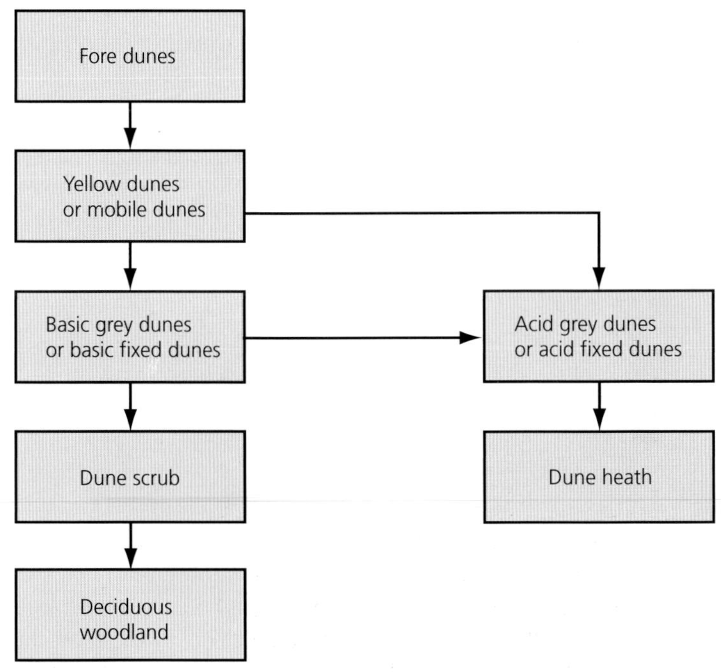

Figure 9.16 Succession in sand dunes.

We will start nearest the sea on a sandy shore, above high tide level. Some plants, such as sea couch grass and lyme grass, are able to colonise and survive in the bare sand. We call species such as sea couch grass and lyme grass **pioneer species** because they are the first plants to colonise the area. Their roots and shoots form a dense network that binds sand particles together and, as a result, sand starts to pile up and form the fore dunes. As the plants grow larger, their aerial parts trap more and more sand. They cannot grow fast enough, however, to avoid being smothered by sand when it is piling up at a rate faster than 30 cm year^{-1}. Under these conditions they are replaced by marram grass.

Marram grass-dominated dunes are called yellow dunes or mobile dunes: yellow because there is very little humus in the sand and it shows yellow through the marram grass cover; mobile because they are continually changing shape as the wind scours the face and blows the sand. This is a very harsh environment and few plants can survive in these conditions. Look at Table 9.6. This table compares the abiotic conditions in these yellow dunes with the conditions in the grey or fixed dunes later in the succession.

Table 9.6 A comparison of some of the abiotic factors in yellow and grey dunes.

Abiotic factor	Yellow dunes	Grey dunes
Mean wind velocity 5 cm above the dune surface/km h^{-1}	12.1	2.4
Organic matter/%	0.3	1.0
Concentration of sodium ions/ppm	8.5	4.2
Concentration of calcium ions/ppm	637.0	297.0
Concentration of nitrate ions/ppm	48.0	380.0

The high wind speed has two important effects. It will pile sand on top of any plants growing there and it will lead to a high rate of water loss by transpiration. There is still very little organic matter, so the sand does not retain water very well and nutrients tend to leach out. Also, there is little material for decomposers to act on and recycle nitrogen-containing compounds. It will dry out rapidly after rain. Important soil nutrients such as nitrates are in short supply, but sea spray results in high concentrations of sodium and calcium ions, so this lowers the water potential of the sand dune, creating osmotic problems for plants. Very few plants can grow here but one that does is marram grass. Marram grass has a number of xerophytic adaptations that enable it to grow in these conditions (Figure 9.17).

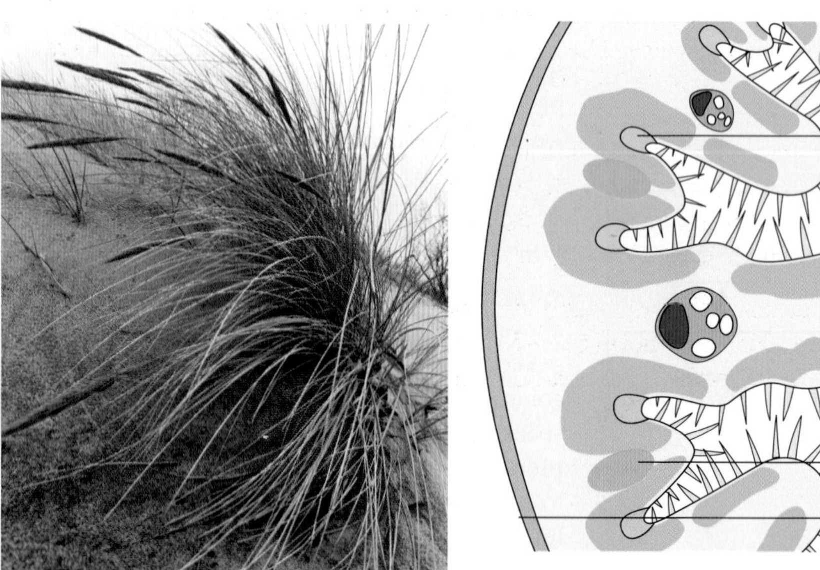

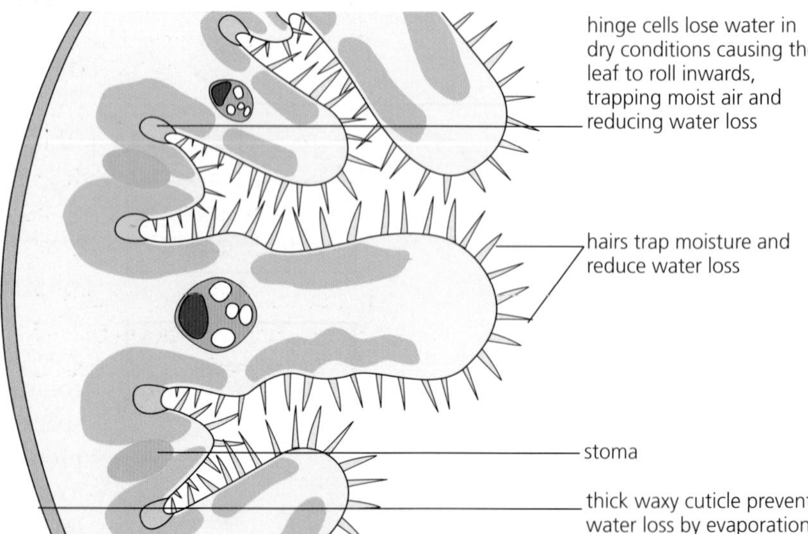
hinge cells lose water in dry conditions causing the leaf to roll inwards, trapping moist air and reducing water loss
hairs trap moisture and reduce water loss
stoma
thick waxy cuticle prevents water loss by evaporation

Figure 9.17 a) On sand dunes, marram grass is an important species. Its leaves grow from a vertical underground stem, so remain above the sand that the wind deposits on the plants. It also has adaptations that enable it to grow in areas where there is often little water available. Some of these adaptations can be seen in the section through a leaf shown in b).

Marram grass illustrates an important biological principle: the more hostile the environment, the fewer the species that are able to survive and the lower the species diversity. In hostile environments, it is generally abiotic factors that determine which species are present.

Now we will go inland to the area of grey dunes or fixed dunes. They are called grey dunes because humus in the sand colours them grey; they also are referred to as fixed dunes because the sand is no longer being blown about, so they are much more stable. Table 9.6 on the previous page shows how abiotic factors change and many of these changes are due to the activity of organisms on the yellow dunes. The roots of marram grass bind the particles of sand and the leaves act as a windbreak, reducing wind-chill and creating shade. The wind velocity is lower so less sand blows about. Dead material falls from the marram grass and is broken down by soil bacteria to form humus. The amount of humus is higher, so the developing soil retains moisture better and the concentration of important soil nutrients such as nitrates rises. Rain is also beginning to leach the soluble sodium and calcium ions from the surface layers of the soil. Other plants can now grow and they gradually replace marram grass because they compete better with marram grass when the soil has more nutrients. We are beginning to arrive at a situation where the environment is less harsh. There are more species and a greater species diversity, creating more complex and therefore more stable food chains and webs. In this environment it is biotic factors, such as competition, that determine whether particular species survive.

The process of succession continues. The species found on these grey dunes may change their environment in such a way that they are replaced by other species. Woody shrubs start to grow. Hawthorn and elder grow and shade out the shorter vegetation. Ultimately woodland develops. We have reached a stage that remains relatively unchanged over long periods of time. This is the climax community. In Britain, it is usually woodland of some sort. If you look again at Figure 9.16 on page 182 and you will see that not all succession on sand dunes follows this pattern. The sand particles that make up the dunes have different origins. Where they originate from rocks such as granite, the soils they produce are more acidic. They support different plants and gradually give way to dune heath dominated by plants such as bracken and heather. Basic soils containing high concentrations of calcium carbonate can also become more acidic as soluble basic ions are gradually leached from them.

The details of the processes of succession in slow-flowing streams and rivers and on bare rock shown in Figure 9.15 are different from those on sand dunes but the principles are very similar. They are summarised in Figure 9.18 overleaf.

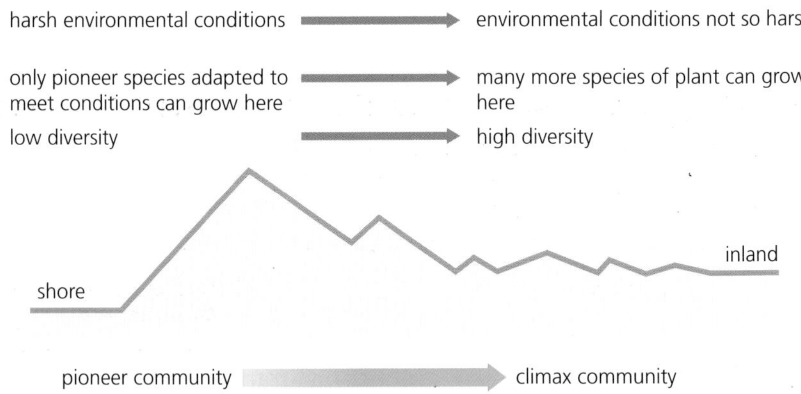

Figure 9.18 A summary of the processes involved in succession.

Managing succession

Succession does not always proceed all the way to a climax community. It can be stopped by various factors often associated with human activity.

There is little, if any, of the landscape of Britain that has not been modified by human activity. Much of our current landscape is the result of centuries of agricultural practice. Chalk grassland is one example. It consists of a diverse mixture of grasses and herbs and can support up to 50 species of flowering plant per square metre (Figure 9.19). There is also a high diversity of invertebrates, such as insects and snails.

Figure 9.19 Orchid-rich chalk grassland is identified as a European conservation priority.

At the beginning of the nineteenth century more than 50% of the South Downs was chalk grassland. Today, the figure stands at just 3%. There are many reasons for this massive reduction. Understanding them requires not only a knowledge of ecology, but also an understanding of other aspects of biology.

- Chalk grassland resulted from sheep grazing the thin, nutrient-poor soils overlying chalk. A massive decline in the number of sheep and the resulting spread of scrub has reduced the amount of chalk grassland. The bushes that form this scrub also spread to remaining areas of chalk grassland. These bushes, however, are also part of the chalk ecosystem and are important in contributing to the overall biodiversity by providing a habitat for many species. A balance needs to be achieved by controlling this process of succession.
- The disease myxomatosis wiped out large numbers of rabbits in the 1950s. Since then rabbits have become resistant to the disease and numbers have increased dramatically. They have little effect in controlling scrub vegetation and their warrens and overgrazing cause serious erosion.
- In order to improve yield, many farmers have added nitrogen-containing fertilisers and selective herbicides to fields. Some species have benefited at the expense of others and there has been a resulting loss in biodiversity.
- Much of the existing chalk grassland is fragmented. This fragmentation causes isolation and makes populations more vulnerable to local extinction from disease and predation, because each population has a limited gene pool and there is limited variation for natural selection to operate on.

Conserving habitats such as chalk grassland requires careful management, to prevent succession taking place. By grazing sheep on chalk grassland, scrub cannot develop, and this maintains a suitable habitat for many different species of wildflowers and insects such as butterflies. Many areas of moorland are mowed to remove tree species and preserve the habitat for moorland birds and other wildlife.

EXAMPLE

Investigating succession when plant cover has been destroyed

In an investigation, ecologists compared control plots with plots on which the plant cover had been partly destroyed by allowing motor cycles to be ridden over it. The results are shown in Figure 9.20.

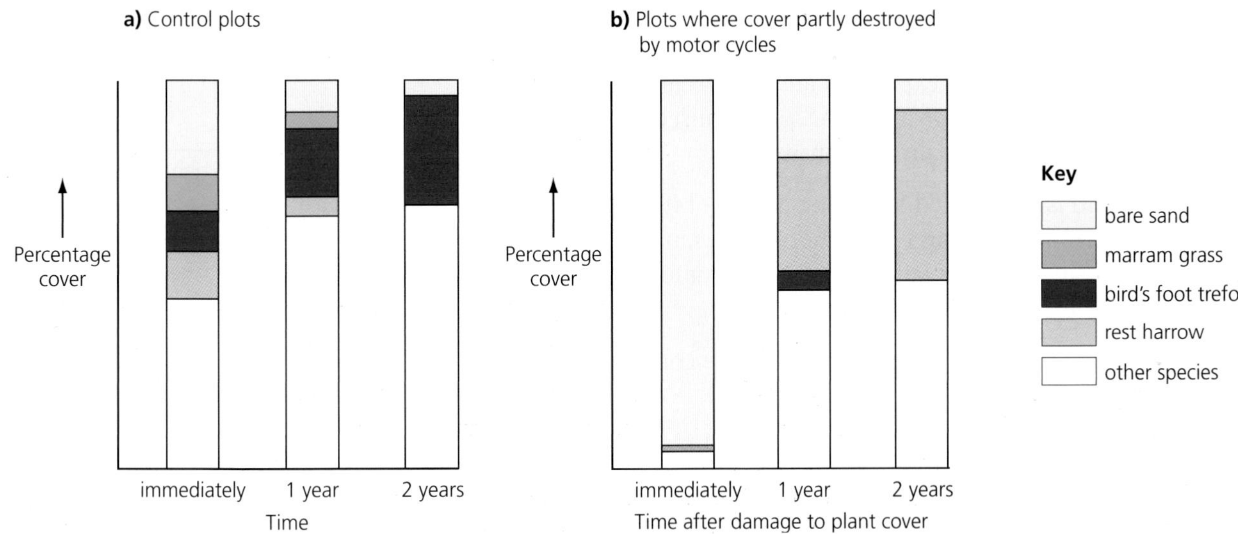

Figure 9.20 A comparison of changes in the vegetation of a) control plots and b) plots in which the plant cover had been partly destroyed by motor cycles.

1 Describe how the control plots should have been treated.
In this case, the important thing was to treat the control plots in the same way as the experimental plots in all aspects except in riding a motor cycle over them.

2 What was the purpose of the control plots?
They provide a comparison with the experimental plots by showing what would happen to similar areas of dune that were not damaged by being ridden over.

3 Describe the process of succession that takes place in the control plots over the period covered by this investigation.
There is a decrease in the amount of bare ground and a change in plant cover. The percentage cover of marram grass and rest harrow decreases and there is an increase in bird's foot trefoil and other species.

This investigation illustrates an important point. When we halt the process of succession, in this case by partly destroying plant cover with a motor cycle, it rarely returns to where it began.

4 Use the results in the bar chart to describe how the process of succession is different after the plant cover has been destroyed. Suggest an explanation for the difference.
In a natural succession, marram grass is present early in succession. In the investigation, we are starting from a position where species other than marram grass are present, although most of the plant cover has been destroyed. The sand in the areas from which the plant cover has been removed is quite different from the sand in which the marram grass became established initially. The sand with reduced plant cover contains more humus and it is not so likely to dry out. It also has a lower concentration of sodium and calcium ions and a higher concentration of nitrate ions. These are conditions in which rest harrow grows better than marram grass. You can see quite clearly how human activity has altered the path of succession.

TEST YOURSELF

11 Use your knowledge of competition to explain why the use of nitrogen-containing fertilisers on chalk grassland has resulted in a loss of biodiversity.

12 Explain two ways in which farming prevents succession occurring in fields.

13 Succession takes place after a fire has destroyed a large area of forest. Suggest how plant species are able to colonise the area.

Practice questions

1 a) What is species richness? *(1)*

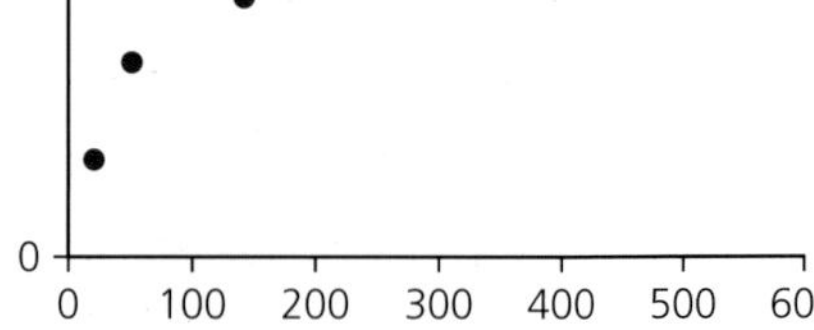

b) The graph shows how species richness changes along a sand dune system.

i) Describe the graph. *(1)*

ii) Explain the results. *(3)*

c) How would you expect the animal species richness to change along the same transect from 0 to 400 m? Explain your answer. *(2)*

2 If a field is abandoned for a long period of time, succession will occur. Over time, changes in the composition of the plant and animal communities will occur until eventually no further change takes place.

a) What name is given to the final stable community at the end of a succession? *(1)*

b) One community is gradually replaced by another during succession. Explain how. *(2)*

c) Explain how the following farming activities interfere with the process of succession:

i) regular grazing by sheep

ii) ploughing fields each year. *(2)*

3 a) What is meant by an ecological niche? *(1)*

b) In a study of one population of badgers, 72 animals were trapped and marked with a harmless dye on their underside. They were then released. One month later, scientists trapped 120 badgers and found that 14 of these had been marked with dye. Use these figures to calculate the size of the badger population. Show your working. *(2)*

c) Suggest two reasons why this answer may not be completely reliable. *(2)*

4 a) Give the meaning of these ecological terms:

i) population

ii) community. *(2)*

b) Some students used the mark-release-recapture technique to estimate the size of a population of woodlice. They collected 97 woodlice and marked them before releasing them back into the same area. Later they collected 86 woodlice, 14 of which were marked. Calculate the number of woodlice in the area under investigation. Show your working. *(2)*

c) Describe how you would use a quadrat to estimate the number of buttercup plants in a field measuring 100 m by 50 m. *(3)*

5 *Paramecium aurelia* and *Paramecium caudatum* are single-celled protoctists that feed on algae. A scientist cultured the organisms in flasks containing a suitable culture medium. He grew them separately and together. The graph shows the results.

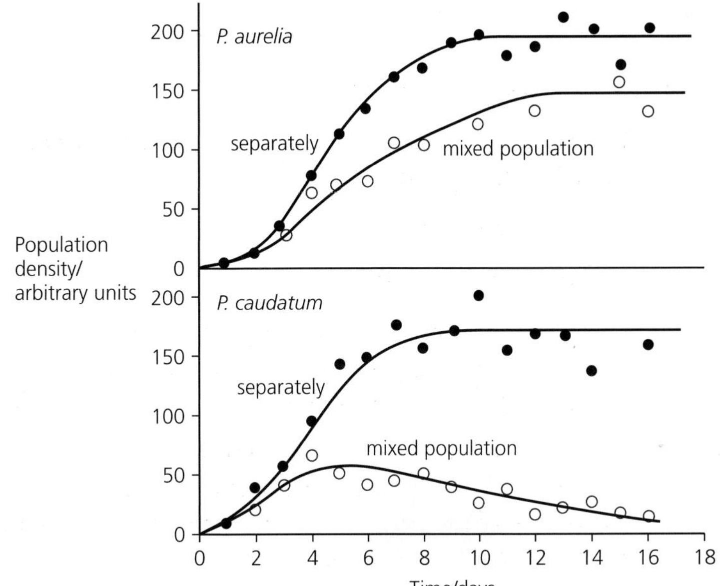

a) Explain why the two species of *Paramecium* were grown separately as well as together. *(1)*

b) Explain the results when both species are grown together. *(2)*

c) Suggest two conditions that the scientist would have kept the same for these results to be reliable. *(1)*

d) Evaluate the benefit of laboratory studies like this in understanding interactions between organisms in a natural environment. *(4)*

6 Scientists investigated the effect of a predatory mite on a herbaceous mite. They grew a population of a herbaceous mite that feeds on oranges in a container. They added the predatory mite that feeds on the herbaceous mite. The population sizes of both mites were measured at regular intervals for a period of just over a year.

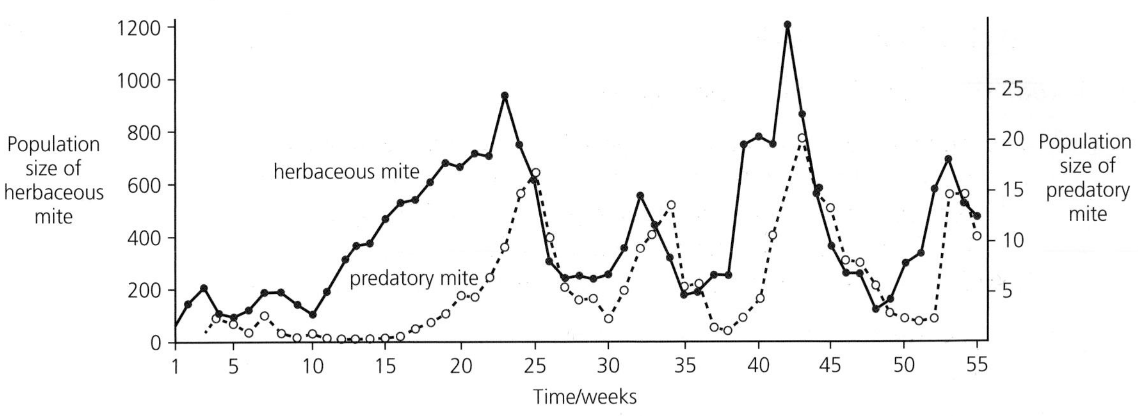

a) Describe and explain the results shown in the graph. *(3)*

b) Explain how you would calculate the percentage increase in the population of the herbaceous mite between weeks 10 and 24. *(1)*

c) Explain two advantages of studying these two mites in understanding the relationship between predators and prey. *(2)*

Stretch and challenge

7 Design an investigation into the effect of an abiotic factor on the distribution of a plant species. Plan your investigation to ensure reliable results, and choose a suitable statistical test to analyse your results.

10 The control of gene expression

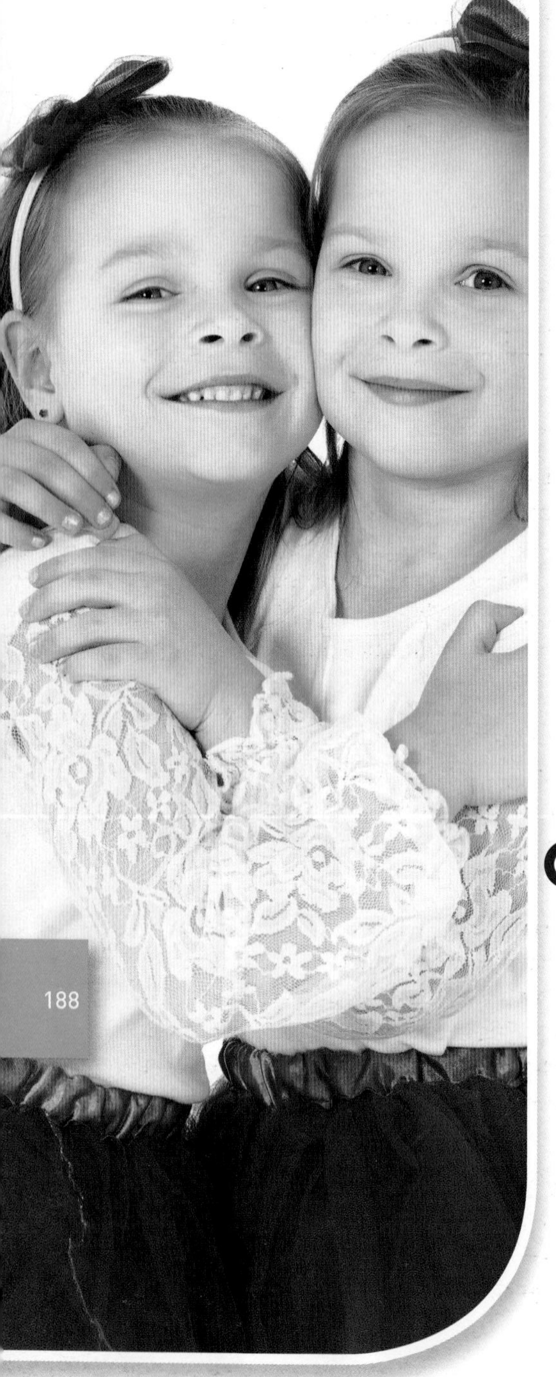

PRIOR KNOWLEDGE

- A gene is a base sequence of DNA that codes for the amino acid sequence of a polypeptide or a functional RNA (including ribosomal RNA and transfer RNA).
- A gene occupies a fixed position, called a locus, on a particular DNA molecule.
- A sequence of three DNA bases, called a triplet, codes for a specific amino acid. The genetic code is universal, non-overlapping and degenerate. Also, three bases in mRNA or tRNA code for one amino acid.
- In eukaryotes, much of the nuclear DNA does not code for polypeptides. There are, for example, non-coding multiple repeats of base sequences between genes. Even within a gene only some sequences, called exons, code for amino acid sequences. Within the gene, these exons are separated by one or more non-coding sequences, called introns.
- The genome is the complete set of genes in a cell and the proteome is the full range of proteins that a cell is able to produce.
- Messenger RNA (mRNA) and transfer RNA (tRNA) have specific structures.
- Transcription is the production of mRNA from DNA. RNA polymerase is used in joining mRNA nucleotides.
- In prokaryotes, transcription results directly in the production of mRNA from DNA.
- In eukaryotes, transcription results in the production of pre-mRNA; this is then spliced to form mRNA.
- Translation is the production of polypeptides from the sequence of codons carried by mRNA.
- Ribosomes, tRNA and ATP are involved in translation.

TEST YOURSELF ON PRIOR KNOWLEDGE

1 Give two similarities and two differences between an RNA and a DNA nucleotide.

2 The coding strand of DNA has the following base sequence: CGGTACGA. What is the base sequence of the mRNA for which it codes?

3 Put the following sentences in order to describe the process of protein synthesis.

A A tRNA molecule brings a specific amino acid to the ribosome.

B RNA polymerase joins the nucleotides together to make a molecule of mRNA.

C A section of DNA (a gene) unwinds and the hydrogen bonds break.

D The amino acids join by a peptide bond.

E The mRNA leaves the nucleus and attaches to a ribosome in the cytoplasm.
F RNA nucleotides attach to the exposed DNA bases by complementary base pairing.
G One strand of the DNA becomes a template.
H The anticodon on the tRNA is complementary to the codon on the mRNA.
I Once the first tRNA has passed on its amino acid, it leaves the ribosome and the ribosome moves along the mRNA, three bases at a time.
J A second tRNA brings its specific amino acid to the ribosome, next to the first tRNA.

4 What is the difference between a polypeptide and a protein?

5 During transcription, the base sequence of only one of the strands of DNA is used to make a molecule of mRNA; the other DNA strand is not used. Give **one** advantage of a molecule of DNA having both strands.

Introduction

You might be familiar with animals that have different body forms during their life cycle. The cabbage white butterfly is a major economic pest in the UK. Its larval stage, the caterpillar, is a voracious leaf-eater and can destroy cabbage crops. The adult butterfly does not eat plant leaves; instead it feeds infrequently on the sugary nectar found in flowers.

Have you ever taken cuttings from a plant, such as the stem, and used them to grow new plants? If so, you have created a plant clone, a group of genetically identical plants. Humans have been doing this for hundreds, if not thousands, of years. Research botanists perform a more sophisticated version of the same process when they isolate individual cells from the root tip of a plant and grow them in tissue culture to produce large numbers of genetically identical plants.

Clone An organism or cell, produced asexually from one ancestor, to which it is genetically identical.

In these examples, we can see different body forms within a single species. Body form is one aspect of an organism's phenotype. Although butterflies have different phenotypes during their life cycle, their genotype is the same at each stage. Similarly, the genotypes of the root and stem used to make plant clones are identical.

Making clones from animals is not as easy as for plants. As you will see in this chapter, one reason for this is that animal cells lose the ability to express many of their genes as they differentiate and mature. However, some animals have been cloned by removing the nucleus from one of their cells and inserting it into an egg cell whose nucleus has been removed. This technology is called somatic cell nuclear transfer (SCNT); we usually refer to it as animal cloning.

You are probably familiar with Dolly, shown in Figure 10.1. She was produced by a team of scientists in Scotland, using SCNT, and is thought to be the first mammal to have been cloned successfully. Dolly was put down when she was only 6 years old. Although she had given birth to a number

Figure 10.1 Dolly the sheep was the first mammal to be produced by somatic cell nuclear transfer. Do you know why she was called Dolly?

Unipotent A cell that can divide to form only one kind of cell.

Pluripotent A cell that can mature into many different kinds of specialised cell.

Totipotent A cell that can mature into any kind of cell type.

of healthy lambs, at the time of her death Dolly showed symptoms seen more commonly in much older sheep. These included arthritis, lung disease and obesity. Some of these symptoms are thought to have resulted from problems associated with cell division and the control of gene expression in her body cells.

In this chapter you will learn how genes are controlled at a cellular level, how they are switched on or off in different cells within an organism. This knowledge will help to explain how different body forms within one life cycle and different organs within one body are possible.

Stem cells

Unlike plants, only a small number of cells in animals retain the ability to divide and give rise to new tissues. Most cells in animal that can divide are only able to produce cells of the same type. These are called unipotent cells. However, some cells retain the ability to divide and are called **stem cells** and can be found in embryonic tissue and in some adult tissues. Whatever their source, stem cells have three general properties.

- They can divide and renew themselves over long periods.
- They are unspecialised.
- They can develop into other specialised cell types.

Human embryonic stem cells are stem cells that exist in all human embryos. They are taken from an embryo that has developed following *in vitro* fertilisation of an egg (in an *in vitro* fertilisation clinic) and been donated for research purposes with the informed consent of the donors. About 30 cells are removed from the inside of a human embryo that is 4–5 days old. These cells are plated into dishes that contain a coating of embryonic mouse skin cells and an appropriate culture medium, where they divide.

Very small numbers of **human adult stem cells** are found in specific tissues, such as bone marrow and the brain. They lie dormant for many years until stimulated to begin to divide, following an injury. In addition to being difficult to isolate, they are also more difficult to grow in tissue culture than embryonic stem cells. Unlike human embryonic stem cells, adult stem cells can only give rise to a limited number of body tissues: they are **multipotent**. In contrast to human adult stem cells, embryonic stem cells are pluripotent; this means, if cultured in appropriate conditions, they can develop into most of the body's cell types. This type of stem call can divide in unlimited numbers and can be used to treat human disorders. Cells taken from a very early embryo, in the first 3 or 4 days, are totipotent and can develop into any kind of cell type.

The use of embryonic stem cells is controversial. In some countries, including the USA, it is currently illegal to use embryonic stem cells, even for research. In other countries, the use of embryonic stem cells is currently legal but is tightly regulated. In April 2008, members of the European Parliament (MEPs) voted to ban across the European Union any research involving embryonic stem cells. The following month, members of the UK parliament, which is a member of the European Union, voted to continue to allow such research.

More recently, scientists have developed **induced pluripotent stem cells** (iPS cells). These are made from somatic (body) cells which have already differentiated. All cells in an organism contain the same genes, but once a cell differentiates some of the genes are 'switched off'. The scientists who developed iPS cells found that there were four genes expressed in a mouse embryo that control pluripotency. They added these four genes to somatic cells and found that the transcription factors produced by the genes made the cells pluripotent. You will learn more about transcription factors later in the chapter. There is a great deal of evidence that these iPS cells are very similar, if not identical, to embryonic stem cells. It may be possible in the near future to use these cells rather than embryonic stem cells, to avoid the ethical issues mentioned above. However, at the moment there is still a great deal of testing to be done, as iPS cells can lead to tumour formation. Scientists need to understand the processes going on in the cells much better before they can be used therapeutically in humans.

Tumour A group of one type of cell that is dividing rapidly and uncontrollably. The formation of a tumour might result from one, or only a few, genetic changes in a cell.

TEST YOURSELF

1 Embryonic mouse skin cells are used in culturing human embryonic stem cells to provide a surface to which the human cells can attach. Suggest one potential danger of this use of mouse cells.

2 Scientists hope that transplanting cultures of stem cells might be used to repair damaged tissues or to replace malfunctioning tissues.
   a) How can transplanted stem cells repair or replace malfunctioning tissues?
   b) Suggest one biological advantage and one biological disadvantage of using embryonic stem cells in such transplants.

Gene mutation

Complementary base pairing is essential if transcription and translation are to work properly. Table 10.1 summarises the pairing relationship between a base in a DNA nucleotide and the bases in complementary nucleotides of mRNA and then of tRNA.

Table 10.1 Complementary base pairing is important during DNA replication, transcription and translation.

Base on DNA nucleotide	Complementary base in the codon of a nucleotide of mRNA	Complementary base in the anticodon of a nucleotide of tRNA
Adenine	Uracil	Adenine
Cytosine	Guanine	Cytosine
Guanine	Cytosine	Guanine
Thymine	Adenine	Uracil

You also learned in *AQA A-level Biology 1 Student's Book* that three bases, called a codon, code for one amino acid. Table 4.6 on page 70 of that book shows the mRNA codons that code for each amino acid.

As you also learned in the first year of your A-level Biology course, DNA base sequences are copied during DNA replication. At a rate of about one in a million, spontaneous errors occur during this copying process.

Some of the errors that can occur are outlined below.

Frame-shift mutation A mutation caused by the addition or deletion of bases so that all of the triplets are changed from the point of mutation.

- **Base deletion**: a base is lost from the DNA sequence. As a result, the whole base sequence following the deleted base changes. This is called a frame-shift mutation and results in a new sequence of amino acids after the deletion.
- **Base addition**: a new base is added, which changes the whole base sequence following this addition. This also results in a frame shift and a new sequence of amino acids after the deletion.
- **Base substitution**: the 'wrong' base is included in the base sequence. This mutation might result in a different amino acid being included in the polypeptide chain. It does not cause a frame shift.
- **Base duplication:** an important source of evolutionary change. This happens when one gene, or part of a gene, is copied, so that there are two copies on one chromosome. The second copy can develop new functions by mutation, while the original copy continues to make the protein, so there is no harmful effect on the organism. This is how the different kinds of haemoglobin – alpha, beta and fetal – are believed to have evolved. Also, in ice fish, which survive sub-zero temperatures in the Arctic Ocean, duplication of a gene coding for a digestive enzyme is believed to have resulted in the second copy of the gene mutating into an antifreeze protein.
- **Base inversion:** this can occur when two breaks in the DNA of a single gene occur. The 'cut' portion can rotate 180° and then re-join the original DNA. As this type of mutation usually affects several amino acids, it generally results in a non-functional protein being produced. However, sometimes it results in a very different protein.
- **Base translocation:** this can occur when part of a gene breaks off and re-attaches to another gene. This usually results in the original gene with a missing section, and the translocated portion, coding for non-functional proteins. However, if an inversion or a translocation occurs in a proto-oncogene or a tumour suppressor gene, this may lead to cancer. For example, a faulty tumour suppressor gene may lead to a cell with faulty DNA continuing to replicate, and translocation of a proto-oncogene to another part of the DNA may cause the gene to be expressed more strongly, or to produce a protein that has oncogenic activity.

Table 10.2 shows the effects that can arise from base deletion, base substitution and base insertion.

Table 10.2 The first two rows of this table show the amino acid sequence encoded by part of a molecule of mRNA. The mRNA sequence is shown as individual codons to help you to read the code. The table demonstrates the effect on the encoded amino acid sequence of a deletion, substitution and insertion of the bases shown in red.

Original base sequence on mRNA	AG**A**	UAC	GCA	CAC	AUG	CGC
Encoded sequence of amino acids	Arg	Tyr	Ala	His	Met	Arg
mRNA base sequence after base deletion	AGU	ACG	CAC	ACA	UGC	GCX
Encoded sequence of amino acids	Ser	Thr	His	Thr	Cys	Ala
mRNA base sequence after base substitution	AGC	**U**AC	GCA	CAC	AUG	CGC
Encoded sequence of amino acids	Ser	Tyr	Ala	His	Met	Arg
mRNA base sequence after base insertion	AGG	**A**UA	CGC	ACA	CAU	GCG
Encoded sequence of amino acids	Arg	Ile	Arg	Thr	His	Ala

TIP

Some mutations do not cause a change in the amino acid sequence at all. For example, for many amino acids, changing the third base in a triplet means the triplet still codes for the same amino acids. This is because the DNA code is degenerate.

Mutagenic agents

Natural mechanisms occur within cells that identify and repair damage to DNA. Many environmental factors increase the rate of mutation. They are called **mutagenic agents** and include:

- toxic chemicals, for example bromine compounds, mustard gas (used in a large number of conflicts to kill soldiers and civilians) and peroxides
- ionising radiation, for example gamma rays and X rays
- high-energy radiation, for example ultraviolet light.

TEST YOURSELF

3 a) Look at the middle two rows of Table 10.2. The final codon in the sequence is incomplete (GCX). Why can we safely identify Ala (alanine) as the encoded amino acid?

b) Use the middle two rows to explain the term 'frame shift'.

c) Use the bottom two rows in Table 10.2 to explain why a base substitution does not cause a frame shift.

4 a) Mutations in some cells are not important. Suggest why a mutation in the gene for haemoglobin is unimportant in a white blood cell.

b) Explain why a gene mutation can have important effects if it is found in:

i) a gamete

ii) a gene controlling cell division.

5 Some people are concerned that the aerials used to transmit mobile phone messages emit radiation that increases the rate of mutation. For this reason they oppose the erection of these aerials in their neighbourhood. Suggest what you would need to measure to investigate whether these aerials increased the rate of mutation leading to cancer.

Control can occur at several stages of gene expression

Gene expression involves the following flow of genetic information: DNA → mRNA → polypeptide. Gene expression can be controlled at any stage in this flow of information, including transcription and translation.

- Control of transcription: only some genes are transcribed at any given time.
- Control of translation: mRNA might be destroyed or its translation by a ribosome blocked.

Control of transcription by specific transcription factors

Every gene has one or more DNA base sequences that control its expression. Figure 10.2 shows one of these, called a **promoter region**. A promoter region is located near the gene it controls, usually about 100 base pairs before the start of its gene. Figure 10.3 shows how a protein, called a **transcription factor**, binds to a gene's promoter region and, in doing so, enables RNA polymerase to attach to the start of the gene and begin its transcription.

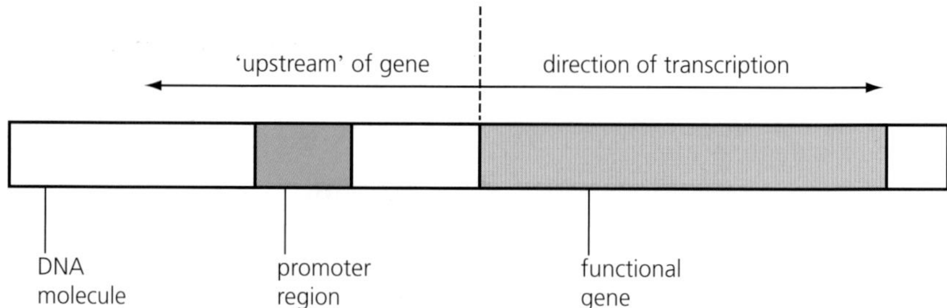

Figure 10.2 Every gene in eukaryotes is controlled by one or more promoter regions. The promoter region lies close to the gene it controls and is 'upstream' of it.

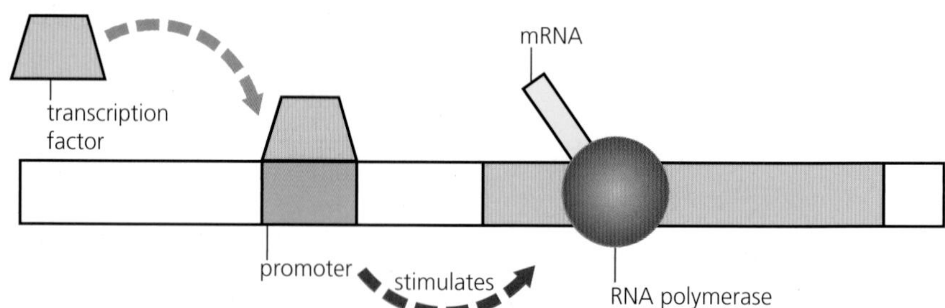

Figure 10.3 The role of a transcription factor and a promoter region in stimulating transcription. The position of a gene encoding a polypeptide is shown in part of a molecule of DNA. In the DNA sequence preceding this gene (referred to as 'upstream' of the gene) is a promoter region. Only if an appropriate transcription factor attaches at this promoter region can RNA polymerase begin to transcribe the gene.

The role of oestrogen in initiating transcription

Oestrogen is a mammalian steroid hormone. It is involved in the control of the mammalian oestrous cycle; it also stimulates sperm production in males. Because it is a small, hydrophobic molecule, oestrogen can diffuse through the plasma membranes of cells. Once in the cytoplasm, oestrogen diffuses into the cell's nucleus, where it binds to a type of oestrogen receptor, called ER alpha (**ERα**).

The ERα oestrogen receptors are transcription factors that can bind to the promoter region of up to 100 different genes. Figure 10.4 (overleaf) shows one way in which these ERα oestrogen receptors work. In the cell, oestrogen receptors are normally held within a protein complex that inhibits their action. When oestrogen binds to an ERα oestrogen receptor it causes the oestrogen receptor to change shape and leave its protein complex. The oestrogen receptor can now attach to the promoter region of one of its target genes, stimulating RNA polymerase to transcribe that target gene.

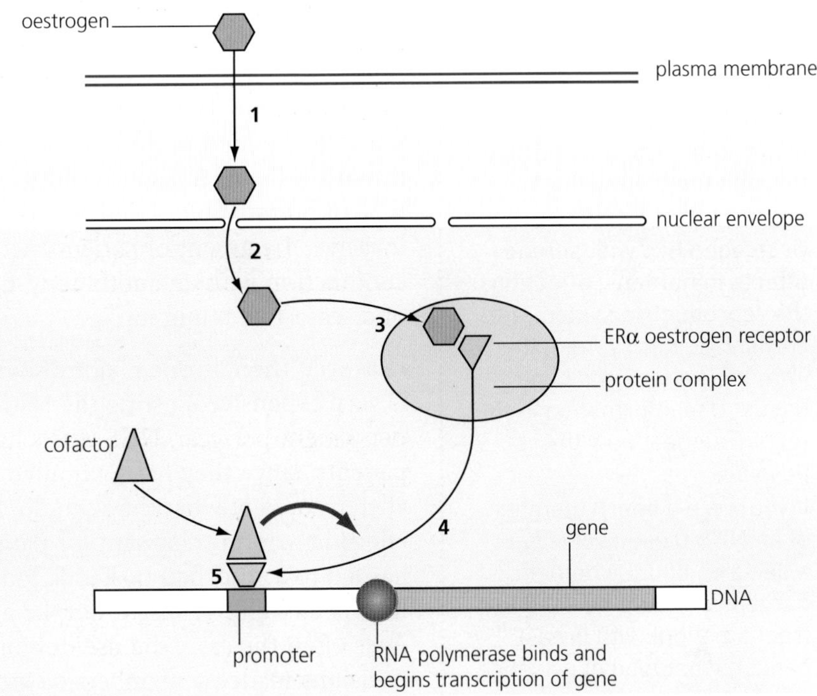

Figure 10.4 A simplified summary of how oestrogen stimulates the transcription of a target gene.
(1) Oestrogen diffuses through the plasma membrane of a target cell and then diffuses into its nucleus.
(2) Here it attaches to an ERα oestrogen receptor that is contained within a protein complex.
(3) This causes the oestrogen receptor to change its shape and leave the protein complex that inhibits its action.
(4) The oestrogen receptor can now attach to the promoter region of a target gene
(5) where it attracts other cofactors to bind with it. The oestrogen receptor, with combined cofactors, enables RNA polymerase to transcribe its target gene.

TIP

You need to know how oestrogen can stimulate gene transcription, but you do not need to know about oestrogen-dependent breast tumours and implications for hospital budgets. However, you might be given information on topics such as this and be asked to evaluate it.

Oestrogen receptors, oestrogen-dependent breast tumours and hospital budgets

Many people in the UK develop breast tumours; the vast majority are women. About 35% of breast tumours are associated with over-stimulation of the gene encoding oestrogen. They are called **oestrogen-dependent breast tumours**.

For over 20 years the main treatment of breast tumours has been a drug called tamoxifen. This drug is effective because it has a chemical shape similar to that of oestrogen (Figure 10.5). This enables tamoxifen to bind permanently to oestrogen receptors in tumour cells. As a result, these tumour cells can no longer bind with oestrogen, which they need in order to grow.

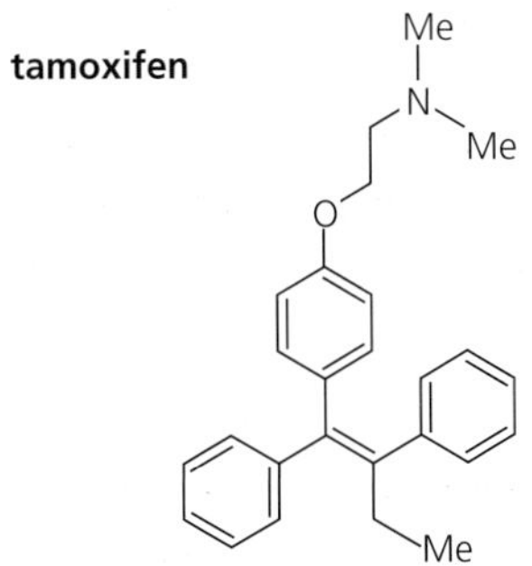

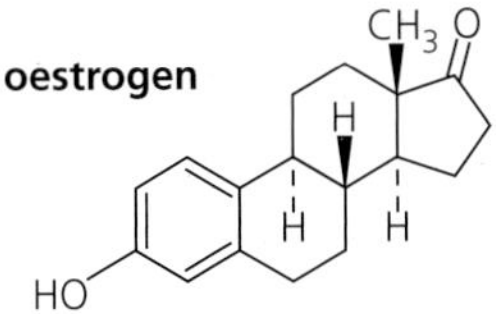

Figure 10.5 Tamoxifen is a drug that is used to treat oestrogen-dependent breast tumours. Tamoxifen has a similar chemical shape to oestrogen.

More recently, new drugs have been developed that inhibit an enzyme, aromatase, which is required for the synthesis of oestrogen. Like tamoxifen, these drugs stop tumour cells responding to oestrogen, so stopping their growth. Large-scale clinical trials suggest that aromatase inhibitors, such as anastrozole and letrozole, can be even more effective in treating breast tumours than tamoxifen.

However, not all breast tumours are stimulated by oestrogen. Tamoxifen and aromatase inhibitors are ineffective against breast tumours that are not oestrogen-dependent. A relatively new drug, called herceptin,

> **TEST YOURSELF**
>
> **6** Explain why having hydrophobic molecules enables oestrogen to diffuse through the cell-surface membrane of a cell.
>
> **7** Oestrogen has widespread effects in humans, affecting the reproductive system, the cardiovascular system, the immune system and bone tissue. Use information in the text to suggest how this is possible.
>
> **8** If you were a board member of an NHS trust involved in a decision about whether herceptin should be used to treat a patient with breast cancer, what type of evidence would you take into account?

is effective against some of these breast tumours because it controls them in a different way. Herceptin is a monoclonal antibody. You learned about monoclonal antibodies in the first year of your A-level Biology course. Herceptin works by binding to a growth factor receptor that is embedded in the surface membranes of some types of breast-tumour cells. As a result, it inhibits the growth of the tumour cells. The type of breast-tumour cell against which herceptin is effective is very invasive. Treatment of patients with this type of breast tumour, used in conjunction with chemotherapy, has been shown to lead to long periods of disease-free remission.

However, there is one major drawback with the use of herceptin. It is very expensive, costing the National Health Service about £22 000 per patient per year. NHS trusts have to consider the cost of treating patients, since they have a limited budget and have to balance the cost-effectiveness of what they can do. Some trusts have not been able to afford herceptin treatment for patients with invasive breast cancers. This has led to a so-called postcode lottery: depending on where a sufferer lives she might, or might not, be able to receive herceptin treatment. As is often the case, the use of biological advances is dependent on decisions made by members of society.

Control of translation by RNA interference

In eukaryotes and some prokaryotes translation of mRNA from target genes can be inhibited by RNA interference, or RNAi.

EXAMPLE

Gene regulation in petunias

> **TIP**
>
> The main idea that you need to remember from this is that interfering RNA binds to mRNa by complementary base pairing.

Petunias are popular plants that are grown in hanging baskets in the UK. The purple colour of the flowers in Figure 10.6 on the next page results from a series of reactions in which a white pigment is converted to a purple pigment. One of the enzymes in this series of reactions is called chalcone synthase.

In 1990, a group of scientists reported their attempts to produce petunia flowers with a very deep purple colour. They used genetic engineering techniques to insert many copies of the gene encoding chalcone synthase into the cells of petunia plants with pale purple flowers.

Figure 10.6 Petunia flowers show a wide range of colours.

1 Use your knowledge of how enzymes work to suggest why the scientists expected the plants with the extra genes for chalcone synthase to produce deep purple flowers.
Enzymes speed up a reaction by combining with molecules of substrate to form enzyme–substrate complexes, which break down to release molecules of the product. We can increase the rate at which the product is formed by adding more enzyme molecules. As a result, more enzyme–substrate complexes will be produced and so the product will be formed faster. The scientists predicted that if they added more genes for chalcone synthase to cells of petunia plants, more mRNA would be transcribed from them and so more molecules of enzyme would be made in the cells of the petunia.

Instead of deep purple flowers, the transformed petunia plants produced flowers that were mottled white. The scientists could not explain this result.

In 2004, in a totally unrelated investigation, a different team of scientists reported their discovery of enzymes called **RNA-dependent RNA polymerases (RDRs)**. These RDRs catalyse the production of **double-stranded RNA (dsRNA)**. They do this by using molecules of mRNA that are in the cytoplasm as a template to synthesise a complementary strand. The two RNA strands are held together by hydrogen bonds between complementary base pairs.

2 What is unusual about the RNA produced by RDRs?
You learned in AQA A-level Biology 1 Student's Book that RNA molecules are single-stranded, but the RNA molecules produced by RDRs are double-stranded (dsRNA).

Figure 10.7 shows what happens when RDRs produce a dsRNA molecule from a molecule of mRNA in the cytoplasm.

3 Look at Figure 10.7. What is the first thing that happens to the dsRNA molecule?
The dsRNA molecule is cut into small fragments. These fragments are about 23 base pairs long and are called ***small interfering RNA (siRNA)****.*

Figure 10.7 shows what happens to the siRNA. In a reaction requiring the hydrolysis of ATP, a protein complex in the cytoplasm takes up one of these siRNA fragments and separates its two RNA strands.

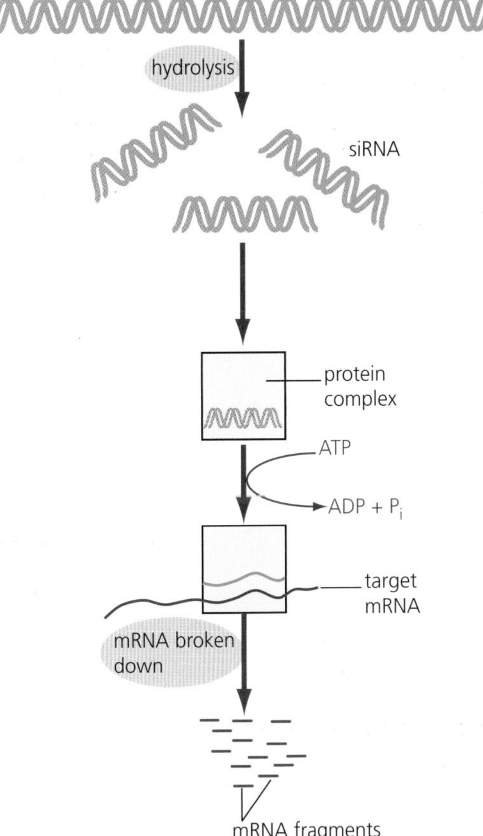

Figure 10.7 Small interfering RNAs (siRNAs) are small double-stranded RNA molecules. They are used by protein complexes in a cell's cytoplasm to break down mRNA. By breaking down the target mRNA, a cell can control the expression of the gene from which the mRNA was transcribed.

4 Suggest why the hydrolysis of ATP is involved in the reaction between the protein complex and the siRNA it takes up.
Like many reactions in metabolism, the reaction between the protein complex and the siRNA requires energy. This energy is released when ATP is hydrolysed to ADP and inorganic phosphate.

Look at the protein complex containing the single strand of RNA in Figure 10.7. The protein complex uses the RNA strand to bind to a molecule of mRNA in the cytoplasm. Once it has attached to the mRNA molecule, the protein complex breaks down the mRNA molecule. This stops the mRNA being translated by the cell's ribosomes.

5 To which type of mRNA will the protein complex attach?

The RDR in Figure 10.7 used a molecule of mRNA to make the dsRNA. One strand of the dsRNA must, therefore, have a complementary base sequence to the mRNA from which it was made. When the protein complex separates the two strands of the siRNA, it uses this complementary base sequence to attach to any of the original mRNA molecules in the cytoplasm. Thus, the protein complex specifically breaks down the mRNA from which the siRNA was made.

6 Will the base sequence of the siRNA be complementary to the whole of the mRNA in Figure 10.7?

The siRNA is a small fragment of the mRNA molecule, so it will only be complementary to part of the base sequence of the mRNA. However, this is enough to enable the protein complex to bind with, and destroy, the target mRNA.

7 The destruction of mRNA by siRNA is stimulated when the concentration in the cytoplasm of one type of mRNA becomes high. Can you use this information to suggest how the discovery of RNA-dependent RNA polymerases and small interfering RNA provided an explanation for the failure of the first team of scientists to produce petunia flowers with a deep purple colour?

The first team of scientists increased the concentration of mRNA encoding chalcone synthase by inserting into petunia cells many copies of the gene for the enzyme. A high concentration of mRNA encoding chalcone synthase stimulated RDRs to produce dsRNA from it. As a result many more siRNA molecules carrying part of the mRNA code for chalcone synthase were produced by the plants, so the mRNA was destroyed by the mechanism in Figure 10.7. Unwittingly, the scientists had stimulated destruction of mRNA encoding chalcone synthase instead of stimulating more of it to be transcribed to produce more enzyme molecules.

Epigenetic control of gene expression in eukaryotes

Epigenetics involves heritable changes in gene function without changes to the base sequence of DNA. Along with environmental influences, it explains why identical twins, whose DNA is identical at fertilisation, become increasingly different from each other as they get older. Figure 10.8 shows two identical twin sisters. They still look very much alike at 78, but the small differences between them are much easier to see now that they are older.

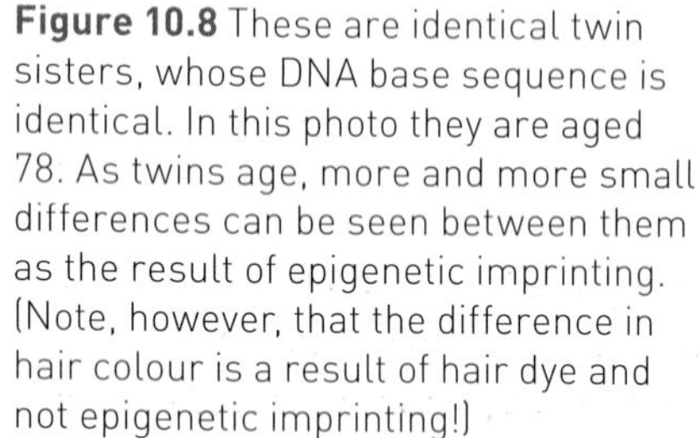

Figure 10.8 These are identical twin sisters, whose DNA base sequence is identical. In this photo they are aged 78. As twins age, more and more small differences can be seen between them as the result of epigenetic imprinting. (Note, however, that the difference in hair colour is a result of hair dye and not epigenetic imprinting!)

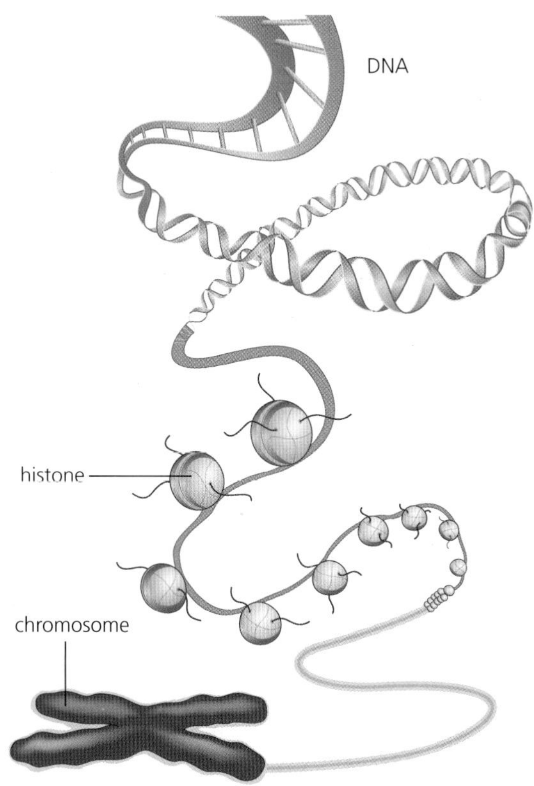

Figure 10.9 DNA winds around histone proteins to form chromosomes. Chemical 'tags' (see Figure 10.12) attached to the DNA and the histones affect how tightly the DNA winds around the histones.

In the cells of eukaryotes, the DNA is wrapped around proteins called histones. Chemical 'tags' can join on to the histone proteins and the DNA (see Figure 10.9). The chemical 'tags' affect how tightly the DNA is wound around the histone proteins. If the DNA is wound tightly round the histone proteins, the gene is effectively 'switched off' but if the DNA is loosely wound, the gene may be 'switched on'. This system is flexible, so the epigenome can be different in one cell type from another.

Acetylation of histones

Histone molecules contain the amino acid lysine. Acetyl groups ($COCH_3$) may be added to these lysine residues, replacing one of their hydrogen ions. This removal of positively charged ions causes the histone proteins to be less tightly wrapped around the DNA. As a result, the enzyme RNA polymerase and other factors needed for transcription can bind to the DNA more easily. You can see this in Figure 10.10. Therefore, in most cases

- adding acetyl groups to the histone proteins (**acetylation**) stimulates transcription
- removing acetyl groups from the histone proteins (**deacetylation**) suppresses transcription.

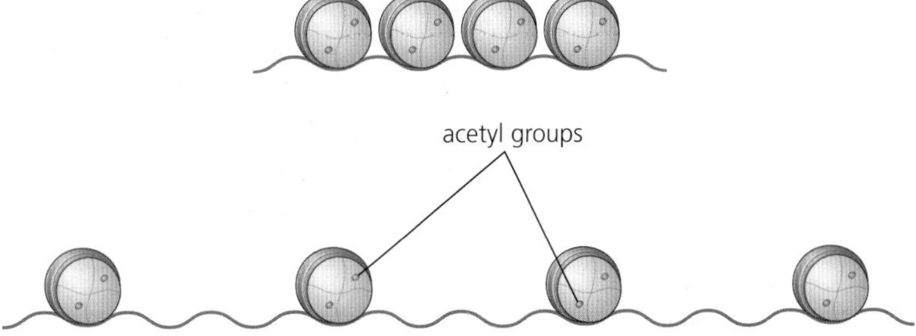

Figure 10.10 Acetylation of the histone proteins makes the DNA less tightly wound round the histone proteins. This allows DNA transcription to occur.

Epigenome The sum of the chemical changes to the histone proteins and the DNA (but not including changes to the base sequence) of an organism.

Methylation of DNA

DNA methylation occurs when methyl groups (CH_3) are added to a DNA molecule, usually to a carbon atom of cytosine bases where they occur in a cytosine–guanine sequence. This methylation suppresses transcription of the affected gene, effectively switching the gene off.

Methyl and acetyl groups are examples of the chemical 'tags' that form part of the epigenome. They can be added to chromosomes as the result of environmental factors, such as diet, stress, smoking or exercise. They can also be the result of signals from neighbouring cells or even from within the same cell. Methylation is an important mechanism for 'switching' genes on and off during embryonic development, when cells are differentiating.

Epigenetic imprinting

You will remember that our body cells have two sets of chromosomes in them, one (paternal) set from our fathers and the other (maternal) set

Epigenetic Describing inherited changes in the DNA that do not involve a change in the DNA base sequence.

from our mothers. During the formation of oocytes and sperms, DNA methylation of certain genes occurs; this process is called **epigenetic imprinting**. This imprinting is reversible. For example, when a woman inherits a chromosome from her father, it will be epigenetically imprinted as 'paternal'. However, when the daughter passes the same chromosome on to her child, it will have become imprinted as 'maternal'.

Recently, scientists have found that the same allele or alleles can have a different effect, depending on which parent it was inherited from. For example, Prader–Willi syndrome is a genetic condition that affects one in every 12 000–15 000 people. It affects both sexes and all ethnic groups. It is caused by inactivation of some of the alleles on chromosome 15.

You can see a pedigree showing the inheritance of Prader–Willi syndrome in one family in Figure 10.11. You will notice that individuals who inherit the defective chromosome from their mother do not develop Prader–Willi syndrome.

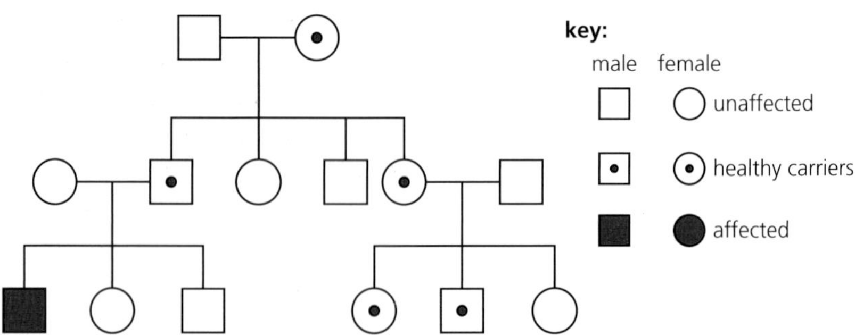

Figure 10.11 Inheritance of Prader–Willi syndrome in one family.

Extension

Prader-Willi syndrome causes a wide range of effects, which may include:

- a constant feeling of hunger, which leads to obesity
- restricted growth
- reduced muscle tone
- no development of secondary sexual characteristics
- learning difficulties and behavioural problems.

Epigenetics is very important in cancer research. The DNA in human tumours shows changes in DNA methylation and histone protein modification. This can cause tumour suppressor genes to be silenced, or oncogenes to be activated. Scientists are developing drugs that treat cancer by reversing epigenetic changes, e.g. by removing acetyl 'tags' on histone proteins or removing methyl groups on DNA.

TEST YOURSELF

9 How does the pedigree in Figure 10.11 show that Prader–Willi syndrome results from a defective chromosome inherited only from the father?

10 The mule, born when a female horse and male donkey breed together, looks very different from a hinny, born to a female donkey and a male horse. Use your understanding of epigenetics to suggest why.

11 In the plant *Arabidopsis*, the *FT* gene causes flowering. This gene is inhibited when the FLC protein is present. The gene that codes for FLC is switched off when environmental temperatures are cold, but the gene is switched on again when the environmental temperature is higher. Suggest

a) the advantage of this process to *Arabidopsis*

b) how the *FT* gene might be switched on or off.

Gene mutations and cancer

Mitosis occurs during the cell cycle of eukaryotic organisms. The rate of mitosis is controlled by two groups of genes

- proto-oncogenes, which control cell division
- tumour-suppressor genes, which slow cell division. These genes also promote programmed cell death (apoptosis) in cells with DNA damage that the cell cannot repair.

Gene mutations can occur in both these two types of gene. A mutated proto-oncogene, called an oncogene, stimulates cells to divide too quickly. Proto-oncogenes often code for proteins (growth factors) that stimulate cell division by binding to receptors in the cell membrane. They may also code for receptors in the cell membrane that control cell division. Mutated proto-oncogenes may result in over-production of these growth factors, or protein receptors in the cell membrane that stimulate cell division even when the growth factor is not present. A mutated tumour-suppressor gene is inactivated, allowing the rate of cell division to increase. Another way in which tumour-suppressor genes may be inactivated is if they undergo epigenetic changes, such as becoming hypermethylated.

ACTIVITY

An investigation into tumour formation in transgenic mice

Clones of laboratory mice are often used in investigations into the causes and effects of tumours.

1 What is a clone?

Scientists genetically transformed mice from a single clone to contain different oncogenes, called *myc* and *ras*. One group of mice contained only the *myc* oncogene, a second contained only the *ras* oncogene and a third contained both the *myc* and the *ras* oncogenes. The scientists then recorded the age at which the mice in each group developed tumours.

2 What is an oncogene?

3 Explain why the scientists used clones in this experiment.

Figure 10.12 shows the percentage of mice in each clone that were free of tumours during the course of the experiment.

4 For how many days were all the mice in this experiment free of tumours?

5 Use the graph to compare the effects of the *myc* and *ras* oncogenes when present alone.

6 Suggest an explanation for the curve showing mice with both the *myc* and *ras* oncogenes.

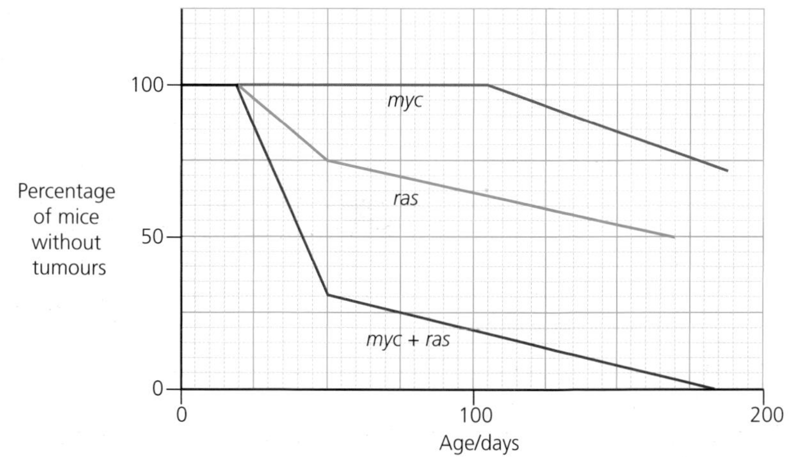

Figure 10.12 Three groups of mice from a single clone were genetically engineered to contain either *myc* oncogenes or *ras* oncogenes or both. The graph shows the percentage of mice in each clone that did *not* have tumours.

Figure 10.12 shows the results of an investigation involving the formation of tumours in mice. A tumour is a group of one type of cell that is dividing rapidly and uncontrollably. The formation of a tumour might result from one, or only a few, genetic changes in a cell. Tumours can be benign or cancerous. Benign tumours are tumours which grow in one place and do not spread. They are not cancerous tumours. However, depending on where they grow, they may still be harmful.

Cancerous tumours have cells that can break off and spread around the body, in a process called metastasis. The cancerous cells invade organs and tissues throughout the body and secondary tumours develop. These are called metastases.

Cancers are tumours in which some cells break away from the group and invade organs and tissues throughout the body. Figure 10.13 shows that the change from tumour cells to cancer cells requires many more genetic changes. Thus, cancer does not usually result from a single gene mutation.

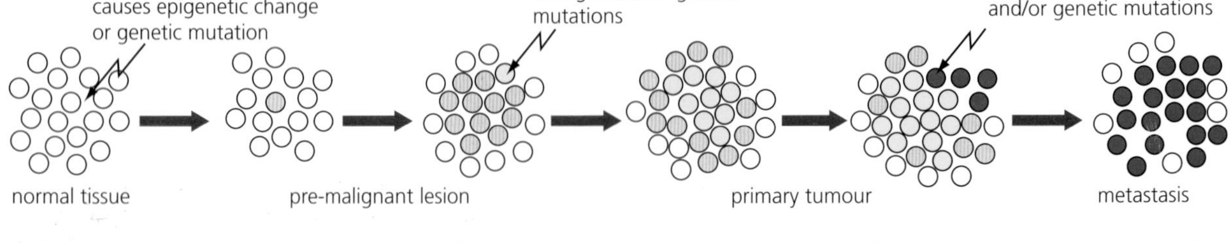

Figure 10.13 Healthy cells become tumour cells as a result of mutation in one, or a few, genes controlling cell division. Many more mutations are required for a tumour cell to become a cancer cell.

EXAMPLE

Meat consumption and cancer

The graph in Figure 10.14 shows the results of a large-scale epidemiological study of Armstrong and Hill (1975). It shows there is a strong link between colon cancer incidence and meat consumption.

1 Describe the relationship shown by the graph.

It shows that the more meat is eaten per person per day, the higher the incidence of colon cancer in women. However, this is not a perfect correlation. For example, meat consumption in Sweden and Hungary is very similar, yet the incidence of colon cancer in women in Sweden is more than double the incidence in Hungary.

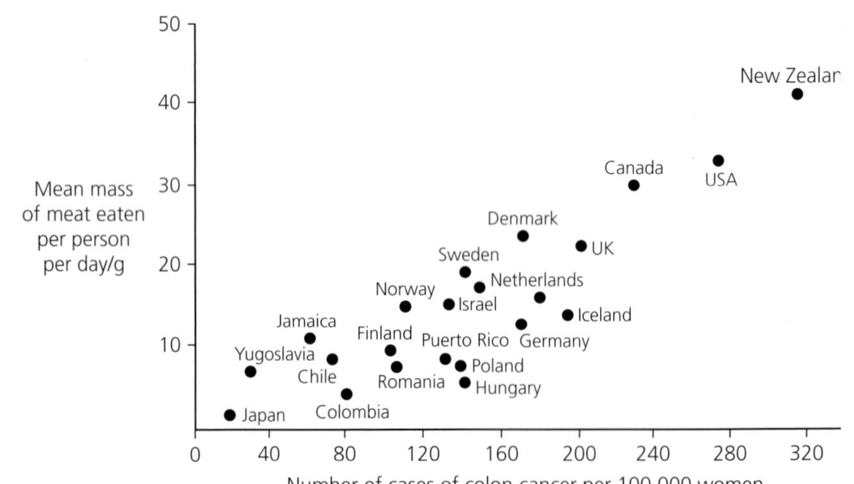

Figure 10.14 Meat consumption and incidence of colon cancer.

2 Explain why the incidence of colon cancer is given per 100 000 women.

Each country has a different size of population, so to give the incidence of colon cancer in each population would be misleading. For example, the population of New Zealand is very much lower than the population of the USA. Giving the incidence per 100 000 means you can compare countries even though their populations are different in size. It is also useful to look at rates of colon cancer among women as the incidence of many cancers varies depending on the sex of the individual.

3 A journalist writing for a vegetarian food magazine wrote an article about these data, under the heading 'Here's the proof that eating meat causes cancer'. Evaluate the journalist's conclusion.

There are many points that you can give here. The word 'evaluate' means 'give the arguments for and against'. You can certainly say that the graph gives evidence that meat-eating increases the incidence of colon cancer, as there is a positive correlation. However, this is a correlation and does not prove cause and effect. It is possible that there is another factor that causes cancer. People who don't eat meat usually eat more vegetables, so it might be that fibre and vegetables protect people from cancer. Countries where people eat a lot of meat tend to be wealthier than countries where people eat very little meat, so there might be a factor connected with affluence that increases the incidence of colon cancer, such as lack of exercise or drinking more alcohol. The rates of cigarette smoking are likely to be very different between these countries. The data does not distinguish between types of meat. It may be that red meat, or processed meat, is more likely to cause cancer than some other kind of meat. The problem with these data is that environmental factors will vary between these countries, so meat-eating is not the only variable.

TIP

Sir Richard Doll was the world's leading epidemiologist in the 20th century and proved the causal relationship between smoking and lung cancer. Look back at *AQA A-level Biology 1 Student's Book* to the section about smoking, cancer, correlations and causal relationships in Chapter 7.

Practice questions

1 Read the following passage.

Scientists think that some of the genetic changes associated with ageing may be the result of epigenetics, which suggests they could be reversed.

Molecules can attach to DNA, enhancing or preventing gene activation without changing the underlying genetic code. Such epigenetic changes are already suspected as factors in psychiatric disorders, diabetes and cancer.

They may also play a role in ageing. A group of scientists looked at the DNA of 86 sets of twin sisters aged 32 to 80, and discovered that 490 genes linked with ageing showed signs of epigenetic change through a process called methylation. They found that these genes were more likely to be methylated in the older twins than the younger twins. This can be triggered by environmental factors.

a) What is epigenetics (line 2)? *(2)*

b) What is methylation (line 10)? *(3)*

c) Name two environmental factors that might lead to methylation. *(1)*

d) Scientists think that some of the genetic changes associated with ageing might be reversed (line 2). Suggest how. *(2)*

2 Read the following passage.

A woman in Japan has received the first medical treatment based on induced pluripotent stem cells, 8 years after they were discovered. The iPS cells were made by taking skin cells from the woman's arm and reprogramming them, so they could be transformed into specialised eye cells. These were used to treat age-related macular degeneration (AMD), a condition that affects millions of elderly people worldwide and often results in blindness.

The woman had a patch of the cells measuring 1.3 × 3 millimetres grafted into her eye. She is one of six people who will receive this treatment, in an attempt to investigate the safety of stem cell treatments.

a) What is the meaning of pluripotent (line 2)? *(1)*

b) Explain why the skin cells must be reprogrammed before using them for this treatment. *(2)*

c) Explain the advantages of using stem cells produced from the woman's own skin cells, rather than cells from an embryo. *(4)*

d) Suggest reasons why it is important to investigate the safety of stem cell treatments such as this. *(4)*

3 Some kinds of breast cancer cells have HER-2 receptors on their cell-surface membrane. These receptors cause the cells to be stimulated by a growth factor so that they divide too much. Trastuzumab is a monoclonal antibody that binds to the HER-2 receptors and stops the cells from dividing.

a) What are monoclonal antibodies? (2)

b) Explain how trastuzumab stops these cells from dividing. (2)

c) Explain why trastuzumab is not effective against all kinds of breast cancer. (2)

d) Trastuzumab is administered by injection into the bloodstream. Suggest why it is not given in the form of a tablet. (2)

4 RNA interference (RNAi) has been proven effective against a human disease, respiratory syncytial virus (RSV), which is harmful to young children but relatively harmless in adults. Eighty-five healthy adults were given a nasal spray containing either a placebo or small interfering RNA (siRNA) designed to silence one of the genes of RSV. The adults used the spray daily for 5 days. On day 2, all the volunteers were infected with live RSV. By day 11, just 44% of those who received the RNAi nasal spray had RSV infections, compared with 71% of the placebo group. The scientists hope to use siRNA on lung transplant patients soon, and then they want to test it on infants.

a) Describe how siRNA can silence a gene. (4)

b) i) What would the placebo contain? (2)

ii) Suggest why the trial was done on adults rather than children. (2)

c) Suggest why siRNA might be useful for people who have had lung transplants. (2)

Stretch and challenge

5 Investigate Angelman syndrome and how it is inherited. Contrast the inheritance of Angelman syndrome with the inheritance of Prader–Willi syndrome.

6 Research the genetic inheritance of tortoiseshell coat colour in cats. Investigate Carbon Copy, the cloned tortoiseshell cat, and explain why she does not look like her clone.

Gene cloning and gene transfer

PRIOR KNOWLEDGE

- The genetic code is universal.
- In transcription, the genetic information in DNA is copied into messenger RNA (mRNA). mRNA then travels to the ribosomes, where proteins are synthesised. This happens in all living organisms.
- In genetic engineering, genes from the chromosomes of humans and other organisms can be 'cut out' using enzymes and transferred to cells of other organisms.
- Genes can also be transferred to the cells of animals, plants or microorganisms at an early stage in their development so that they develop with desired characteristics.
- New genes can be transferred to crop plants. Crops that have had their genes modified in this way are called genetically modified crops (GM crops), and examples of such crops include ones that are resistant to insect attack or to herbicides. GM crops generally show increased yields.
- Concerns about GM crops include the effect on populations of wild flowers and insects, and uncertainty about the effects of eating GM crops on human health.

TEST YOURSELF ON PRIOR KNOWLEDGE

1 The DNA code is universal. Explain what this means, and how this makes genetic engineering possible.
2 What is a gene?
3 What is the name for three bases in DNA that code for an amino acid?

Do you intend to study biology at university? If so, are you planning to use your degree to make a career? Many graduate biologists do, for example your biology teachers. Some biologists are employed by large pharmaceutical companies, where they are employed to develop new drugs.

In the 1980s, pharmaceutical companies were attracted by new developments in DNA technology, including the genetic manipulation of organisms. Some molecular biologists even set up their own companies to capitalise on their research. The potential profits from this biotechnology industry attracted people to invest their money in these companies. They believed that buying shares in the companies would give them a better return on their money than other types of investment. This sudden growth in ownership of shares in the biotechnology industry was called the 'biotechnology bubble'. Biotechnology companies became very valuable in a short space of time and some molecular biologists who had set up their own companies became very wealthy on paper.

Although the potential benefits of biotechnology were promising, there were many difficulties in converting research methods into profit-earning products. Many of these problems were technical, but some resulted from negative public opinion. As a result, the price of shares in these companies fell dramatically in 1999; the biotechnology bubble burst and investors lost a good deal of the money they had invested.

Recombinant DNA technology

Entire molecules of DNA are, in general, too large to be used in gene technology. Instead, molecular biologists use fragments of DNA, which contain the gene or genes they are interested in. They can produce these fragments using three different methods. They use:

- a reverse transcriptase to convert mRNA to cDNA
- a restriction endonuclease to cut DNA fragments out of a large DNA molecule
- a 'gene machine' to synthesise the required piece of DNA.

We will look at each of these methods in turn.

Using a reverse transcriptase to convert mRNA to cDNA

You learned in Chapter 10 that most cells do not transcribe all their genes. Thus, the cytoplasm of a specialised cell contains only mRNA transcribed from some of the genes in its nucleus. This is because specialised cells make a lot of a limited number of different proteins. It is often easier to extract mRNA from the cytoplasm of a cell than to find the gene from which it was transcribed. For example, mRNA encoding the mammalian hormone insulin will be found in high concentrations in the cytoplasm of β-cells in the islets of Langerhans of the pancreas, but in no other cell.

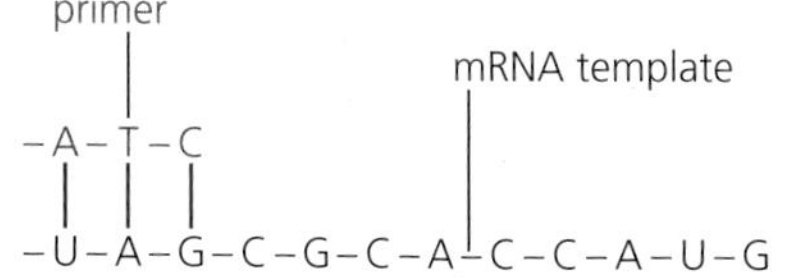

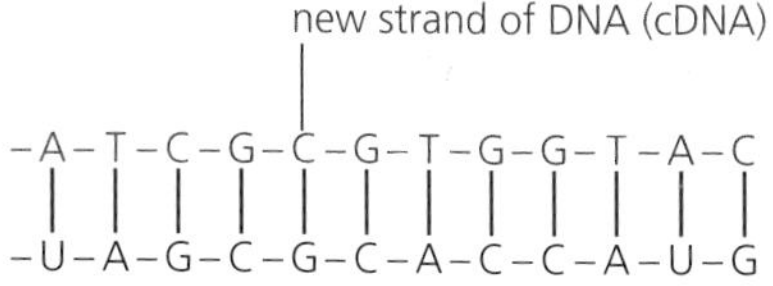

Figure 11.1 Complementary DNA (cDNA) is made using reverse transcriptase to copy the base sequence of mRNA.

Molecular biologists can extract this mRNA and make a DNA copy from it (Figure 11.1). Purified mRNA is mixed with free DNA nucleotides and an enzyme called **reverse transcriptase**. You learned about reverse transcriptase when you studied HIV in the first year of your course (see *AQA A-level Biology 1 Student's Book* Chapter 6). This enzyme catalyses a process that is the reverse of transcription: DNA is made from mRNA. The DNA formed in this way is called **complementary DNA (cDNA)**. The cDNA is single stranded, but it can be made into double-stranded DNA using DNA polymerase.

Using a restriction endonuclease to cut fragments out of a large molecule of DNA

Nucleases are enzymes that break the bonds linking one nucleotide to the next in a DNA strand. An **exonuclease** removes nucleotides, one at a time, from the end of a DNA molecule. We are more interested in **endonucleases**. Figure 11.2 shows how an endonuclease hydrolyses bonds within the DNA molecule, producing fragments of DNA.

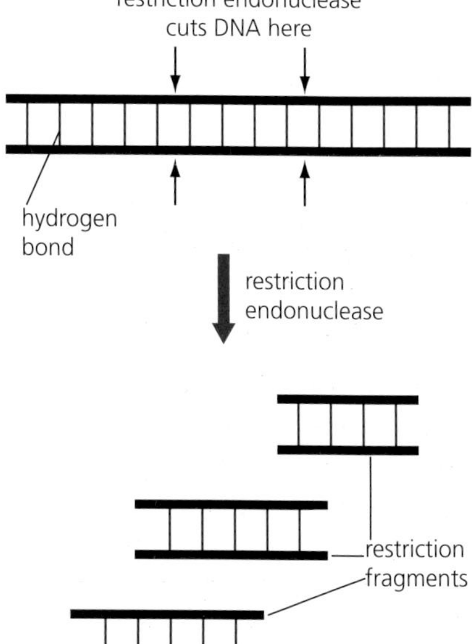

Figure 11.2 A restriction endonuclease cuts DNA into double-stranded fragments, called restriction fragments.

Restriction endonucleases (or restriction enzymes, for short) are enzymes that break bonds in the sugar–phosphate backbones of *both* strands in a DNA molecule. As a result, they produce double-stranded fragments of DNA, like the ones shown in Figure 11.2. Each restriction enzyme cuts DNA only at a particular sequence of bases, called the **recognition sequence** of the enzyme.

A restriction endonuclease can cut a DNA molecule in one of two ways. Look at Figure 11.3a. This shows how one restriction enzyme makes a simple cut right across the middle of its recognition sequence. You can see that this results in DNA fragments with **blunt ends**. Figure 11.3b shows how another restriction enzyme makes a staggered cut across its recognition sequence. This produces fragments with short single-stranded overhangs. Since these overhangs can form base pairs with other complementary sequences, they are called **sticky ends**.

a) Production of blunt ends by the restriction enzyme *Alu*1

–B–B–A–G–C–T–B–B
–B–B–T–C–G–A–B–B
restriction enzyme *Alu*1
–B–B–A–G–   C–T–B–B
–B–B–T–C–   G–A–B–B
blunt ends

b) Production of sticky ends by the restriction enzyme *Eco*R1

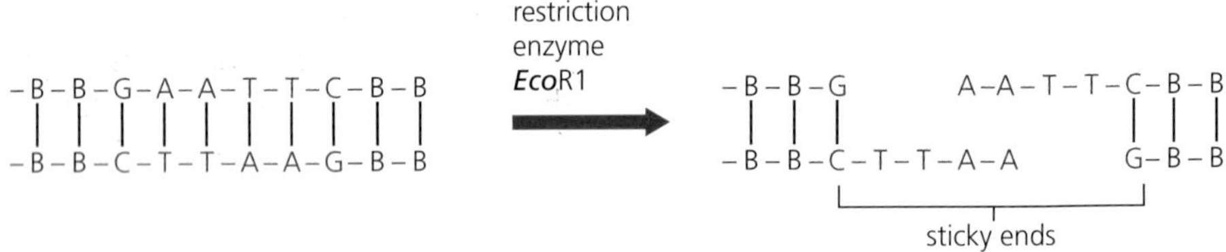

Figure 11.3 A restriction endonuclease produces double-stranded fragments of DNA by breaking DNA molecules at a specific recognition sequence. The bases involved in the recognition sequence are shown in red. A is adenine, C is cytosine, G is guanine and T is thymine. The symbol B represents other organic bases that are not involved in the recognition sequence.

EXAMPLE

The properties of recognition sequences

1 What does Figure 11.3 tell us about the recognition sequence of the restriction enzymes from these two bacteria?
The recognition sequence of Alu1 is different from the recognition sequence of EcoR1. In fact, every restriction enzyme has a unique, specific recognition sequence where it cuts a DNA molecule.

2 Can you use your knowledge of how enzymes work to explain why each restriction enzyme has a specific recognition sequence?
You should recall that enzymes are specific. An enzyme is specific because only one type of substrate can fit into its active site. The active sites of restriction endonucleases are specific to particular sequences of bases; that is, to specific recognition sequences.

3 How long are the recognition sequences in Figure 11.3?
In Figure 11.3a, the recognition sequence is only four base pairs (4 bp) long. In Figure 11.3b, the sequence is only 6 bp long. Although some restriction enzymes have recognition sequences that are longer than this, most have recognition sequences between 4 bp and 8 bp long.

4 Look carefully at the sequence of bases in the upper and lower strands of each DNA fragment shown in Figure 11.3a and b. What is the common feature about their patterns of bases?
In Figure 11.3a, the upper strand of the recognition sequence reads AGCT. If you read the lower strand from right to left, it also reads AGCT. Look again at Figure 11.3b. The upper strand of the recognition sequence reads GAATTC, which is the same as the lower strand when read from right to left. The order of bases in one strand of a recognition sequence is always the same as the complementary strand when read backwards. Something that reads the same forwards and backwards is called a palindrome. A simple palindrome in English is 'never odd or even'. Figure 11.3 shows us that, in addition to being short, recognition sequences are palindromic. We can now define the recognition sequence of a restriction enzyme as a short, specific, palindromic base sequence.

5 How could molecular biologists ensure that they only cut a specific gene out of DNA?

Provided they know the base sequences at each end of a gene, biologists can use restriction endonucleases with base sequences that are complementary to them. As a result, the two restriction enzymes will cut the DNA, one at each end of the desired gene. Of course, this will only be helpful if the recognition sequences do not also occur within the gene.

6 Bacteria can be infected by viruses that inject their nucleic acid into the host cell. Suggest one natural function of restriction endonucleases in bacteria.

A bacterium uses its restriction endonucleases to hydrolyse the nucleic acid injected by a virus before it can be transcribed. In this way, bacteria protect themselves against viral infection. The bacterium's own DNA is not broken down because, unlike the DNA of the virus, bacterial DNA contains methyl groups ($-CH_3$) that protect it from the action of its own restriction endonucleases.

Creating a gene in a 'gene machine'

If the primary structure of a protein is known, then it is possible to synthesise the gene required to produce the protein using a 'gene machine'. The amino acid sequence required is entered into a computer. The triplet code for each amino acid is known (see Table 4.6 in *AQA A-level Biology 1 Student's Book*, Chapter 4, Page 70) so the DNA sequence that will produce that protein is worked out. The computer then controls the machine and the required DNA sequence is made. The advantage of synthesising a gene in this way is that it does not contain introns, so the gene can be transcribed and translated in prokaryotic cells.

Analysing restriction fragments

Since they are produced by restriction endonucleases, the DNA fragments produced by these enzymes are called **restriction fragments**. DNA is digested by restriction endonucleases into a number of restriction fragments of different lengths. We can find out the number and size of the restriction fragments using **gel electrophoresis** followed by visualisation of the DNA. This is important in many genetic analyses, such as genetic fingerprinting (see page 235) or analysing DNA to find out whether the DNA contains a specific allele or not.

Separating restriction fragments using electrophoresis

Electrophoresis uses a slab of gel made of agarose or polyacrylamide. An electrode is placed at each end of the gel and an electric current is passed through it. A sample containing restriction fragments is placed in a well that is cut in the gel near the negative electrode. Since DNA has a negative charge, the restriction fragments will migrate through the pores in the gel towards the positive electrode. The smaller fragments will move faster through the pores in the gel than the larger molecules and so will move the furthest from the well. Figure 11.4 shows a cross section through a gel with restriction fragments migrating through it. The fragments produce bands of restriction fragments of different sizes.

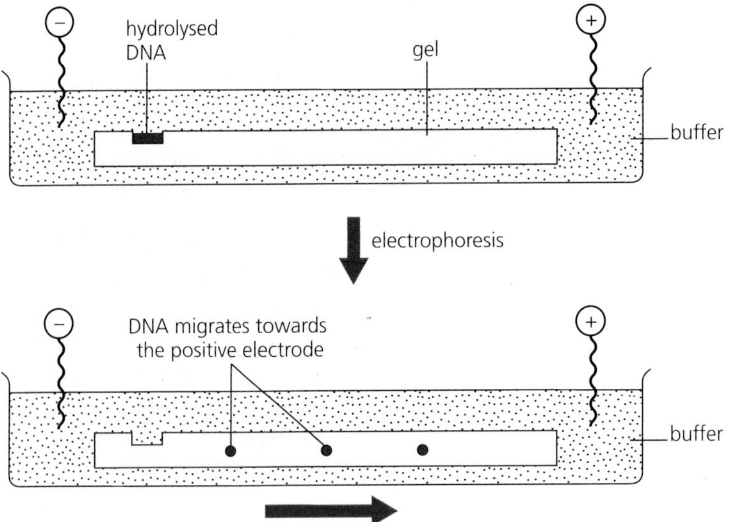

Figure 11.4 Restriction fragments migrate through a gel towards the positive electrode of an electric field. The smaller fragments move through pores in the gel faster than the larger fragments. Thus, bands of restriction fragments are formed in the gel.

Visualisation of DNA bands

DNA is colourless, so we must treat the DNA in such a way that we can see the bands in the gel after electrophoresis. This can be done using a suitable stain. The result is a series of coloured bands in the gel. An alternative is to treat the DNA with a **radioactive marker** before it is digested by restriction endonucleases. The fragments can then be seen by placing the gel in contact with a photographic film that is sensitive to X-rays. The developed film, called an **autoradiograph**, shows the DNA bands. Figure 11.5 shows a developed film showing labelled restriction fragments. Alternatively, the DNA may be visualised by adding fluorescent probes that bind to the DNA (see page 232).

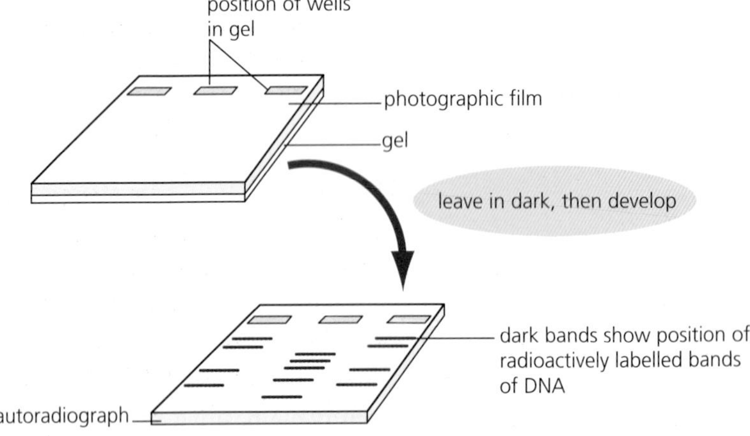

Figure 11.5 Radioactively labelled DNA has been separated using gel electrophoresis. After separation by electrophoresis, the gel has been placed in contact with a photographic film. After development, the different restriction fragments appear as dark bands on the film, called an autoradiograph.

TEST YOURSELF

1 What will be the base sequence of cDNA made from a section of mRNA with the base sequence: ACG CGA UCA UGA?

2 Reverse transcriptase can only begin to copy mRNA in the presence of a primer. Use Figure 11.1 to suggest what a primer is.

3 Reverse transcriptase is found in some viruses such as HIV. What is its role in HIV?

4 The recognition sequence of a certain restriction endonuclease occurs six times along a molecule of DNA. How many fragments of DNA will be produced by the action of the restriction enzyme?

5 Why does DNA have a negative charge?

6 The agarose gel used in electrophoresis must be very pure and uniform. Suggest why.

Gene cloning

You learned in Chapter 10 that a clone is a group of cells, or organisms, that are genetically identical to each other. **Gene cloning** involves making identical copies of a gene. There are two major technologies that enable molecular biologists to clone genes (Figure 11.6).

One form of gene cloning or DNA amplification is the **polymerase chain reaction (PCR)**. Because the PCR occurs in a test tube, this method of gene cloning is called *in vitro* gene cloning (literally 'in glassware'). You are probably familiar with the term *in vitro* in the context of *in vitro* fertilisation (IVF). In an IVF clinic, the fertilisation of an egg cell by a sperm cell occurs in glassware.

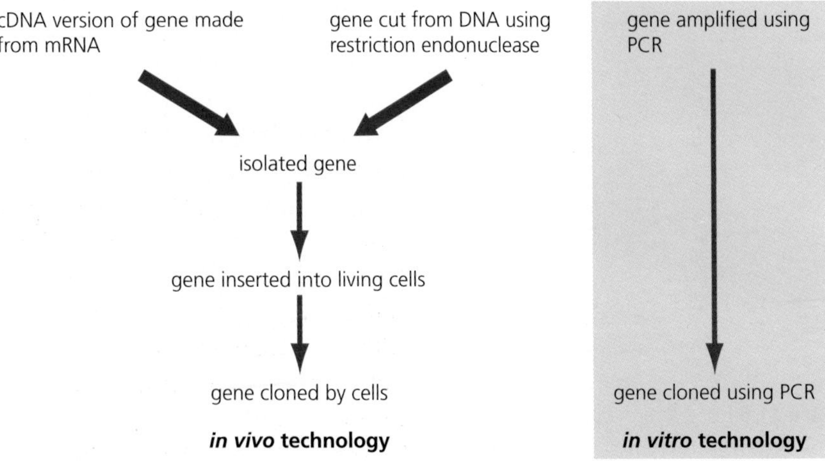

Figure 11.6 Gene cloning involves making identical copies of a gene. This can be done using two different technologies: *in vivo* and *in vitro*.

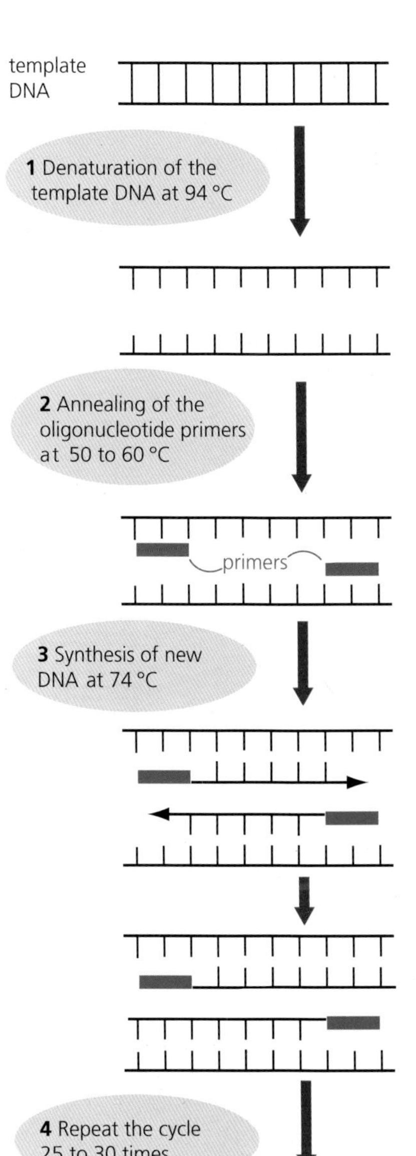

Figure 11.7 The basic steps in the polymerase chain reaction (PCR).

An alternative way of cloning a gene involves isolating the gene and inserting it into a suitable host cell. As the host cell divides, it replicates the inserted gene along with its own DNA. Since this method of gene cloning requires the use of living cells, it is called *in vivo* gene cloning (*in vivo* means 'in life').

Using the PCR to amplify part of a DNA molecule

According to scientific folklore, the idea of the PCR occurred as a brainwave to its inventor one evening as he drove along the coast of California. What is true is that the PCR has since become one of the most useful tools in molecular biology. It is highly likely that you have seen the PCR 'performed' in TV crime programmes. As long as molecular biologists can design primers that mark the beginning and end of a section of DNA, the PCR can be used to amplify a gene or any length of DNA or DNA fragment.

PCR involves mixing the DNA to be copied (**template DNA**) with a set of reagents in a test tube and placing the tube in a thermal cycler. The reagents include an enzyme, short lengths of single-stranded DNA (called **oligonucleotides**) and free DNA nucleotides, each with an adenine, cytosine, guanine or thymine base. The thermal cycler is an automated machine that changes the temperature at which the test tube is incubated in a pre-programmed sequence. Figure 11.7 shows the three basic steps in one PCR cycle, which we will now describe in more detail.

- **Step 1**: denaturing of the template DNA. The thermal cycler heats the mixture to 94 °C. This causes the hydrogen bonds that hold together the two strands of the template DNA to break down. As a result, the two individual DNA strands separate and can now act as templates for building complementary strands.
- **Step 2**: annealing the primers. The cycler cools the mixture to between 50 °C and 60 °C. This allows hydrogen bonds to re-form. In theory, the two strands of template DNA could re-join at this temperature. Most do not because the oligonucleotides bind to the template DNA strands instead. This binding is called **annealing** and occurs only at a site on the template DNA that has a base sequence complementary to that of the oligonucleotides. These oligonucleotides are called **primers** because the enzyme involved in the next step must attach to one of them before it can start to work.
- **Step 3**: synthesis of new DNA. The cycler raises the temperature to 74 °C. This is the optimum temperature of the **DNA polymerase** used in this step of the PCR. The enzyme attaches to one end of each primer and synthesises new strands that are complementary to the template DNA. At the end of this stage the original molecule of template DNA has been copied and two molecules are in the mixture. The cycler now raises the temperature back to 94 °C and the cycle occurs all over again.

The PCR is usually repeated about 25 times. As a result, over 50 million copies of the template DNA are formed. The result of a small number of cycles is shown in Figure 11.8.

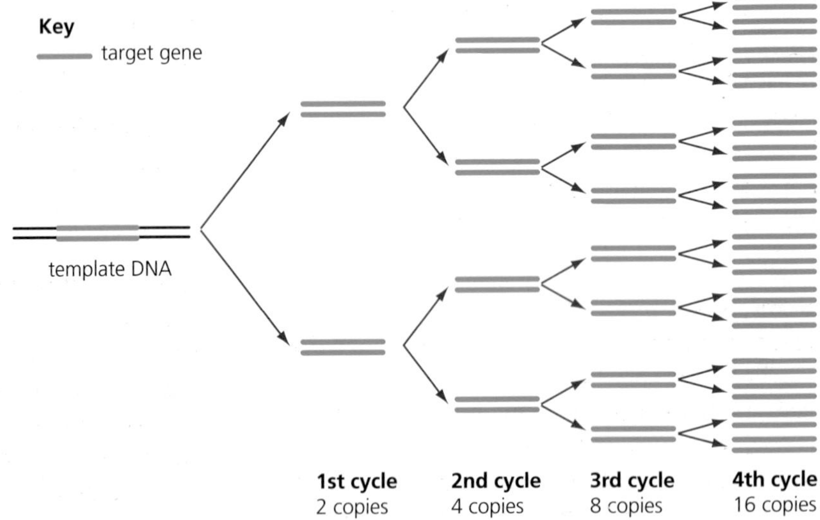

Figure 11.8 For each cycle, the PCR produces two molecules of DNA from a template molecule of DNA. In a short time, many copies of the template DNA can be made.

TEST YOURSELF

7 The enzyme most commonly used in the PCR is *Taq* DNA polymerase. This enzyme is extracted from a bacterium, *Thermus aquaticus*, that lives in hot-water springs. Many of this bacterium's enzymes, including *Taq* DNA polymerase, are thermostable.

a) What is the advantage to the bacterium of having thermostable enzymes?

b) Explain why it is important that the DNA polymerase used in the PCR is thermostable.

8 You have learned already about the importance of molecular collisions in explaining the action of competitive inhibitors. Use this understanding to suggest why it is important to the success of Step 2 that the PCR mixture contains a high concentration of primer.

Extension

How do scientists use the PCR to amplify specific fragments of DNA, including genes?

So far, our account of the PCR suggests that the whole of a DNA molecule will be copied in each cycle. In eukaryotes, this would involve copying entire chromosomes, which would not be helpful. A whole chromosome contains too much information to transfer or to store. Molecular biologists usually want to amplify only part of a DNA molecule, for example a gene that they wish to transfer from one organism to another or a gene that they wish to store in a gene library.

During the PCR, DNA polymerase starts replication where it binds to a primer. It will stop when it reaches another primer on the same strand of DNA. The correct design of primers is critical to the success of amplifying fragments of DNA.

The primer is a sequence of nucleotides that attaches to template DNA wherever it meets a base sequence that is complementary to its own. It would be possible to design a primer that contained only two nucleotides, but not very useful. A sequence of only two bases, for example CT, is likely to occur a large number of times within a short length of template DNA. DNA polymerase starts copying when it attaches to a primer, then moves along the template DNA but stops when it reaches another primer. With primers attached only short distances apart, this would result in a large number of very small sections of amplified DNA.

It is easy to estimate how often base sequences of different lengths are likely to occur along template DNA. For example, a particular sequence of four bases is likely to occur once in every $4^4 = 256$ nucleotides, a particular sequence of eight bases is likely to occur once in every $4^8 = 65\,536$ nucleotides and a particular sequence of 16 bases is likely to occur once in every $4^{16} = 4\,294\,967\,296$ nucleotides.

You will have spotted that a particular sequence of *n* bases is likely to occur every 4^n nucleotides in a DNA molecule. This means that a primer with only two bases is likely to attach to template DNA every $4^2 = 16$ nucleotides and our PCR would amplify fragments of DNA only 16 nucleotides long. These would be too short to be useful. Using primers that are too short would also make it almost impossible to amplify an entire gene during gene cloning.

As we have seen, short primers attach at short distances along a DNA molecule. Therefore, it is highly likely that short primers would attach to a DNA molecule at several points within a gene. As DNA polymerase copies DNA from one primer to the next, only fragments of the gene would be copied, not the whole gene.

By using a variety of primers, molecular biologists could collect an array of DNA fragments. It is much easier for them to find a gene in a DNA fragment than in the entire genome of an organism. It is also easier for them to sequence an organism's genome fragment by fragment. For an increasing number of organisms, molecular biologists have been able to determine the base sequence of genes and of the DNA at either end of genes.

If molecular biologists know the base sequences at each end of a gene, they can design primers that will attach to them. Provided the primers do not attach within the gene, DNA polymerase will amplify the entire gene plus any bases to either side. Genes are controlled by promoter regions that lie close to them. If they wished to transfer a human gene into a bacterium, molecular biologists would need to amplify the gene and its promoter region.

If molecular biologists wish to transfer a human gene successfully into a bacterial cell they would need to amplify the gene and its promoter and terminator regions. Bacteria are prokaryotic cells whereas human cells are eukaryotic. Prokaryotic cells have a mechanism for controlling expression of their genes that is different from the mechanism in eukaryotes. Without its promoter region, the human gene might not be expressed in the bacterial cell. Also, if the human gene is cut out of the human DNA using a restriction enzyme, it will still contain its introns. If this gene is transferred to bacteria, the correct protein will not be made as bacteria do not splice introns out of the pre-mRNA before translation. Therefore any human gene that is transferred to bacteria needs its introns removed.

An overview of *in vivo* gene cloning using bacteria

Because they are easy to culture and increase in number at a very fast rate, bacteria are often used to clone genes. This involves the following steps:

- **Step 1**: a fragment of DNA containing one or more genes is obtained by one of the three methods described on pages 214–216.
- **Step 2**: the DNA fragment is inserted into a vector which will transfer it into a bacterium. In this case, the vector is a small circular DNA molecule called a **plasmid**.
- **Step 3**: the vector is transported into a bacterial cell.
- **Step 4**: the bacterium is allowed to multiply.
- **Step 5**: in reality, only some bacteria will have successfully taken up the plasmid containing the target gene(s). In Step 5, these bacteria are identified so that they can be cultured. Any that have not taken up the target gene(s) are destroyed.

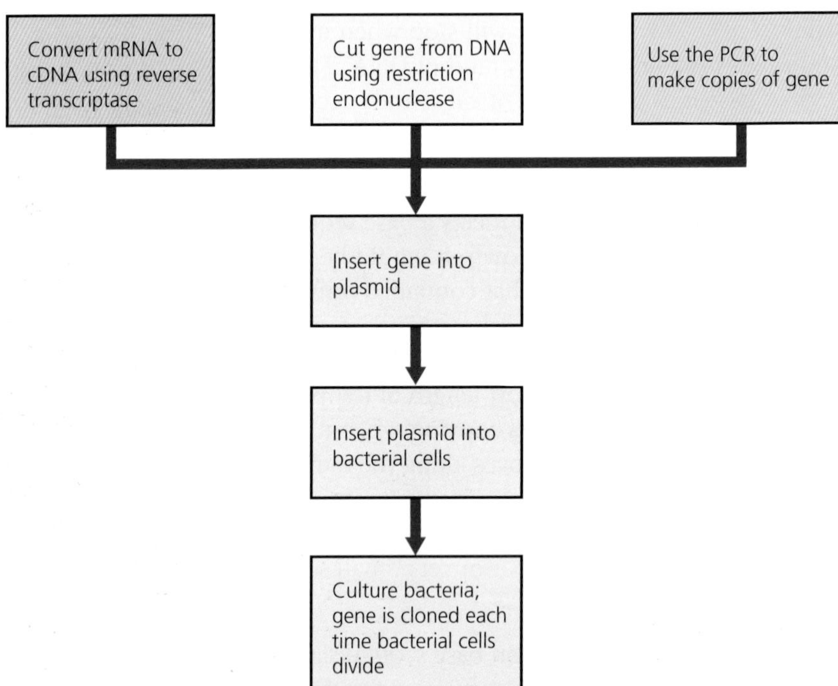

Figure 11.9 The basic steps in *in vivo* gene cloning. Although other cells can be used, bacteria are the most common host cells for this technology.

Inserting DNA fragments into vectors

Fragments of DNA can be produced by any of the three methods shown in Figure 11.9. Each method has been described in some detail earlier in this chapter. Having obtained an appropriate DNA fragment, a promoter and a terminator sequence need to be added. You learned about promoters in Chapter 10, and will realise that these are needed for transcription to start. A terminator is a sequence of DNA that acts as a signal for transcription to stop at the end of the gene sequence. The DNA fragment, with its promoter at one end and the terminator at the other end, must be inserted into a **vector** that will carry it into a target cell. Viruses, including those that infect bacteria (bacteriophages) can be used, but plasmids are the most commonly used vectors. **Plasmids** are small, circular DNA molecules that lead an independent existence in many bacterial cells. Each plasmid carries a small number of genes that are expressed in the phenotype of the bacterium. For example,

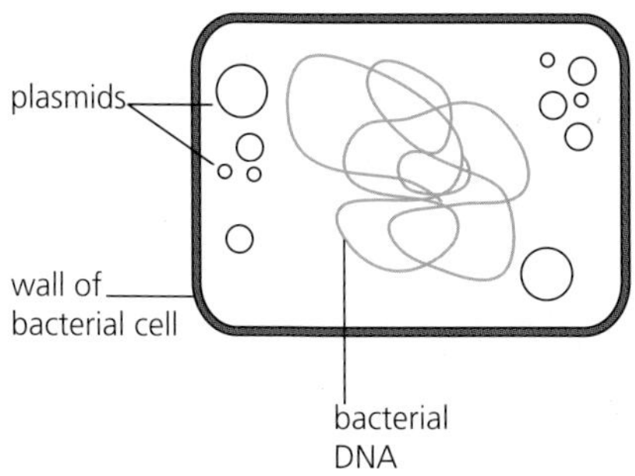

Figure 11.10 This bacterial cell contains many plasmids. They vary in size but they are smaller than the bacterial 'chromosome'.

the ability to survive toxic concentrations of antibiotics, such as ampicillin, chloramphenicol and kanamycin, is often due to the presence inside the bacterium of a plasmid carrying genes for antibiotic resistance. Figure 11.10 shows that a single bacterial cell can contain many plasmids, each of which is much smaller than the circular DNA that makes up the main bacterial DNA.

Before it can be used as a vector, a plasmid must be cut open. This is done using a restriction endonuclease. The DNA fragment can now be inserted into the cut plasmid. Figure 11.11 shows how, under the right conditions, the ends of the DNA fragment and of the plasmid join together. This process is called **ligation** and is catalysed by a **ligase** enzyme. Some ligases join together DNA fragments and plasmids that have blunt ends. However, ligation is much more efficient if the plasmid and the DNA fragment have sticky ends (page 209). This allows complementary base pairing to hold the two ends of DNA together while the ligase joins them up. If either the DNA fragment or the plasmid does not have sticky ends, they can be added before ligation occurs.

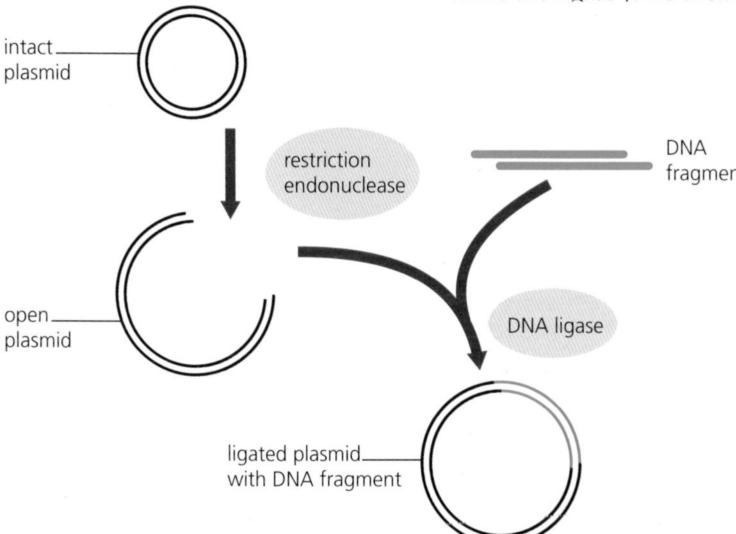

Figure 11.11 The cut ends of a plasmid and of a DNA fragment are joined together by a ligase enzyme to form recombinant DNA. This process is much more efficient if the plasmid and DNA fragments have sticky ends rather than blunt ends.

If successful, the result of this process is a plasmid that contains a fragment of foreign DNA. Because it contains DNA from two sources, the DNA of this plasmid is referred to as **recombinant DNA**.

Transferring vectors into host bacteria

Most species of bacteria are able to take up plasmids. There are several methods for getting vectors into bacterial cells. Most were found by trial and error and examiners will not expect you to recall any of them. One of the earliest methods involves soaking bacterial cells, together with plasmids, in an ice-cold solution of calcium chloride followed by a brief heat shock, during which the temperature is raised to 42 °C for 2 minutes. Quite why this works is poorly understood, but the treatment increases the uptake of plasmids into the bacteria. Bacteria that have taken up plasmids are said to be **transformed** and, because they contain a gene from another organism, they are also said to be **transgenic**.

Identification of transformed bacteria

When presented in a diagram, such as Figure 11.11, the process for producing transformed bacteria seems simple enough. In reality, biologists cannot see the bacterial cells or the plasmids. They carry out the procedures described above not knowing whether:

- any plasmids have recombinant DNA or whether they have just joined back on themselves (**self-ligated**)
- any bacterium has taken up a plasmid
- any bacterium that has taken up a plasmid has taken up a plasmid containing recombinant DNA. Only about 0.01% of them do.

Biologists need methods to identify which bacteria have been transformed, so that they can grow these, and only these, in a culture medium. There are many ways that they can identify transformed bacteria. Most involve the use of **marker genes** on the plasmids.

TIP

You need to be able to interpret information given to you about any method of identifying transformed bacteria but you do not need to recall any particular method.

EXAMPLE

Finding transformed bacteria

Escherichia coli bacteria are often used in gene cloning experiments. This bacterium is able to hydrolyse lactose into its constituent monosaccharides.

1 Name the monosaccharides formed by the hydrolysis of lactose.

Glucose and galactose.

E. coli uses a series of enzyme-controlled reactions to hydrolyse lactose. One of the enzymes in this series is β-galactosidase. The enzyme is normally encoded by the bacterial gene *lacZ*. The normal *lacZ* gene contains segments that encode different peptide portions of the β-galactosidase molecule. Some strains of *E. coli* have a *lacZ* gene which lacks a segment, called *lacZ′*, that encodes the α-peptide portion of the enzyme.

2 Does β-galactosidase have a quaternary structure? Justify your answer.

Yes, because it consists of more than one chain of amino acids.

Figure 11.12a shows a plasmid, called pUC8, which is commonly used as a vector. This plasmid contains two genes that are of interest to us: one confers resistance to ampicillin and the other is the *lacZ′* gene. If a cell of *E. coli* that lacks the *lacZ′* gene contains a copy of the normal pUC8 plasmid shown in Figure 11.12a, it can produce normal β-galactosidase. Figure 11.12b shows where a foreign gene can be inserted into the pUC8 plasmid using a restriction enzyme called *Bam*HI. Note that the recognition sequence of *Bam*HI is in the middle of the *lacZ′* gene in the plasmid.

a) Normal vector molecule produces α-peptide

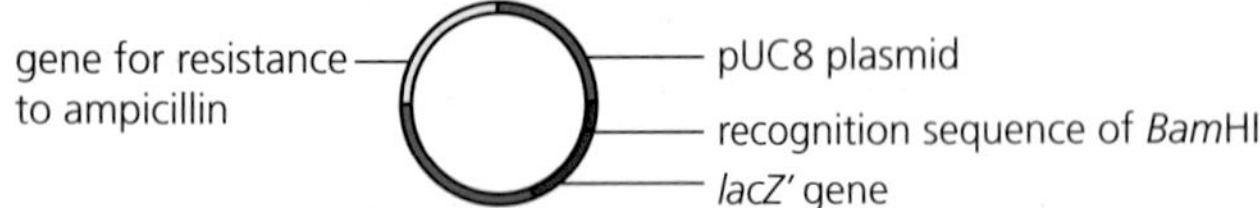

b) Recombinant vector molecule does not produce α-peptide

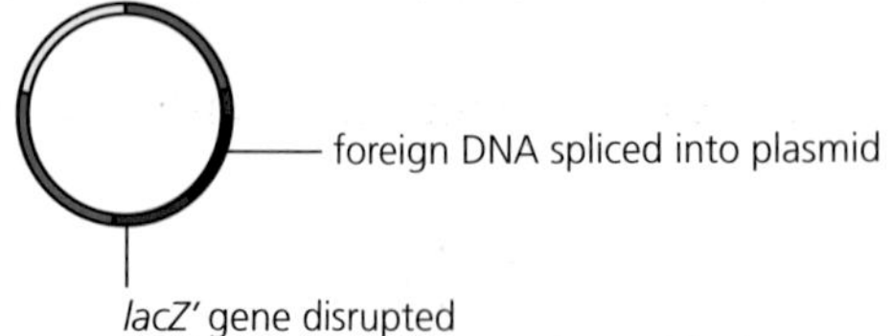

Figure 11.12 The plasmid pUC8 is commonly used as a vector in gene cloning experiments. a) The normal plasmid with two of its genes indicated; b) a recombinant plasmid, containing a foreign gene.

3 Explain why a cell of *E. coli* that lacks the *lacZ′* gene is able to make β-galactosidase if it also contains a copy of the normal pUC8 plasmid, shown in Figure 11.12a.

The normal pUC8 plasmid contains a copy of the lacZ′ gene.

Having followed the protocol to splice foreign DNA into pUC8 plasmids and then insert the plasmids into cells of *E. coli*, molecular biologists cultured the *E. coli* in a suitable medium.

4 What is a protocol?

A protocol is a detailed plan for a scientific procedure.

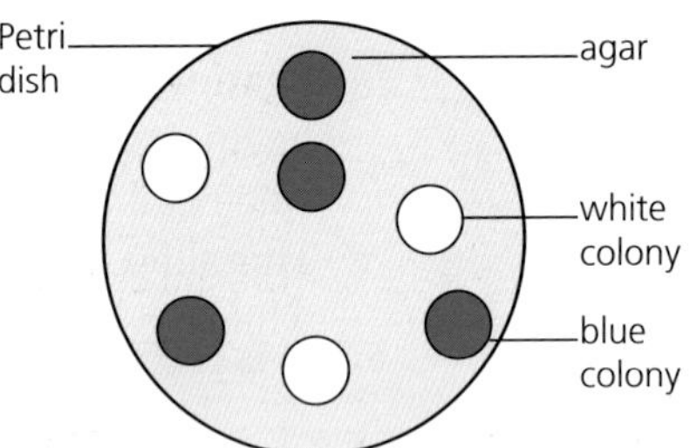

Figure 11.13 This agar plate was inoculated with bacteria. The circles represent colonies of bacteria that have grown on the agar plate. The agar contains the antibiotic ampicillin and X-gal. This helps us to identify which bacteria have taken up a transformed pUC8 plasmid.

After some time, the biologists inoculated the bacteria on to agar plates containing ampicillin and X-gal. Figure 11.13 shows the appearance of one agar plate after it had been incubated for 48 hours. The circles represent *E. coli* colonies that had grown on the agar. Each colony is a clone of a single *E. coli* cell that was able to grow on the agar.

5 All the colonies in Figure 11.13 are clones of cells that had taken up the pUC8 plasmid. Explain why they are clones.

Because they have all grown from one original cell by binary fission.

6 Before inoculating the bacteria on the agar plates, the scientists cultured the bacteria in a suitable medium. Use your knowledge of gene expression to suggest why.

Some of the bacteria would be unable to hydrolyse lactose so they need to be cultured in a medium that contains a suitable carbon source such as glucose.

In addition to ampicillin, the agar contained X-gal. This is a lactose analogue. It is white but is broken down by β-galactosidase to produce a blue-coloured product.

7 Use your knowledge of enzyme action to suggest the meaning of 'lactose analogue'.

This means it is similar in shape to lactose, so can fit into the active site of β-galactosidase and will break down to produce a blue product.

8 Which of the colonies shown in Figure 11.13 contain bacteria that can produce β-galactosidase?

The blue colonies.

9 Use Figure 11.13 and your answer to Question 8 above to explain how we can identify bacteria that have taken up plasmids containing the foreign gene.

These are the white colonies as they are resistant to ampicillin but they cannot break down X-gal so they do not have a functional lacZ′ gene.

Relative advantages of *in vivo* and *in vitro* gene cloning

We have looked at two methods of cloning genes: the *in vivo* method, using bacteria, and the *in vitro* method, using the PCR.

Each cycle of the PCR takes between 3 and 5 minutes. Normally, 25–30 cycles are carried out in any one experiment. Consequently, a PCR experiment takes only a few hours to produce a large gene clone. In contrast, the *in vivo* method takes several weeks. This gives the PCR such a time advantage that you might wonder why scientists continue to use the *in vivo* method at all. Table 11.1 compares the relative advantages of the PCR and of the *in vivo* method and explains why the *in vivo* method continues to be used.

Table 11.1 The relative advantages of the PCR and cell-based methods of cloning genes.

In vitro method, using the PCR	*In vivo* method, using bacteria
The PCR can copy DNA that has been partly broken down. This makes it useful in forensic science.	Partly broken down DNA is not copied.
The PCR is very sensitive. Even minute amounts of DNA, such as that contained in a single cell, can be copied.	*In vivo* methods are less sensitive, so large amounts of sample DNA are needed. This makes *in vivo* methods less useful in forensic work.
DNA that is embedded in other material can be copied. This makes the PCR useful for analysing DNA in formalin-fixed tissues or in archaeological remains.	Unless DNA can be isolated from the medium in which it is embedded, it cannot be copied by *in vivo* methods.
The cloned genes are produced in solution. They cannot be used directly to manufacture the protein that they encode.	The cloned genes are already inside cells that will manufacture the protein that they encode.
DNA polymerase will only copy a gene if it is marked at each end by complementary primers. If we do not know the base sequences at each end of the gene, we cannot make appropriate primers. Thus, we cannot use the PCR to copy genes that have not been studied before.	Once incorporated into its host DNA, the gene will be copied by the host cell. This method can be used to copy genes that have not been studied before.
The PCR becomes unreliable when we use it to copy DNA fragments longer than about 1000 base pairs (1 kbp).	*In vivo* methods reliably copy genes up to about 2 Mbp long.
In vitro methods lack error-correcting mechanisms, so the error rate is higher than in cell-based methods.	Cells have mechanisms for correcting any errors that are made when copying genes. This reduces the error rate of gene copying.

TEST YOURSELF

9 Extracts from bacterial cells will contain the DNA of the bacterium as well as plasmids. Suggest one property of plasmids that will allow them to be separated from bacterial DNA for use later as vectors.

10 Complete the table with the names of the enzymes that carry out the processes described.

Enzyme	Process
	Cuts DNA at a specific base sequence
	Makes a single-stranded DNA copy of an RNA sequence
	Joins a new piece of DNA into a plasmid

Transgenic organisms can be useful to humans

A transgenic organism is one that contains a gene from another organism. You have seen how genes can be transferred into bacterial cells, producing transgenic bacteria. When a gene is successfully transferred into another organism, the transgenic organism will produce the protein encoded by the transferred gene. This happens because the genetic code is universal, so a gene from any organism can be transcribed and translated by any other organism.

Collecting human proteins from transgenic organisms

Many human diseases result from an inability to produce a vital protein. For example, Type I diabetes is caused by failure to produce insulin and haemophilia is caused by failure to produce an essential blood-clotting protein. If the gene encoding one of these proteins is successfully transferred into a transgenic organism, the protein it produces can be extracted and used to treat patients with the relevant disease. Table 11.2 shows a range of organisms that have been genetically modified to produce proteins useful to humans on a commercial scale.

Table 11.2 The range of transgenic organisms that have been used in the commercial production of human proteins. Many more are at the pre-clinical trial stage of production.

Transgenic organism	Human protein	Clinical use of protein
Bacteria	Alpha-1-antitrypsin	Treatment of emphysema
	Growth hormone	Treatment of dwarfism
	Insulin	Treatment of Type 1 diabetes
Yeast	Alpha-1-antitrypsin	Treatment of emphysema
Plants	Collagen	Reconstructive surgery
	Growth hormone	Treatment of dwarfism
	Interleukins	Treatment of cancer
	Vaccine	Prevention of measles
Cattle	Fibrinogen	Treatment of blood clotting disorders
Goats	Anti-thrombin	Treatment of deep-vein thrombosis
Pigs	Factor VIII	Treatment of haemophilia

Figure 11.14 These large industrial fermenters contain a culture of transgenic bacteria that produce a human protein. Since the bacteria secrete the human protein, it is relatively simple to extract it from the culture medium.

Extraction of the human protein is much easier if the transgenic organism secretes it into the fermentation medium in which it is growing, or in its milk if it is a mammal. This explains the presence of microorganisms and mammals listed in Table 11.2. Plants do not secrete large volumes of protein; extraction of a human protein from plants is easier if we know where in the plant it is produced. Scientists can ensure that a human protein is secreted in milk or deposited only in the seeds of plants by transferring the human gene into a specific part of the genome of the transgenic organism. By placing it downstream of a promoter controlling lactation in mammals, or of a promoter controlling seed production in plants, scientists control where in the transgenic organism the human protein is made. Producing pharmaceutical products using transgenic farm animals (sometimes called 'pharming') and transgenic plants is an increasingly important source of new drugs.

Transgenic plants produce genetically modified (GM) food

In addition to producing human proteins, research into gene transfer in plants has focused on improving plant productivity. In particular, research has focused on two plant characteristics: resistance to insect pests and resistance to herbicides. One species of soil bacterium, *Agrobacterium tumefaciens*, has been especially important in transferring new genes into crop plants. Figure 11.15 explains why this bacterium has been so important.

Figure 11.15 The damage to this plant is called a gall. It is caused by a bacterium, *Agrobacterium tumefaciens*, which has infected the plant. When *A. tumefaciens* infects a plant cell, it injects its own plasmid, a T-plasmid, which the plant cell incorporates into its own DNA. By inserting genes into T-plasmids, biologists have been able to transfer foreign genes into plants.

A. tumefaciens produces chemicals that make the bacterium resistant to attack by insects. The genes encoding the production of these **natural insecticides** have been successfully transferred from bacteria to crop plants, reducing the need to spray crops with insecticides. In 1995, maize was the first plant to be marketed that had been genetically modified in this way, with potato plants and cotton plants soon following them on to the market. Today, there are many competing agrochemical companies around the world, marketing many types of transgenic insecticide-producing crop plants.

No one has yet produced a plant that is resistant to weeds growing around it. Instead, transgenic plants have been produced that are resistant to the most commonly used herbicide, **glyphosate** (sold commercially as Roundup). Glyphosate inhibits an enzyme (EPSP synthase) involved in the pathway, which plants use to make essential amino acids. Without a functional EPSP synthase, weed plants cannot make some of the proteins they need and die. Unfortunately, the same happens to crop plants that have been sprayed to get rid of weeds. The solution was to transfer into crop plants a bacterial gene that confers resistance to glyphosate. Such herbicide-resistant crop plants (called Roundup Ready) are routinely grown in many countries.

Figure 11.16 Some people take extreme measures in support of their opinions. These activists are destroying what they believe to be a crop of transgenic plants (GM plants).

The food produced by transgenic plants is commonly called genetically modified food (**GM food**). It was hoped that transgenic plants would increase crop production and so help to reduce starvation in the world. You would probably support this aim, to reduce world starvation, but are you in favour of genetically modified food? The issue has caused great debate in the UK. The growth of transgenic crops in the UK is restricted by government legislation and any food produced by transgenic plants must be clearly labelled as GM food. Many people refuse to buy GM food and are angry when they find that the constituents of processed food include GM products. Some people, like the activists in Figure 11.16, take extreme action and destroy experimental GM crops where they discover them.

People's concerns about GM foods tend to fall into one of three categories.

1 **Does eating GM foods harm us?** Some people are worried about eating GM foods, and the major UK supermarkets do not sell them. However, some people are concerned that meat and animal products, such as milk and eggs, can be sourced from animals that have been fed GM animal feed.

2 **Do GM plants adversely affect the environment?** Many people are concerned about the damage that pesticides might do to natural communities. Some people point out that the use of herbicides seems to have increased where herbicide-resistant GM crops have been grown.

Others are concerned that **horizontal gene transfer**, where pollen from one species cross-pollinates another plant species, might result in plants other than the GM crop gaining the transferred genes for herbicide resistance.

3 **Does globalisation disrupt local enterprise?** These concerns are about the dominance of large, multinational companies and the effect of their behaviour on small farmers, rather than about the science of genetic modification. The seeds of genetically modified plants are produced by a small number of multinational companies. They sell these seeds as part of a package; the farmer must agree to buy seeds, fertiliser and pesticides from the same company. Tying in to a single provider increases the risk to the farmers of changes in company policy or of changes in political relations between countries.

You might like to discuss the GM issue in class, including why people in some countries appear to be less concerned about GM crops and GM foods than people in the UK.

TIP

You need to be able to discuss aspects of genetic engineering, but you do not need to recall any particular examples.

TEST YOURSELF

11 Unlike eukaryotic cells, bacterial DNA does not contain introns.

a) Suggest the advantage of transforming bacteria by inserting complementary DNA (cDNA) rather than a copy of the original gene.

b) Suggest one advantage of using transgenic yeast rather than transgenic bacteria to produce human proteins.

12 In addition to the DNA found in plant chromosomes, chloroplasts contain their own DNA. Genes conferring resistance to insecticides and herbicides are commonly inserted into the DNA of chloroplasts, rather than into the transgenic plant's chromosomes. Suggest why this technique reduces the risk of foreign genes passing from one plant species to another by cross-pollination.

Gene therapy may be used to overcome the effect of defective genes

Many people object to the concept of producing transgenic humans, yet this is what gene therapy does. The aim of gene therapy is to treat an inherited disease by providing the sufferer with a corrected copy of their defective gene. The process of putting a corrected gene into a cell is called **transfection** and the cell that has received the new gene is said to have been **transfected**. There are two approaches to gene therapy: somatic cell therapy and germ cell therapy.

Extension

Somatic cell therapy

During **somatic cell therapy**, copies of a functional gene are inserted directly into the body cells of sufferers. In some cases, body cells are removed, copies of the functional gene inserted into them and the transfected cells put back in the same patient's body. This works well with blood diseases, such as leukaemia, where cells from the patient's bone marrow are extracted, transfected and replaced. In other cases, where large numbers of body cells cannot be removed safely, the genes are inserted directly into the affected tissues. This technique has been used to treat lung diseases, such as cystic fibrosis.

Vectors are needed to transfer genes into the cells of a sufferer. **Retroviruses** are often used as vectors for bone marrow because they have been found to transfect a large number of stem cells successfully. Viruses, known as adenoviruses, have also been used as vectors for treating cystic fibrosis, as have **liposomes**: small droplets of lipid that fuse with the surface membranes of epithelial cells in the lungs (Figure 11.17).

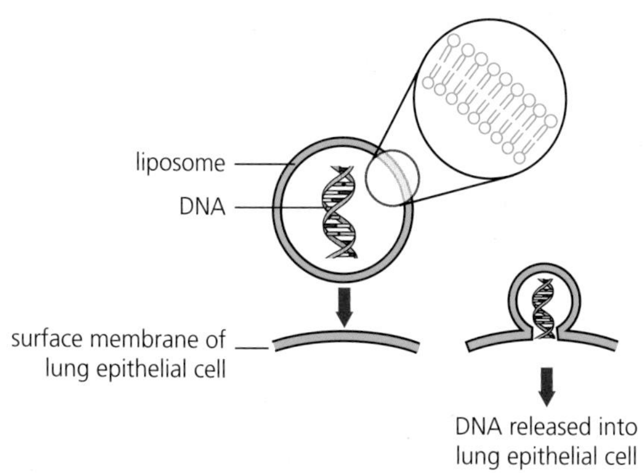

Figure 11.17 The corrected gene for cystic fibrosis can be carried into the epithelial cells of the lungs by small droplets of lipid, called liposomes. A liposome is able to fuse with the surface membrane of a lung epithelial cell, releasing its DNA into the cell.

When stem cells from bone marrow are transfected and replaced in the marrow of patients, all the different types of white blood cells produced from them contain the added gene. This provides a long-term treatment for blood diseases, such as leukaemia. Where stem cells cannot be used, the effects of the corrected gene last only as long as the transfected cells remain alive. This explains why gene therapy to treat cystic fibrosis must be repeated every few weeks.

Attempts have been made to use gene therapy to treat cancer. As you know from Chapter 10, cancers can result from activation of **oncogenes** or from inactivation of tumour-suppressor genes. Gene therapy has been used to insert corrected copies of active tumour-suppressor genes into cancer cells and to attempt to prevent the expression of an oncogene in cancer cells. The success of this research is currently hampered by a lack of vectors that will ensure the genes are taken up by cancer cells.

Germ cell therapy

In **germ cell therapy**, a corrected gene is inserted into an egg cell that has been fertilised using *in vitro* fertilisation (IVF) techniques. Once fertilised, the embryo is allowed to develop before being re-implanted into the mother's uterus. If gene transfer is successful, mitosis will ensure that every cell in the embryo contains a copy of the corrected gene. As an adult, the transfected individual will pass on copies of the corrected gene to her or his offspring. Consequently, germ cell therapy provides a treatment for an inherited condition that not only lasts the lifetime of the sufferer but also crosses generations. Germ cell therapy is currently illegal in humans.

Moral and ethical issues

As a biology student, you should be able to recognise and discuss some of the moral and ethical issues raised by DNA technology. It would be wrong for this book to suggest that there is a 'correct' response. You must make up your own mind.

However, you will be expected to discuss the controversial issues raised by gene cloning and DNA technology using appropriate scientific terminology and understanding. Three examples of how you could show your biological knowledge and understanding are given below.

You might express concern that producing transgenic animals causes the animals to suffer.

- As a biologist, you should be able distinguish between the effects of a foreign gene in an animal and the procedures involved in producing transgenic animals. Although the presence of a foreign gene does not appear to cause suffering, the manipulations involved in producing a transgenic animal might do so. For example, animals produced by

TEST YOURSELF

13 Somatic cell therapy is more successful in treating diseases caused by a recessive allele of a gene than those caused by a dominant allele. Suggest why.

14 Some people are concerned that the techniques of germ cell therapy could be misused. Suggest one way in which the techniques could be misused.

15 How could the techniques of germ cell therapy be used to produce cattle that secrete a human protein in their milk?

somatic cell nuclear transfer (page 190) suffer a relatively high rate of birth defects and even the healthy animals appear to suffer from premature ageing.

- You might use Dolly the sheep as an example of premature ageing. If so, you should make it clear that she was *not* a transgenic animal. Her premature ageing was associated with the somatic cell nuclear transfer technique rather than any foreign gene.

You might express concern about the use of genetically modified plants to improve crop productivity.

- As a biologist, you should not justify this concern using statements such as, 'this harms the environment'. You can show knowledge and understanding of biological concepts and principles by giving a specific example of how a GM crop could cause damage. For example, you could say that horizontal gene transfer between bacteria could result in the foreign gene for herbicide resistance being transferred from the GM crop plants to weed plants and that this would cause the weeds to become an even greater pest. You might also recall that foreign genes are often inserted into the DNA of chloroplasts to reduce the likelihood of their horizontal transfer to other plants via pollen.

You might express concern about the use of gene therapy to cure human diseases. There is no easy answer to problems such as this, but you can give a balanced argument that includes biological knowledge. For example, you might argue that:

- it is unreasonable to object to the management of cystic fibrosis using liposomes in respiratory inhalers to insert functional versions of the gene into the lung cells of sufferers
- if bone marrow transplants are acceptable, it is difficult to reject the correction of blood disorders by inserting functional genes into blood stem cells
- germ cell therapy has been beneficial in producing cows that yield milk with less fat and more protein. However, since this type of genetic manipulation involves changing the genotype of this animal and of future generations of this animal in a directed way, it would be unacceptable to use it with humans.

TIP

It is *never* a good idea to use the expression 'playing at God', since this shows no biological knowledge or understanding. Remember that a good scientist uses evidence from reliable sources, and not unsubstantiated opinions, when formulating an argument.

Practice questions

1 Scientists have isolated a gene that codes for the protein that a type of spider uses for making a web. They have inserted the gene into a very early goat embryo. The modified goat produces the spider web protein in its milk. The protein is very useful for medical applications, such as treating nerve damage and repairing wounds.

a) Describe how the following are used in genetic engineering:

i) a vector *(1)*

ii) restriction endonuclease. *(2)*

b) i) Why is it important that the gene is inserted into a very early embryo? *(2)*

ii) Describe how the DNA inserted into the goat cells produces the spider web protein. *(6)*

c) One animal rights organisation described this example of genetic engineering as 'an unethical use of farm animals'. Do you agree with this statement? Give reasons for your answer. *(4)*

2 Genetically engineered insulin can be used to treat diabetes. The figure shows some of the stages in genetically engineering insulin.

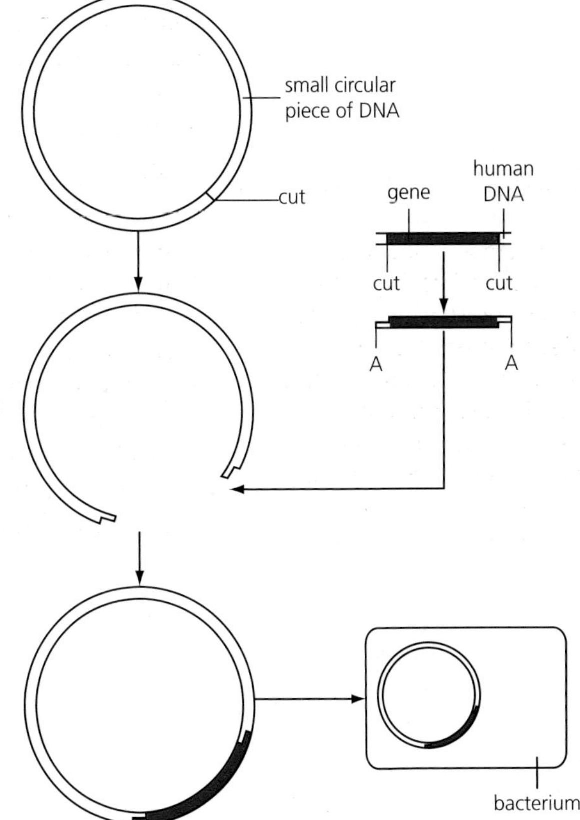

a) Complete the table to give the names of the items being described. *(4)*

Item	Description
	The term used for the small circular piece of DNA
	The name for the structure labelled A
	The enzyme used to cut the DNA
	The enzyme used to attach the human gene to the circular DNA

b) What is a marker gene? *(1)*

c) Erythropoietin is a glycoprotein used to treat some medical conditions. It is produced by genetically modified animal cells.

i) Give two differences between an animal cell and a bacterial cell. *(2)*

ii) Explain why a glycoprotein, such as erythropoietin, cannot be produced using bacterial cells. *(2)*

3 Some scientists wanted to carry out the PCR on a sample of DNA. They put the DNA they wanted to copy in a test tube together with two types of primer.

a) What else should be added to the tube for PCR to be successful? *(1)*

b) Why are two different primers needed? *(1)*

c) Starting with just one DNA molecule, what is the maximum number of DNA molecules that would be present after six cycles of PCR? *(1)*

d) One cycle of PCR involves heating the mixture and then cooling it again. What happens during the heating stage of the cycle? *(2)*

4 Some scientists have devised a test to detect an animal parasite, *Sarcocystis*, in meat. This parasite can cause harm to humans if it is present in undercooked meat that is then eaten by a human. The test involves PCR, using primers that bind to certain genes in *Sarcocystis*.

a) Meat will contain DNA from the animal and may also contain bacterial DNA. Explain why this kit will detect *Sarcocystis* DNA, and not any other DNA present. *(1)*

b) i) The PCR uses DNA polymerase enzyme. Why is this enzyme needed? *(1)*

ii) The DNA polymerase enzyme is heat-stable. Explain the advantage of this. *(2)*

5 The Mediterranean fruit fly is a serious agricultural pest that causes extensive damage to crops. It can be controlled by insecticides and biological control but these are not always effective.

Now scientists have genetically engineered flies by adding a female-specific gene into the insects. This stops females developing to reproductive maturity. The male flies are released into the environment, where they mate with wild females. However, they pass on the female-specific gene to their offspring, ensuring that only males develop into adulthood.

a) Suggest how this new gene could be inserted into an insect zygote. *(4)*

b) If a chemical repressor is added to the food supply of the genetically modified fruit flies, the female-specific gene is inactivated. Suggest the advantage of this. *(2)*

c) Evaluate this use of genetically engineered Mediterranean fruit flies. *(4)*

Stretch and challenge

6 Discuss the concerns that many people in the UK have about genetic modification. To what extent are these concerns justified?

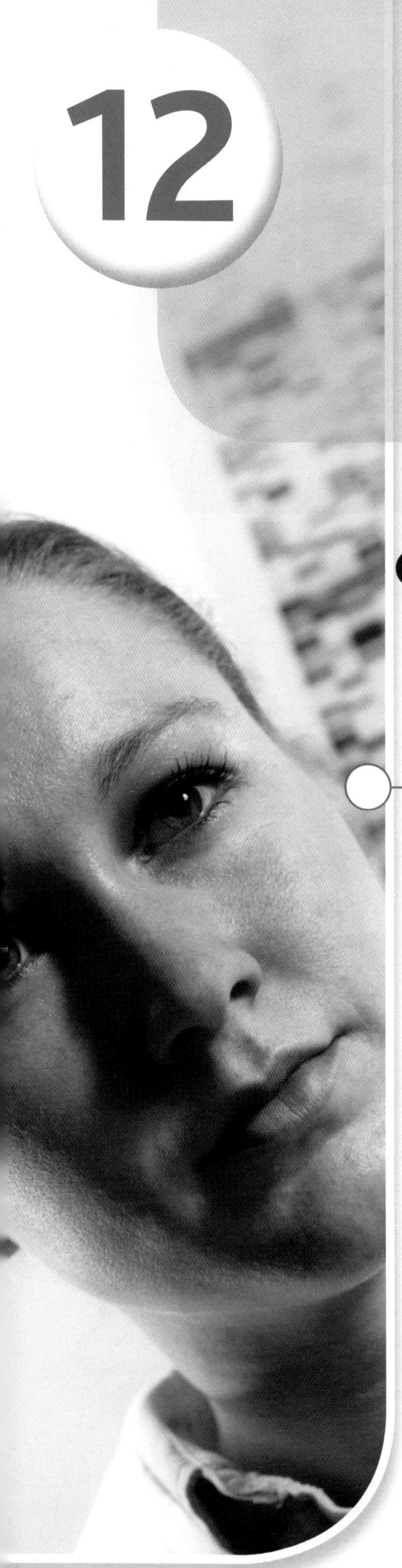

12 Using gene technology

PRIOR KNOWLEDGE

- Restriction enzymes cut DNA at specific base sequences.
- The polymerase chain reaction can be used to make many copies of a sequence of DNA.
- A mixture of pieces of DNA of different lengths can be separated using gel electrophoresis.

TEST YOURSELF ON PRIOR KNOWLEDGE

1. Fragments of DNA move towards the positive electrode in gel electrophoresis. Explain why.
2. Name the enzyme used in the PCR.
3. What are primers used for in the PCR?

Introduction

Genome is the term for all the genes in an organism. The Human Genome Project (HGP) was an international, collaborative research programme that aimed to completely map and understand all the genes in a human being.

The Human Genome Project was launched in 1990. It started in the USA with collaborators from several other countries. In the UK, scientists in Cambridge had been working for several years on mapping the genome of a nematode worm that is widely used in research. Although it was not complete, they recognised that they had the technology to be able to achieve this. In the UK, the Sanger Institute near Cambridge provided the British contribution to the Human Genome Project. Laboratories all over the world were allocated different sections of different chromosomes to sequence.

One key decision made by these scientists was that all the data generated would be shared publicly before being published. This allowed data to be shared as quickly as possible. Because this was a publicly funded project, they wanted to make sure that all information discovered could lead to as many benefits for humans as possible without commercial interests being involved.

On 26 June 2000 a draft sequence of the human genome was published. In 2003, the International Human Genome Sequencing Consortium announced they had completed the detailed reference human genome with 99.99% accuracy. This work involved huge numbers of scientists from 20 institutions all over the world.

Figure 12.1 The Wellcome Trust funded the building of the Sanger Institute near Cambridge to provide the British contribution to the Human Genome Project.

Genome projects

The finished sequence of the human genome was over 3 billion letters long. Scientists were surprised to find that the human genome contains only about 25 000 genes, which was a great deal fewer than they had expected. However, the scientists still needed to find out the **proteome**. This is the full range of proteins that a cell is able to produce.

Applications of the Human Genome Project

Scientists also need to work out how to use this sequence to improve human health and cure diseases. In the human genome, scientists have identified a number of genes that can increase the likelihood of people developing certain types of cancer and other diseases, such as Alzheimer's disease. It is hoped that, in time, it will be possible to understand the causes of inherited diseases, and to find effective treatments.

Sequencing of genomes

Scientists at the Sanger Institute near Cambridge, as well as sequencing one-third of the human genome, have sequenced the genomes of many other organisms. These include organisms that are used in genetic research, such as zebra fish, roundworms and the plant *Arabidopsis*, and organisms that cause disease, including parasites such as *Plasmodium*, which causes malaria.

One of the most severe forms of malaria is caused by *Plasmodium falciparum*. Scientists have sequenced hundreds of different *P. falciparum* parasites. These genomes are then examined by computers to find differences between the sequences. Scientists can then identify the genes that show the most variation between parasites. This indicates the genes that are under the greatest selection pressure, so these can be investigated further to find out whether they code for proteins that function as antigens that could be targeted for a vaccine. Similarly, scientists are investigating whether specific gene variants are associated with drug or insecticide resistance. They are also trying to find human genes associated with protection against severe malaria.

TIP

Remember that you learned about the immune response in the first year of your course (see *AQA A-level Biology 1 Student's Book* Chapter 6).

Identifying genes that might be a suitable target for a vaccine is just the first stage. Scientists need to work out the protein that the gene codes for. Then they need to inject that protein into people living in areas where malaria is common, to see whether they produce antibodies against the protein. Scientists at the Sanger Institute are working with centres in Africa and Asia to carry out these tests. It may be some years before an effective anti-malaria vaccine is readily available, but sequencing DNA enables scientists to focus their efforts on the most likely targets.

One reason why scientists can sequence so many genomes is that sequencing methods are improving rapidly. Most of the processes involved are automated. Data from sequences can be fed into huge banks of computers that are programmed to compare sequences from different organisms.

TEST YOURSELF

1. The scientists involved in the Human Genome Project decided that all the data generated would be shared publicly before being published. Evaluate this decision.
2. Explain why a gene for an antigen that shows a great deal of variation is likely to be under a high selection pressure.

Extension

DNA sequencing

DNA sequencing finds the base sequence of genes that have not been studied before. The methods for doing this are all based on the use of **dideoxyribonucleotides**. Look at Figure 12.2, which shows molecules that you might recognise. Molecule a) is ribose, the five-carbon sugar found in every RNA nucleotide. Molecule b) is deoxyribose, the five-carbon sugar found in every DNA nucleotide. Note that it has one oxygen atom less than ribose, hence the name *deoxy*ribose (see Chapter 4 in *AQA A-level Biology 1 Student's Book*). Now look at molecule c). Can you see how this differs from ribose and deoxyribose? It has one less oxygen atom than deoxyribose and two less than ribose: this is *di*deoxyribose.

Just like ribose and deoxyribose, a molecule of dideoxyribose can join up with phosphate and an organic base to form a nucleotide. During DNA replication, a nucleotide containing dideoxyribose (a **dideoxynucleotide**) can pair with a deoxynucleotide on the template strand that has a complementary base. However, DNA polymerase stops replicating DNA when it encounters a nucleotide containing dideoxyribose on the developing strand. This is the basis of the **chain-termination technique** for sequencing DNA.

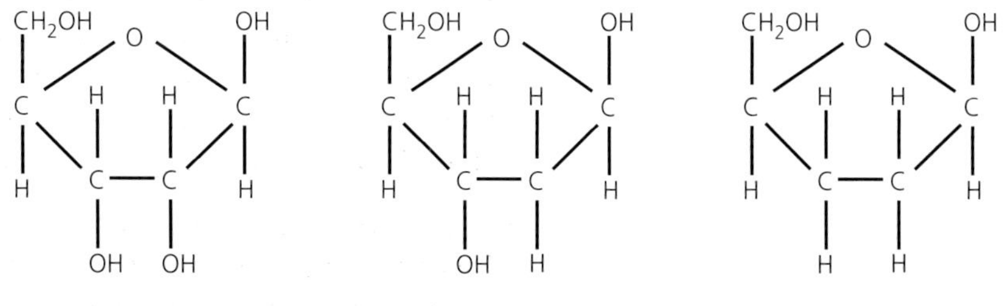

Figure 12.2 A molecule of a) ribose, b) deoxyribose and c) dideoxyribose.

Figure 12.3 The principles involved in the use of an automated DNA sequencer. a) The length of single-stranded DNA to be sequenced is cloned into a vector and a primer added. b) DNA polymerase attaches to the primer and begins DNA replication, adding nucleotides with bases that are complementary to the bases on the recombinant DNA. c) The DNA polymerase inserts a dideoxynucleotide and replication of the chain terminates. d) Provided the ratio of normal (deoxynucleotides) to dideoxynucleotides is high enough, chains of one to several hundred nucleotides are produced.

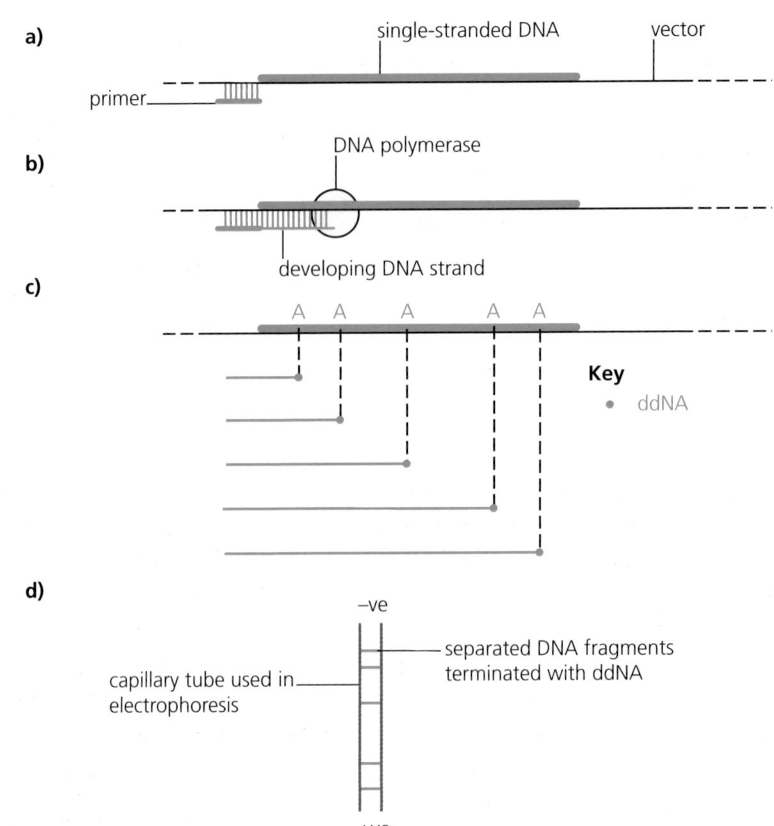

Automated DNA sequencing

Figure 12.3a shows a single-stranded version of the DNA we wish to sequence. It has been cloned into a vector, which in this technique is usually the DNA of a bacteriophage called M13. A short oligonucleotide has been annealed to the DNA to act as a primer.

DNA polymerase is added to the recombinant DNA, together with a mixture of deoxynucleotides containing each of the four bases: adenine, cytosine, guanine and thymine. Under normal circumstances, the DNA polymerase attaches to the primer and begins the process of replicating the recombinant DNA. The new strand starts to grow as free nucleotides form hydrogen bonds with complementary bases in the starting strand (Figure 12.3b).

However, also present in the mixture of nucleotides are some dideoxynucleotides. When, by chance, DNA polymerase inserts one of these instead of a deoxynucleotide, DNA replication stops. If the ratio of deoxynucleotides to dideoxynucleotides is high enough, we end up with complementary chains of DNA varying in length from a single nucleotide to several hundred nucleotides. Figure 12.3c shows how this is possible, using just one of the four types of dideoxynucleotide (ddNA). Whenever ddNA forms a complementary base pair with thymine on the recombinant DNA strand, replication stops.

Each type of dideoxynucleotide is labelled with a different fluorescent dye. The dideoxynucleotides carrying adenine (ddNA) are labelled with a green fluorescent dye, ddNC with a blue dye, ddNG with a yellow dye and ddNT with a red dye.

At the end of the incubation period, the newly formed DNA fragments are detached from the template DNA and these new single-stranded chains are separated by size, using a special type of electrophoresis that takes place inside a capillary tube (Figure 12.3d). The resolution of this technique is so high that it separates strands that are different in length by a single nucleotide.

When illuminated by a laser beam, each of the four dideoxynucleotides fluoresces. The colour and position of the fluorescence is read by a detector which then feeds this information into a computer, which either stores the data for future analysis or produces a printout, like the one in Figure 12.4, overleaf. An automated sequencer can read almost 100 different DNA sequences in a period of 2 hours.

You will remember that the genome is the DNA sequence of an organism. Scientists have found the DNA base sequence of many different organisms, including humans. You will also remember that genes code for proteins, and we already know the triplet sequences (codons) that code for each amino acid.

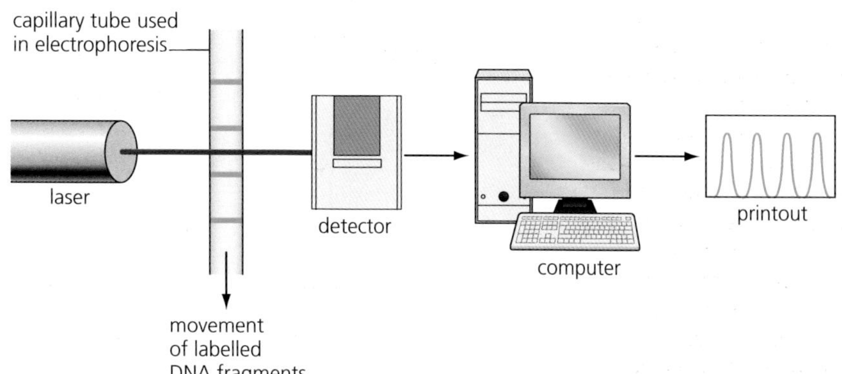

Figure 12.4 As the replicated fragments of DNA are separated, they pass through a laser and fluoresce. The fluorescence detector 'reads' the fluorescence and sends the data to a computer for analysis. The coloured printout shows the base sequence of the DNA sample.

This means we can work out the primary structure of the proteins that the genes code for. The structure of the proteins that a cell's DNA codes for is called its **proteome**. Knowing the proteome can be very useful, for example in identifying potential antigens to use as targets for vaccines. Scientists trying to develop a vaccine against malaria have sequenced the genome of many different malaria parasites. They compare the results. The genes that show the greatest variation must be those that are subjected to the greatest selection pressure, so these are the best targets for a vaccine.

However, it is not easy to work out the entire proteome of an organism from sequencing the DNA. You will remember that complex organisms such as humans have a great deal of non-coding DNA and regulatory genes, and it is not always easy to identify these from the DNA that codes for proteins.

Interpreting the results of manual DNA sequencing

The manual method is similar to the automated method above, except that:

- a separate run is made for each type of dideoxynucleotide – ddNA, ddNC, ddNG and ddNT
- after incubation, the four mixtures are placed in separate wells of a gel and separated by gel electrophoresis
- the dideoxynucleotides are labelled with radioactivity, rather than with fluorescent dyes
- a Southern transfer of the gel is made and an autoradiograph made from it.

Reading the DNA sequence is relatively easy. Figure 12.5 shows part of an autoradiograph. For each dideoxynucleotide base, it shows bands that are DNA fragments for which replication was terminated. As you know, the band that has moved the furthest is the smallest fragment. This will be a fragment where replication stopped at the first base.

In Figure 12.5 the fragment that has moved furthest is in the ddNA track. This is the smallest fragment that was produced when a dideoxyribonucleotide stopped DNA polymerase forming a new strand of DNA. The smallest fragment of DNA that could be formed is a single nucleotide. By chance, a dideoxynucleotide was the first to be inserted into the new DNA strand and its presence stopped DNA polymerase inserting any more nucleotides. We have now found the first nucleotide that is incorporated into the new DNA strand. It carries adenine.

The next smallest fragment is in the ddNC track. This is a fragment of DNA that is only two nucleotides long and it carries cytosine. The first two bases in the test DNA are adenine and cytosine. The fragment which has moved the third furthest is in the ddNG track. This fragment stopped DNA replication because it carries the ddG nucleotide. So our third base is guanine and the base sequence of the test DNA becomes ACG. By repeating what we have done so far, you should find that the base sequence is ACGCATGTTC.

Figure 12.5 The wells at the top of this autoradiograph contained a sample of a reaction mixture that had been incubated with dideoxynucleotides containing adenine (ddNA), cytosine (ddNC), guanine (ddNG) or thymine (ddNT), respectively. Below each well is the track of the separated DNA fragments produced by DNA replication in the presence of each dideoxynucleotide. The DNA sequence of the newly synthesised DNA is read by identifying the distance each fragment has moved, starting with the one that has moved furthest.

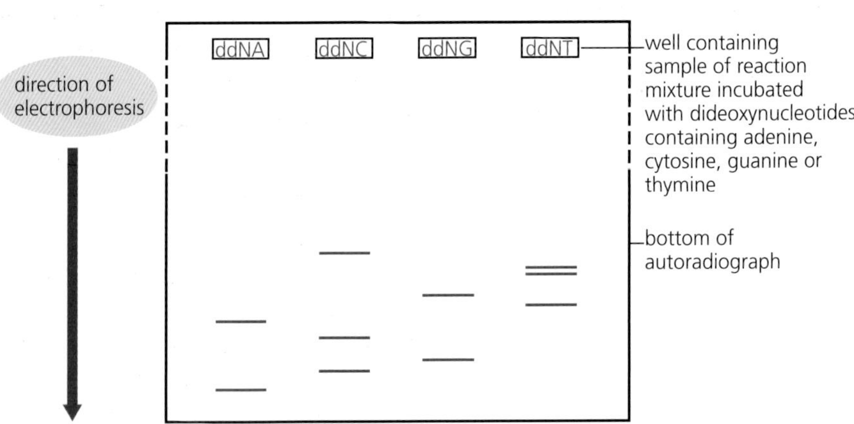

Using labelled DNA probes and DNA hybridisation to locate specific alleles of genes

DNA hybridisation The process of combining two complementary single-stranded DNA molecules and allowing them to form a single double-stranded molecule through base pairing.

If we know at least part of the base sequence of a harmful allele of a gene, we can use **DNA hybridisation** to find out if this allele is present in DNA samples. Before hybridisation, the test DNA is treated using techniques you learned about in Chapter 11. Let's revise these techniques.

First, the DNA is extracted from the sample of cells taken from a patient. After purification, the test DNA is amplified using the polymerase chain reaction (PCR). The test sample will only contain a few cells, for example cheek cells, cells from the amniotic fluid or cells from the umbilical cord. These cells only yield a small amount of DNA. We need a large amount of DNA for analysis, so the PCR is used to produce this extra DNA. Of course, the PCR produces DNA that is identical to the test DNA.

The amplified test DNA is then digested using a restriction endonuclease. Whole DNA molecules are too long to be analysed successfully in one go. This is why the test DNA is digested using restriction endonuclease.

The restriction fragments are then separated using gel electrophoresis. Since DNA has a negative charge, DNA fragments migrate to the positive electrode of an electric field. The gel through which they move has fine pores. Small fragments of DNA move faster through these pores than large fragments. As a result, the fragments are separated, with the smallest nearer the positive electrode, forming bands in the gel.

In the final preparatory stage the DNA fragments on the nylon membrane are treated to break the hydrogen bonds holding the DNA together. Hydrogen bonds between complementary base pairs hold

DNA probe A piece of single-stranded DNA that has a specific base sequence that is complementary to a specific base sequence of a specific allele (of a gene).

together the two polynucleotide chains in a DNA molecule. When these bonds are broken, the nylon membrane contains bands of single-stranded DNA fragments. These fragments anneal with any strands of nucleotides that have complementary base sequences. This is where labelled DNA probes come in. A DNA probe is a piece of single-stranded DNA that is complementary to a specific base sequence. The DNA probe is attached to a fluorescent or radioactive label. This is called **labelling** the probe.

At least part of the base sequence of the harmful gene is known. This enables us to produce single-stranded DNA probes with a complementary sequence using a gene machine. These probes will anneal to any complementary DNA fragment on the nylon film. This is the process of DNA hybridisation. If the DNA probe is labelled, its position can be found. Figure 12.6 shows how a DNA probe will attach to a complementary base sequence if it is present in the test DNA and how, if labelled, we can find the hybridised DNA after electrophoresis of the restriction fragments.

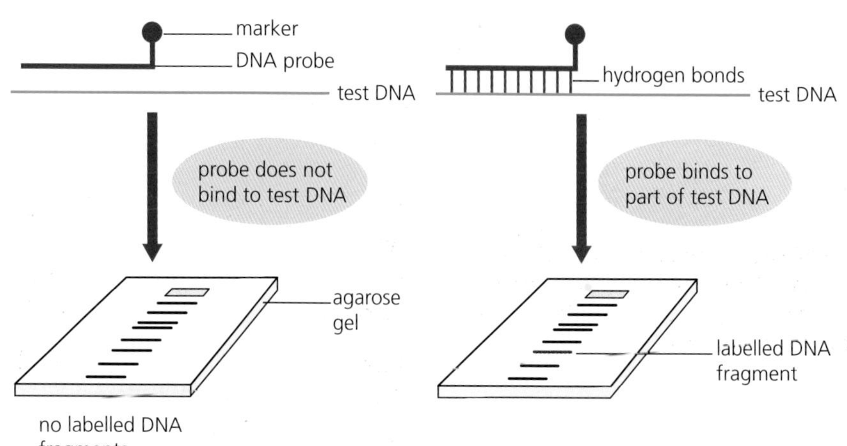

Figure 12.6 A labelled DNA probe can be made which has a sequence that is complementary to part of the harmful allele but not to the normal allele of a gene. Once the DNA probe has attached to a fragment of the harmful allele we can find it in the separated restriction fragments.

We can use Figure 12.6 to explain how DNA hybridisation enables biologists to locate the harmful allele of a gene in a sample of DNA taken from, for example, cells in amniotic fluid. Since the restriction fragments were made from the entire DNA in the sampled cells, one of them will contain all, or part, of the gene in question, for example the gene for β-globulin. The sequence of bases that is different in the harmful allele is known. The labelled DNA probe has a base sequence that is complementary to that part of the harmful allele. If the label attaches to any of the bands of restriction fragments on the nylon membrane, the DNA fragment in that position must be from the harmful allele. If no label attaches to any band, the DNA does not contain the harmful allele of the gene.

TEST YOURSELF

3 Use your knowledge of natural selection to suggest why the genes in a pathogen that show most variation are good targets for a vaccine.

4 Suggest how DNA probes are made.

5 Before a DNA probe is added to DNA on a nylon membrane, the DNA on the membrane is made single-stranded. Explain why.

6 If the DNA probe, complementary to the base sequence found in a harmful allele, does not bind to a person's DNA on a membrane, this is not 100% proof that the person does not carry a harmful allele. Explain why.

Medical screening and genetic counselling

DNA sequencing, DNA probes and DNA hybridisation are important techniques used in DNA analysis. Many human diseases result from mutations of genes, which give rise to alleles that are useful in one context but not in another. Sickle-cell anaemia results from an allele of the gene encoding β-globulin, one of the two types of polypeptide in a haemoglobin molecule. In areas of the world where malaria is endemic, heterozygotes for this gene are at a selective advantage because they are resistant to malaria. In countries where there is currently no malaria, for example the UK, heterozygotes are at a disadvantage because some of their children might be homozygous for the sickle-cell allele. These children will suffer uncomfortable and life-threatening symptoms.

Gene probes can also be used to test DNA from cancer cells to detect certain alleles. This can lead to targeted therapy for that specific mutation. Some cancer drugs only work on cancer caused by a specific mutation. Another use of gene probes is to find whether a person has a gene that makes them more likely to develop a particular kind of cancer. For example, women with the *BRCA1* and *BRCA2* alleles are at higher risk of developing breast cancer than other women.

A couple intending to have children might be concerned if they are aware that there is an inherited disease, such as sickle-cell anaemia, in their family. They might take medical advice about the risk of having a baby with the inherited condition. This advice is called **genetic counselling**. In the past, genetic counsellors were able to use the sort of information in Chapter 7 to work out the chances of an affected baby being born. For example, you can use your understanding of monohybrid inheritance to work out that the chances of two people who are both heterozygous for the sickle-cell allele having a baby with sickle-cell anaemia is 1 in 4 (a probability of 0.25).

Samples of the amniotic fluid that surrounds a developing fetus may be taken. This fluid contains cells from the fetus. Fetal DNA can be analysed using gene probes to find out whether the fetus has a specific genetic condition. A couple who are at risk of having a child with a genetic condition, such as cystic fibrosis, may opt for *in vitro* fertilisation (IVF). This means creating a number of embryos, which can be tested using gene probes. Healthy embryos can be selected and implanted into the mother's uterus.

ACTIVITY

Breast cancer and ovarian cancer

Each year in the UK about 200 000 cases of breast cancer and about 25 000 cases of ovarian cancer are diagnosed. Research has shown that about 10% of these cases are inherited. At least two genes are involved in the development of breast cancer: *BRCA1* (BReast CAncer 1) and *BRCA2*. Normally, the *BRCA1* and *BRCA2* genes control cell growth but mutations in these genes increase the risk of breast and ovarian cancers.

1 What type of gene is *BRCA1* and *BRCA2*?

We each have two copies of the *BRCA1* and *BRCA2* genes, inheriting one from each parent. Someone who carries one of the mutant alleles of either gene has an increased risk of developing breast cancer.

2 What does this tell you about the nature of the mutant alleles of the *BRCA1* and *BRCA2* genes?

Table 12.1 shows data about the risk of a woman developing breast cancer and ovarian cancer in the UK.

3 The risk of developing cancer in the middle column of Table 12.1 is shown as a range, whereas the risk in the right-hand column is shown as a single value. Suggest why.
4 Give two conclusions about the effect of carrying a mutant allele of a *BRCA* gene on the risk of developing breast cancer.
5 Suggest why the risk of developing breast cancer is higher than the risk of developing ovarian cancer.

Table 12.1 The risk of developing cancer among women who are carriers of one mutant allele of the *BRCA1* gene or *BRCA2* gene compared with that of the general population.

Risk of developing	Carrier of one mutant allele of *BRCA1* or *BRCA2* gene/%	General population/%
Breast cancer by age 50	33–50	2
Breast cancer by age 70	56–87	7
Ovarian cancer by age 70	27–44	<2

Genetic counsellors not only advise people about their chances of having a baby with an inherited disease or about their chances of developing an inherited disease, such as cancer, later in life. They also provide emotional and medical support to the affected families. This is needed because the results of screening tests provoke difficult decisions. For example, a woman who finds that she is carrying a child with an inherited disease might have to make a decision about carrying her pregnancy to full term or having an abortion. Similarly, a woman who finds she has an increased risk of developing breast cancer might have to decide to have her healthy breasts surgically removed so they cannot later develop cancer. None of these decisions is easy to make and affects not only the person making the decision but other members of their families as well.

It is likely that shortly after you were born, a drop of blood was taken from your heel. This 'pin prick' procedure collected blood that could be used to test for the presence of common inherited disorders, such as phenylketonuria. This represents a simple form of medical screening. Advances in DNA technology have resulted in much more sophisticated medical screening procedures. Using the techniques you learned about in Chapter 11 it is now possible to analyse DNA taken from scrapings of cells from inside your cheek, cells taken from the amniotic fluid surrounding an embryo in its mother's uterus or cells from the umbilical cord connecting the embryo to its placenta.

TEST YOURSELF

7 Is it always useful to be able to screen a person for alleles that might lead to health problems?
8 What are the arguments for and against screening men for the *BRCA1* and *BRCA2* genes if there is a family history of breast cancer?

Genetic fingerprints

You have learned that any one of our chromosomes carries the same genes in the same order. Consequently, you might expect that we would get the same pattern of restriction fragments if we used the same restriction endonucleases to digest copies of the same chromosome from two different people. In fact, we would not; we would get different patterns of restriction fragments. The reason for this is that, in the non-coding regions of our DNA, each of us has different repeated sequences of bases. These repeated base sequences include **variable number tandem repeats (VNTRs)**. Because the length of these repeated base sequences is different in different people, we can use them to identify the source of DNA in tissue samples. This is the basis of classical **genetic fingerprinting**. VNTRs are chromosomal regions in which a short DNA sequence (such as AGCT) is repeated a variable number of times end-to-end at a single location (tandem repeat). Everyone has these tandem repeats, but a variable number of repeats. However, the number of tandem repeats that a person has is inherited, and so they could be used to identify biological parents.

Figure 12.7 shows how the distance between the recognition sites of a restriction endonuclease on two samples of DNA varies if the number of repeated sequences varies. As a result of these differences, DNA fragments of different lengths would be produced by digestion of these two DNA molecules with the restriction endonuclease. Figure 12.8 shows how these different lengths of DNA would produce different patterns of restriction fragments after electrophoresis. The important idea to grasp is that only identical twins have the same pattern of non-coding, repeated base sequences and so only they would produce the same pattern of restriction fragments.

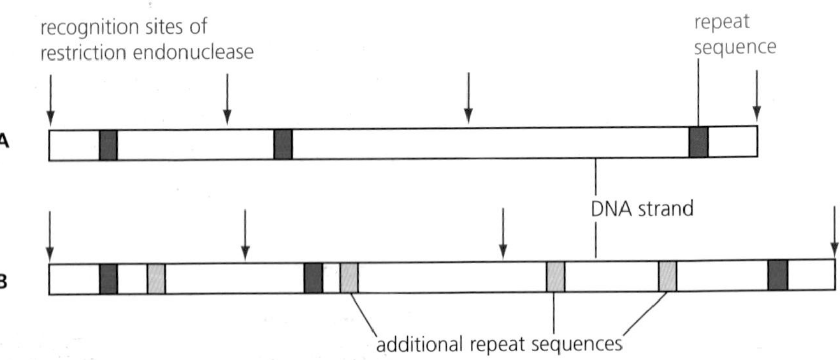

Figure 12.7 Each band represents part of a DNA molecule that has been taken from the same chromosome location from two different people, A and B. The recognition sites of an endonuclease are shown on each DNA molecule. The recognition sites are further apart in B because the non-coding regions of this DNA contain more repeated base sequences.

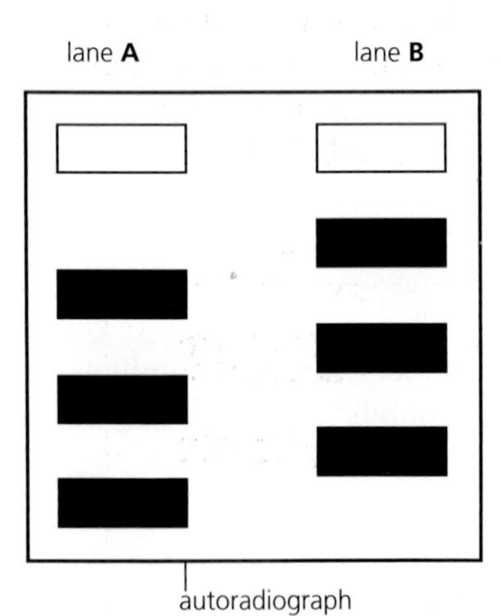

Figure 12.8 The banding patterns produced after the two sets of DNA shown in Figure 12.7 have been digested using the same restriction endonuclease. Lane B contains longer DNA fragments than lane A because the DNA of B contained more repeated base sequences than the DNA of A.

Genetic fingerprinting by hybridisation with labelled probes

This is the classical method for genetic fingerprinting. You have come across the techniques involved earlier in this chapter. If the sample of DNA is very small, more copies are made using the PCR. A sample of DNA is digested into restriction fragments using a restriction endonuclease. The restriction fragments are of different lengths, depending on the number of repeat sequences within them, and are unique to each individual. These fragments are then separated using gel electrophoresis and a Southern blot is prepared from it. Radioactively labelled probes with the complementary base sequence to the repeat sequence are added to the blot. The probes reveal a labelled band wherever they hybridise with a restriction fragment containing the repeat sequences. Thus, each band in the fingerprint represents a DNA fragment that contains the repeat sequence. Figure 12.8 shows the genetic fingerprints of two people. The pattern of bands is different because the location of the repeat sequences was different in the two sets of DNA.

The very first repeated sequence to be used in classical genetic fingerprinting had the base sequence GGGCAGGABG, where B represents any base. This sequence was repeated a different number of times in the non-coding regions of the DNA samples used.

Figure 12.9 shows the results of the very first analysis that led to a criminal receiving the death sentence in the USA. A young couple had been murdered while they slept in their car. Their bodies were discovered the next day. A post mortem showed they had both died of gunshot wounds and that the woman had been raped. One man was later arrested driving the couple's stolen car. Under police questioning he identified a friend who had been with him on the night of the murders. DNA from the semen found in the woman's body and a DNA sample from each suspect was digested using the same restriction enzyme, and the restriction fragments separated using gel electrophoresis. If you look at Figure 12.9, you will see that the genetic fingerprint of the semen does not have the same pattern of restriction fragments as suspect 1 but does have the same pattern as suspect 2. On the basis of this evidence, suspect 2 was found guilty of rape and murder and given a double death sentence. The jury at the time was told that the chance of an innocent person showing the same pattern of restriction fragments was about 1 in 9 billion. At the time of the trial, the human population of the world was less than 6 billion.

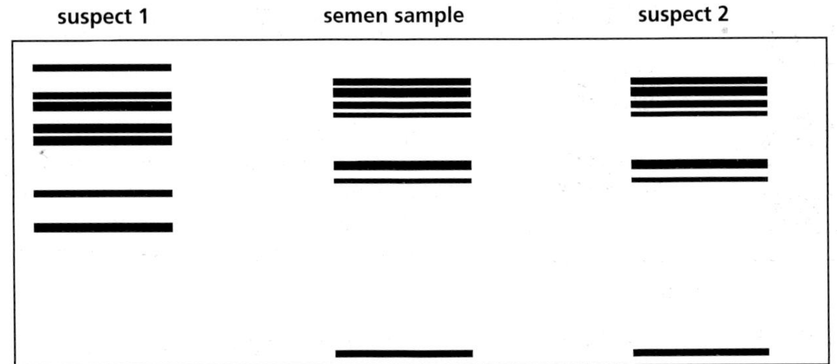

Figure 12.9 The results of the first repeated sequence analysis which led to a criminal receiving the death sentence. The genetic fingerprint of suspect 2 and of the semen taken from a murdered woman show that he was the rapist. He was convicted of rape and murder and received a double death sentence.

EXAMPLE

The Bains family

Mr Bains first settled in England in 1990. Two years later, he applied to bring his wife and four children to join him. The immigration authorities needed to be sure that the children were entitled to come into England and so they carried out genetic fingerprinting on the whole family. Figure 12.10 shows the results.

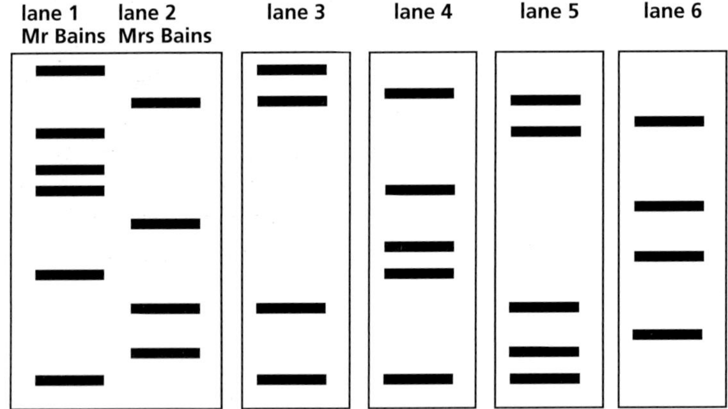

Figure 12.10 The genetic fingerprints of the Bains family members.

Mr Bains had been married before. Sadly, his first wife had died giving birth to their son. Some years later, Mr Bains remarried and he and his new wife had two daughters. They also adopted another son.

1 Look at lanes 1 and 2 in Figure 12.10. They show the separated restriction fragments of Mr and Mrs Bains. Explain why the two patterns of restriction fragments are different.
This is because the different DNA samples have different numbers of repeats of the sequence.

2 Lane 6 of Figure 12.10 shows the restriction fragments from Mr and Mrs Bains' adopted son. Explain why this conclusion is valid.
His bands don't match either of Mr or Mrs Bains's bands, so he cannot be their biological child.

3 Use Figure 12.10 to identify Mrs Bains' stepson. Explain your answer.
All the bands in a child's DNA must match those from either the mother or the father. This is true for the bands in lanes 3 and 5 showing these contained the DNA from Mr and Mrs Bains' daughters. Some of the bands in lane 4 match those of Mr Bains, suggesting he is the biological father. The bands that do not match those of Mr Bains do not match those of Mrs Bains either. These bands must represent DNA from Mr Bains' first wife. So lane 4 is the DNA from Mrs Bains' stepson.

DNA profiling by PCR of variable number tandem repeats (VNTRs)

A more powerful technique has replaced the classical fingerprinting technique. This involves DNA profiling.

As we have seen, variable number tandem repeats (VNTRs) are found in the non-coding regions of our DNA. VNTRs are repeated blocks of base sequences such as GCAT or GC. In our own genome, the most common repeat is a sequence of two bases, cytosine and adenine, repeated between 5 and 20 times, for example CACACACACACACA. The number of repeats in any one VNTR is variable; in a single population, there might be as many as ten different versions of one VNTR. In DNA profiling, the different versions of a selected number of VNTRs are determined.

Using primers that anneal to the DNA to each side of a particular repeat sequence, the PCR is used to make many copies of the DNA containing these VNTRs. After amplification by the PCR, the DNA fragments are separated using gel electrophoresis. Normally, DNA profiling is carried out using several selected VNTRs, so that many more bands are present in the gel.

Just as with the automated sequencing technique described earlier, the VNTR fragments can be labelled with fluorescent dyes and separated using capillary electrophoresis. Scanners 'read' the sequences and feed the data into a computer which compares them with sample material. This technique is much more sensitive than the standard test and is now commonly used by teams of forensic scientists. Figure 12.11 shows the type of printout from an automated scanner.

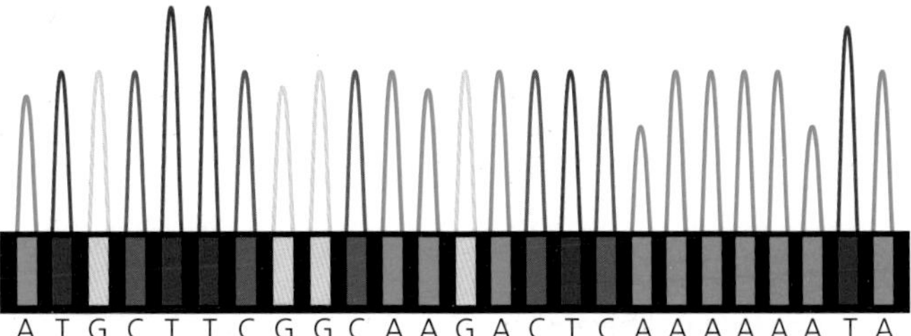

Figure 12.11 The printout from an automated DNA profiler. You might have seen this type of DNA fingerprint in a film or TV programme about crime scene investigators.

TEST YOURSELF

9 Why must the DNA probes used be radioactive or fluorescent?

Practice questions

1 Scientists in South Africa have used genetic engineering to create a vaccine for farm animals that protects them against several different diseases. They have taken a gene that codes for an antigen of Rift Valley virus and inserted it into a smallpox virus. Then they added a gene that codes for an antigen of lumpy skin virus and inserted that into the smallpox virus.

a) i) What is a vaccine? *(1)*

ii) This vaccine will produce immunity to three different diseases. Explain how. *(5)*

b) Suggest the advantage of adding the genes for these antigens into a live virus, rather than injecting the animal with the antigens alone. *(2)*

2 a) Describe how a genetic fingerprint is produced. *(6)*

Some blood was found on a broken window at the scene of a crime. The police found four suspects. They carried out genetic fingerprinting on the crime-scene blood, plus blood samples from each of four suspects. The results are shown in the diagram.

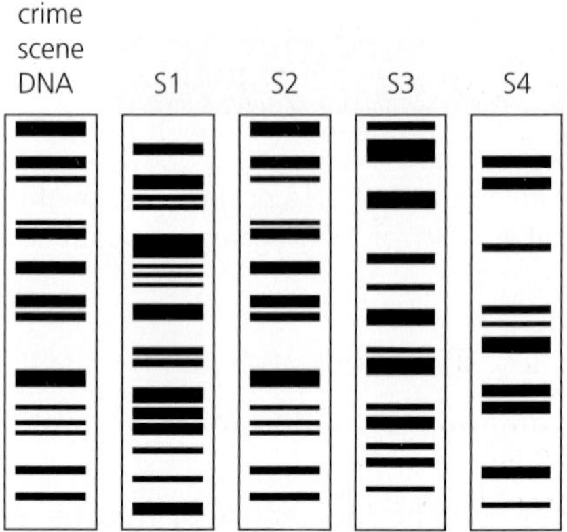

b) Which suspect does the crime-scene DNA come from? Explain your answer. *(2)*

c) It has been suggested that a blood sample should be taken from every person in the country, and the person's genetic fingerprint found. This could be stored on a national database. Any crime-scene genetic fingerprint could then be compared to this database. Do you think this is a good idea? Give reasons for your answer. *(4)*

3 A test can be carried out to find out whether a person's DNA carries a gene mutation that can cause tumour formation. This mutation is recessive.

DNA is taken from the patient and the section where the gene is found is copied many times using the PCR. The DNA is incubated with a restriction enzyme. If the mutated allele is present it is cut by the restriction enzyme. If the normal allele is present it is not cut by the restriction enzyme. The diagram shows possible results.

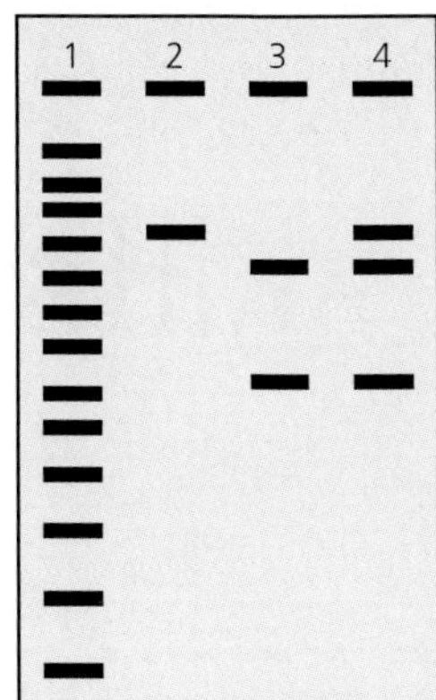

Lane 1 Standard ladder
Lane 2 Control or normal DNA
Lane 3 Mutated DNA for tumour cells
Lane 4 Carrier for the mutation (one normal DNA and one mutated DNA)

a) i) How is it possible to amplify only the section of DNA that carries the gene being investigated? *(2)*

ii) Lane 1 contains DNA fragments of different known lengths. Suggest why this is used. *(2)*

iii) Draw a circle around the shortest piece of DNA on the gel. *(1)*

b) Lane 2 contains DNA from a person who does not carry the mutated allele. Explain the results. *(2)*

c) Lane 3 contains DNA from a person who has a tumour caused by the mutated allele. Explain the results. *(2)*

d) Explain the results for lane 4. *(2)*

Stretch and challenge

4 It has been suggested that a child's genome should be sequenced at birth. This would enable the parents and the child to be aware of harmful alleles, for example those that predispose to cancer, and to make lifestyle choices to reduce exposure to risk factors. Evaluate this proposal, giving reasons to support your point of view.

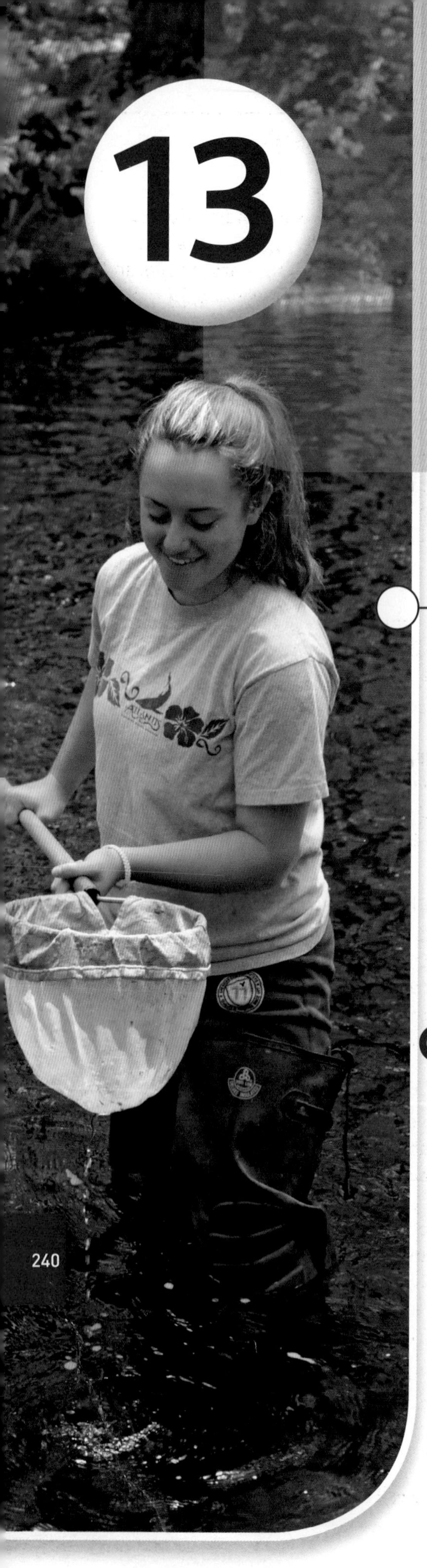

13 Developing mathematical skills

During the second year of your Biology A-level course you will be practising and developing the skills you learned in your first year, as well as learning some new ones.

Using units

In any measurements or calculations, using units correctly is critical. Just as for the first year of the course, you need to be familiar with the common units used in biology and be confident converting between units, for example from millimetres to micrometres or from cubic centimetres to cubic millimetres.

There are a thousand millimetres (mm) in a metre (m), a thousand micrometres (μm) in a millimetre and a thousand nanometres (nm) in a micrometre. There are also a thousand cubic millimetres (mm^3) in a cubic centimetre (cm^3) and a thousand cubic centimetres in a cubic decimetre (dm^3).

Since these are the units you are most likely to need to convert between, the 'rule of thousands' is a helpful one. But there are some important exceptions such as the centimetre. There are, of course, only 100 cm in a metre and 10 mm in a centimetre.

TIP

Avoid using mixed units such as minutes and seconds. Convert the whole time to seconds.

Sometimes units are combined, for example in a rate such as how fast oxygen is being produced ($mm^3\,minute^{-1}$) or how fast an animal is moving ($cm\,second^{-1}$). It is easy to forget to combine the units when you give an answer following a rate calculation. Sixty is the important number to remember for converting units of time. There are 60 minutes in an hour and 60 seconds in a minute.

For some rates, the first unit is a count such as the number of individuals in a quadrat or the number of beats in a heart rate. In this case, the correct units are number of beats $minute^{-1}$.

Sometimes three units are combined, such as when giving figures for gross and net primary production ($kJ\,m^{-2}\,day^{-1}$ or $kJ\,m^{-3}\,day^{-1}$ depending on whether it is a terrestrial or aquatic habitat).

Calculating areas, volumes and circumferences

You need to be confident with arithmetic calculations for finding the size and surface area of biological structures such as cells, organs and whole organisms. You also need to be able to give answers in the correct units and with the appropriate number of decimal places. It is also very useful to be able to recognise when your answer is outside of the expected range so that you realise when you may have made a mistake.

Calculating areas

Areas of rectangles are calculated by multiplying the lengths of the two sides whereas areas of circles are calculated by multiplying the square of the radius by π. Area calculations could be used to find the area of the field of view of a microscope, the area of a disc cut out from a leaf or the size of the area in which some random quadrat samples have been taken.

Decimal places and rounding

When you carry out calculations like finding the area of a disc cut out from a leaf, the answer shown on your calculator will often include a number of **decimal places**. This is the number of figures after the decimal point.

TIP
Rounding should be done using the rule that below half is rounded down and half or more is rounded up, so 1.24 becomes 1.2 and 1.25 becomes 1.3.

Using significant figures

You will remember from *AQA A-level Biology 1 Student's Book* that significant figures in a number include all the non-zero digits, any zeros between non-zero digits and, in numbers containing a decimal point, all zeros written to the right of the digits. The number of significant figures in a measurement gives an indication of its uncertainty.

TIP
When you do calculations, you should not round any numbers used in the calculation, only round the answer.

If you were calculating the area of part of a meadow in which you had carried out some quadrat sampling, you might measure two sides of the area with a tape measure. Let's assume you measured one side as 40.50 m and the other as 35.75 m. When you multiply the two to get 1447.875 m^2 on your calculator, you should only give the area as 1448 m^2 because the two original measurements were both to four significant figures.

Calculating areas of cubes and spheres

Different sized cubes are often used to compare imaginary organisms of different sizes. The surface area of a cube is calculated by finding the area of one side and multiplying that by six, because it has six sides. So a cube of side length 8 mm has a surface area of $(8 \times 8) \times 6 = 384$ mm^2. A cube of side length 16 mm has a surface area of $(16 \times 16) \times 6 = 1536$ mm^2. Notice that although the side length has been doubled, the surface area has increased four times. You should always estimate roughly what the answer should be for a calculation like this so that you can spot if your answer is different from what you are expecting and look for your mistake.

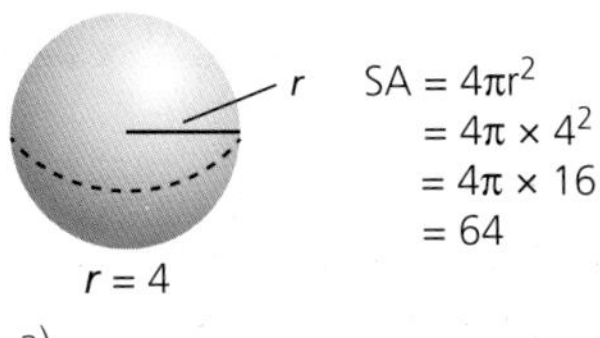

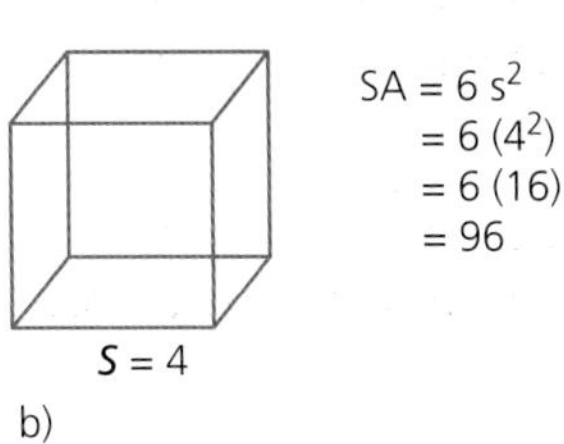

Figure 13.1 Calculating the surface area of a) a sphere and b) a cube.

In the same way that cubes can represent different sized organisms, spheres can be used to represent cells of different sizes. You need to remember that

Area of a sphere = $4 \times \pi \times \text{radius}^2$

Again, this means that if the radius of a cell is doubled, the surface area increases by a factor of four.

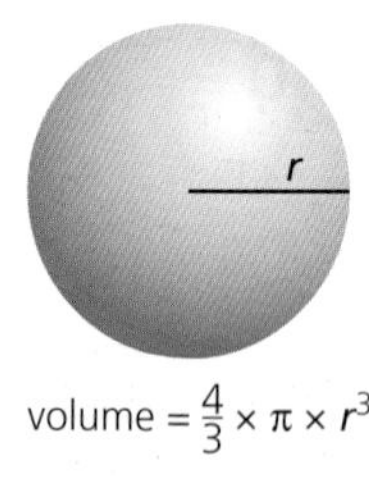

a)

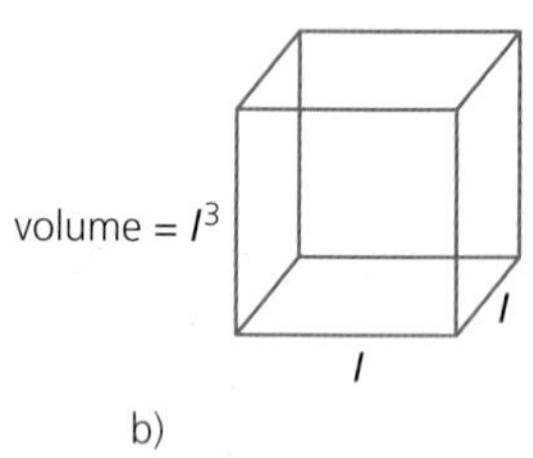

b)

Figure 13.2 Calculating the volume of a) a sphere and b) a cube.

Calculating volumes

Being able to calculate the volumes of cubes and spheres is also useful.

Volume of a cube = breadth × width × height

$$\text{Volume of a sphere} = \frac{4}{3} \times \pi \times r^3$$

Apart from calculating the volumes of cubes and spheres, you should also be able to calculate the volume of a cylinder. This is the cross-sectional area of the cylinder multiplied by its height.

$$\text{Volume of a cylinder} = (\pi \times r^2) \times h$$

This is useful when using a respirometer. The volume of gas taken up or produced is measured by finding how far a bead of coloured liquid moves along a capillary tube. The gas in the tube forms a cylinder shape, so the volume of gas taken up or produced is equal to the distance moved by the bead of liquid multiplied by the cross sectional area of the capillary tube **lumen** (*AQA A-level Biology 1 Student's Book*, Chapter 14, page 242).

The cross-sectional area of the lumen is $\pi \times r^2$. So to find the volume of gas taken up or produced you need to know the radius of the capillary tube (or its diameter so you can halve it). The radius of the capillary tube will most probably be in millimetres, and you should also measure the distance the bead of coloured liquid moves in millimetres so the answer will be in mm^3. If you have timed how long the bead of liquid took to move this distance, you could give the answer as a rate by dividing the distance by time in minutes. The appropriate units would then be $mm^3\,minute^{-1}$.

TIP
You should keep all your units the same in calculations like this to avoid decimal place errors.

Calculating circumferences

Remember that the circumference of a circle can be found by multiplying the diameter by π.

Using ratios and percentages

Standard form

You will remember from *AQA A-level Biology 1 Student's Book* that standard form is a way of using powers of 10 to describe very large or very small numbers. Standard form for small numbers moves the decimal place to the right. If you measured the length of a resting skeletal muscle sarcomere as 2.2 µm, that is 0.0022 mm. In standard form, this becomes 2.2×10^{-3} mm.

Standard form for large numbers moves the decimal place in the opposite direction. If you found that a growing population of yeast cells contained 240 000 cells per cubic centimetre, 240 000 in ordinary form becomes 2.4×10^5 in standard form; in other words, the decimal place is actually five places to the right.

Standard form avoids having to write out all the zeros in small or large numbers. However, you must be very careful to keep the same number of significant figures when converting numbers to and from standard form. A value of 0.0040 indicates that the value is exactly 0.004 rather than a rounded value. The last zero is a significant figure as well as the figure four, so 0.0040 has two significant figures.

TIP
When using standard form, there can only be **one** digit to the left of the decimal point so 240 000 must be written as 2.4×10^5, **not** as 24×10^4 or 240×10^3.

If you were converting 0.0040 to standard form, you would need to include the last zero by writing it as 4.0×10^{-3}. Both significant figures are shown in the standard form.

Ratios

> **TIP**
> You need to understand that it matters which way round the numbers are written in a ratio.

AQA A-level Biology 1 Student's Book described how **ratios** can be used to relate one attribute of something to another. For example, **C : N ratios** are helpful in describing the composition of organic material. Organisms that decompose organic material use carbon-containing biological molecules as their respiratory substrates. They use nitrogen-containing biological molecules for protein and nucleic acid synthesis.

Microorganisms use far more carbon-containing compounds as respiratory substrates than they use for protein and nucleic acid synthesis. Since they use about 30 parts of carbon for each part of nitrogen, a C : N ratio of 30 : 1 promotes rapid composting. If the C : N ratio is greater than this, it means there is more carbon and less nitrogen. If the C : N ratio is lower than this it means there is less carbon and more nitrogen so the soil microorganisms have more nitrogen than they can use.

Surface area to volume ratios are used when comparing the rate of heat loss to their surroundings in mammals of different sizes generally. Smaller mammals have a high surface area relative to their volume so lose heat more quickly.

A sphere of radius 10 mm has

Volume equal to $\frac{4}{3} \times \pi \times 10^3 = 4188.79\,\text{mm}^3$

A surface area equal to $4 \times \pi \times 10^2 = 1256.64\,\text{mm}^2$.

This means it has a surface area to volume ratio of 1257 : 4189, which simplifies to 1 : 3.33. The ratio is **simplified** by making the first number 1, so the second number is $\frac{1}{1257} \times 4189 = 3.33$.

A larger sphere of radius 20 mm has

Volume equal to $\frac{4}{3} \times \pi \times 20^3 = 33\,510.32\,\text{mm}^3$

Surface area equal to $4 \times \pi \times 20^2 = 5026.54\,\text{mm}^2$.

It has a surface area to volume ratio of 5027 : 33 510, which simplifies to 1 : 6.67 by dividing both numbers in the ratio by the first one. So the larger sphere has a smaller surface area to volume ratio.

In genetics, you will come across **phenotypic ratios**. These relate the proportion of one phenotype to the proportion of another. For example, if a pair of fruit flies had 160 offspring and 120 had red eyes and 40 had white eyes, the ratio of red : white would be 3 : 1.

Fractions

Fractions are used to describe the portion of a value that fits into different categories. For example, you might find that $\frac{1}{2}$ of the sunlight hitting some leaves is reflected by the shiny surface or $\frac{1}{3}$ of the woodlice you are watching have moved to a more humid area of a choice chamber after 5 minutes. Fractions are useful for description, but they are often converted into decimals or a percentage to make them easier to work with. You probably already realise that $\frac{1}{3}$ = 0.33 as a decimal, or 33% as a percentage. You may need to do this for genetic crosses. If $\frac{1}{4}$ of the offspring are a particular phenotype, that is the same as 25%.

Percentages

You will come across **percentages** and especially the idea of a **percentage change**, quite often in biology. You will remember from *AQA A-level Biology 1 Student's Book* that converting data to percentage values allows a valid comparison of data from populations of different size. This year you will also come across the idea of percentage cover when sampling with quadrats (see page 270).

Calculating percentage change

Percentage change is often used when considering changes in population size. An increase in the number of individuals in a population in a year may be expressed as a percentage increase. This is the additional individuals as a percentage of the number in the original population. If a population had 1300 individuals in one year and a year later contained 1576, the percentage increase is $\frac{1576 - 1300}{1300} \times 100 = 21.2\%$.

Remember that you should always try to estimate an answer to check that your calculated value is appropriate, rather than simply accepting it. In this case, the increase in the population is 276 compared to an original population of 1300 so an answer of roughly a fifth, or 20%, seems realistic.

Percentage error

When you make measurements in practical work, the uncertainty in the measurement can be expressed as a percentage error. The percentage error in the measurements obtained from different pieces of apparatus can be calculated by dividing the uncertainty by the measured value and multiplying by 100.

If a volume of 25 cm^3 measured with a measuring cylinder has an error of ±0.5 cm^3, then the percentage error for this piece of apparatus is $\frac{0.5}{25} \times 100 = 2\%$.

If a mass of 0.120 g measured using a balance has an error of ±0.001 g then the percentage error is $\frac{0.001}{0.120} \times 100 = 0.833\%$.

Dealing with orders of magnitude

Drawings and images of small structures such as cells and organelles are often larger than reality. Micrographs (photographs taken using a microscope) are magnified thousands of times. The magnification of the micrograph is usually given as a magnification factor such as ×10 000. This means that the image is 10 000 times larger than the object really is. You might be asked to calculate the actual size of something in the image, such as the width of a mitochondrion or the diameter of a cell. The method for doing this is always the same, regardless of the object or the magnification.

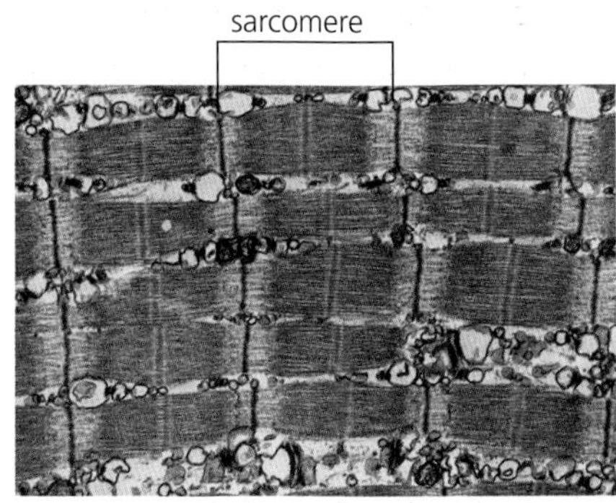

Figure 13.3 Electron micrograph of myofibrils (magnification approximately × 10 000).

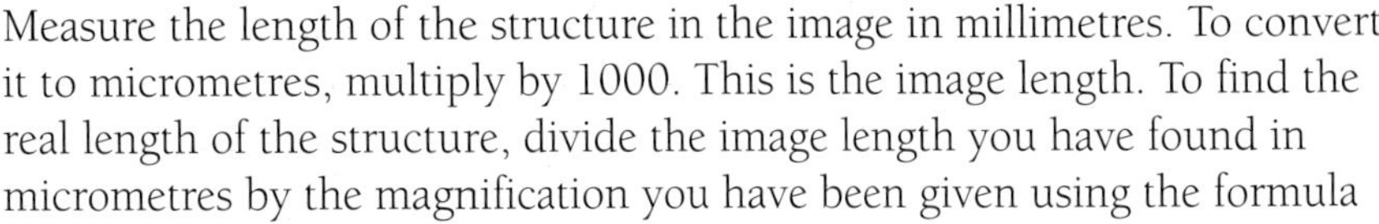

Measure the length of the structure in the image in millimetres. To convert it to micrometres, multiply by 1000. This is the image length. To find the real length of the structure, divide the image length you have found in micrometres by the magnification you have been given using the formula

$$\text{Size of real object} = \frac{\text{size of image}}{\text{magnification}}$$

This will give you a real length in micrometres. Try this for the length of the sarcomere in Figure 13.3. For microscopic structures, micrometres are the appropriate unit to use for the answer.

TIP

Always convert your measured length in millimetres to micrometres by multiplying by a thousand before doing anything else. Do not measure in centimetres. Always measure the image in millimetres. This helps to avoid decimal place errors.

To calculate the magnification of a structure, use the formula

$$\text{Magnification} = \frac{\text{size of image}}{\text{size of real object}}$$

You should realise that magnification is how big something looks (its image length) compared with how big it really is (its real length). The answer should be given in the form ×2000 rather than just the number.

Figure 13.4 Electron micrograph of a Pacinian corpuscle with scale bar.

If there is a **scale bar** on the micrograph, this is the real length (see Figure 13.4). Measure it with your ruler in millimetres and convert to micrometres. Then divide your measurement by the real length, the value on the scale bar. This gives the magnification factor.

Using logarithms to deal with orders of magnitude

Sometimes you may be faced with a very large range of numbers to plot on a graph. If you have a mixture of very small numbers and very large numbers, it is difficult to work out a suitable scale for your axis. To fit the large numbers on, you need a scale that means the small numbers are impossible to plot accurately.

In this sort of situation, using a logarithmic, or log, scale on the graph is a useful approach. You may see graphs like this used for plotting the numbers of yeast in a culture over several hours. The numbers increase very rapidly from a small number of yeast cells in the first place.

Figure 13.5 shows a graph that uses a log scale. You can see that the numbers on the y-axis are not evenly spaced. The number of cells at each time interval has been converted to a logarithm before being plotted. This means that the wide range of values more easily fit onto the same scale.

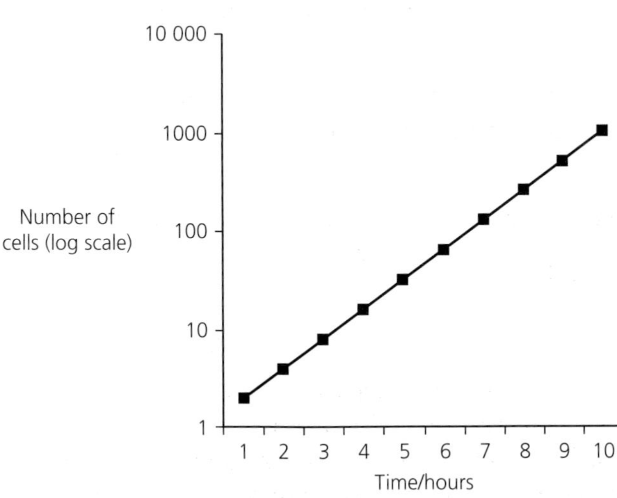

Figure 13.5 Yeast cell numbers plotted on a logarithmic scale.

Using symbols and equations

Table 13.1 The meaning of some symbols used in equations.

Symbol	Meaning
=	is equal to
<	is less than
<<	is much less than
≤	is less than or equal to
>	is greater than
>>	is much greater than
≥	is more than or equal to
∝	is proportional to
≈	is roughly similar to

You need to know the meaning of the mathematical symbols shown in Table 13.1 and when to use them.

In questions sometimes you will need to rearrange an equation in order to find the answer. This was done with the magnification factor equation in the previous section. Magnification = size of image/size of real object can be rearranged as Size of real object = size of image/magnification depending on what you already know and what you need to find.

Two equations you will have seen in Chapter 1 and Chapter 8 are

$$N = I - (F + R)$$

or

Net production of consumers = chemical energy in ingested food − (chemical energy lost in faeces + heat lost to the environment)

and

$$p^2 + 2pq + q^2 = 1$$

or

Frequency of homozygous dominant phenotypes + frequency of heterozygous phenotypes + frequency of homozygous recessive phenotypes = 1

These can be both rearranged depending on which values you know and which you need to find out. If you know the net production, the chemical energy lost in faeces and the heat lost to the environment, you can find the chemical energy in ingested food (e.g. $I = N + F + R$).

Using the Hardy–Weinberg equation, if you are given the frequency of black rabbits where black is the dominant allele, you are being given $p^2 + 2pq$. The frequency of white rabbits is $q^2 = 1 - (p^2 + 2pq)$.

There are a few equations that you need to be able to rearrange. Often it is only necessary to put the known values into the equation to calculate an answer. An example is the mark-release-recapture equation (see page 180), $N = n_1 \times n_2/n_3$ where n_1 is the number initially marked, n_2 is the number caught in the second sample and n_3 is the number re-caught.

Another equation you may need to use is the equation to calculate pH from the hydrogen ion concentration of a solution.

$$\text{pH} = -\log_{10}[\text{H}^+]$$

In this equation, square brackets mean concentration. You would need to find the logarithm of the value you are given for the concentration using a calculator. Reversing the sign would give you the pH. You may also have to use your calculator to calculate values such as x^y where y is a **power**. In this case, it might be to calculate how many bacteria would be present in a culture after a certain time. The number of bacteria doubles every generation. If there were 12 000 bacteria in a culture and they divide every 15 minutes, you could find the number of bacteria you would expect after 8 hours:

Number of generations = 8 × 60/15 = 32

Your calculator will have a power function, usually a key called x^y, which you can use to do these calculations. For this example you would type 2 x^y 32 which would return the answer of 4.29×10^9, which is the number of cells one cell would give after 32 generations. You would then multiply this by 12 000. The number of bacteria after 8 hours would be 5.15×10^{13}.

Plotting and using graphs

Drawing a graph helps you to see the relationship between two variables much more clearly than looking at data in a table. The two types of graph you will see most frequently are scattergrams (also called scattergraphs) and line graphs.

Scattergrams

You will already know about scattergrams from your first year. Scattergrams are used to look for a correlation between two variables (see *AQA A-level Biology 1 Student's Book*, Chapter 14, page 248). However, a scattergram can only indicate if there is a correlation. It cannot tell you if the correlation is significant or not. A Spearman's rank correlation test is required for that (see Chapter 14, page 254). You will know that when you draw a scattergram you should ensure that you label the axes and include units. Someone else should be able to look at the scattergram and know exactly what it shows without any further explanation. Plot each point as a dot with circle round it (⊙) or a cross (×). Draw (or imagine) a line of best fit if asked to.

TIP

If you are drawing your own scattergram, you should try to use a similar scale on each axis so that the results are spread out equally in both directions.

Correlation

If your line of best fit slopes upwards, you can say that there is a **positive correlation** between the two variables. In other words, as one variable increases, so does the other. If your line of best fit slopes downwards, you can say that there is a **negative correlation**. In this case, as one variable increases, the other decreases. Sometimes the line of best fit is horizontal, sometimes it is vertical and sometimes it is completely impossible to draw a line of best fit. In these cases, all you can say is that there is no correlation.

TIP

You should always remember that just because two variables are correlated, it doesn't mean that one *causes* the other.

Line graphs

You will have come across a lot of line graphs by now. You use a line graph to show measured results for a dependent variable when you alter an independent variable in an experiment. You might draw line graphs for the growth rate of a population or for the rate of photosynthesis at different light intensities.

When you draw a line graph the independent variable is plotted on the *x*-axis, and the dependent variable is plotted on the *y*-axis. The axes should be labelled and you should include units. Someone should be able to look at the graph and know exactly what it shows without any further explanation.

To choose a suitable scale, make sure all the points you need to plot will fit on the graph. Avoid a scale which involves parts of grid squares on the paper. Using whole squares is better. This makes it easier to plot points accurately. You should plot the individual points as clearly and accurately as possible. Use either a dot with a circle around it (⊙) or a cross (×). You can draw either a smooth curve or straight lines joining the points.

> **TIP**
> You should only draw a smooth curve if you feel that you can confidently predict intermediate values. Otherwise, join the individual points with straight lines.

You should have learned to recognise the other sorts of curves you usually see in biology. You will already have seen some curves that are **linear**, others that reach a **plateau** and some that are **exponential**.

> **TIP**
> - Lines on line graphs are called curves even if they are straight lines.
> - You should not draw lines beyond the range of the data collected.

Linear relationships

An example of a linear relationship is that between the rate of a reaction and substrate concentration when the enzyme is in excess. Provided the enzyme concentration remains in excess, the line will continue as a straight line. Graphs like this show a **linear relationship** which is described by the equation

$$y = mx + c$$

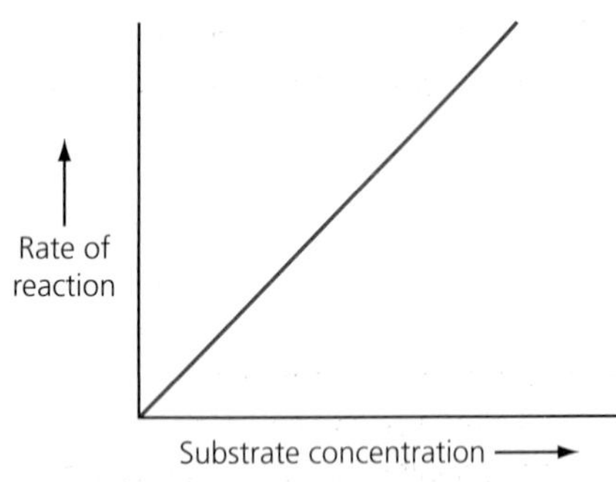

Figure 13.6 Relationship between the rate of reaction and substrate concentration when the enzyme is in excess.

where m is the **gradient** and c is the **intercept** on the y-axis. Figure 13.6 shows a line graph with a linear relationship like this. In this case the intercept value is zero.

Sometimes, if it is not zero, you will need to read off the intercept value. The intercept value is where the line reaches or crosses the y-axis.

An example of a graph with intercepts is shown in Figure 13.7. The graph shows the effect of light intensity on the rate of carbon dioxide uptake by two different plants. You need to be able to see that the y intercept for the linear or straight part of each line is –2 arbitrary units.

On this graph there are also intercepts on the x-axis. When the light intensity becomes so low that the rates of photosynthesis and respiration are balanced, the plant is at the **compensation point**. There is no net uptake of carbon dioxide. You could find the light intensity at the intercept, in other words where the line crosses the x-axis.

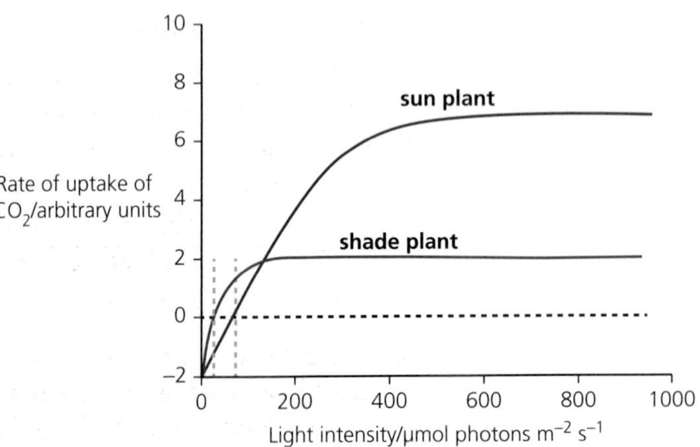

Figure 13.7 Rate of carbon dioxide uptake at different light intensities for sun and shade plants.

Notice that the intercept is different for shade-tolerant species and species that prefer full sun.

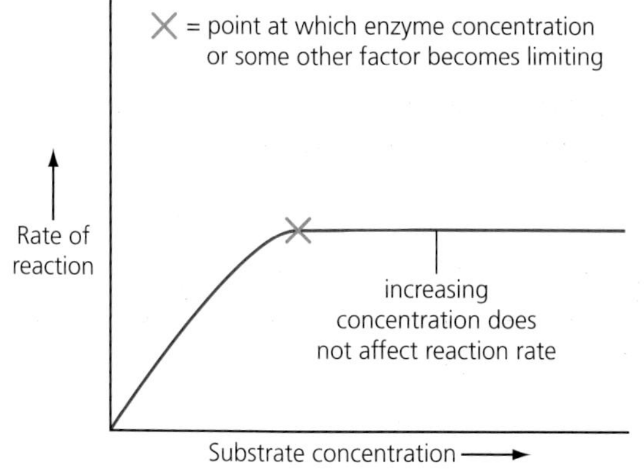

Figure 13.8 When the enzyme concentration becomes the limiting factor, the curve flattens off.

Measuring rate of change

You will already have done some practical work involving enzymes. You will know that if the enzyme concentration becomes the limiting factor, the curve will reach a plateau. An example of this type of curve is shown in Figure 13.8.

Another line graph in which the curve flattens off is shown in Figure 13.9. Here, note that the x-axis shows time rather than substrate concentration and the y-axis shows the mass of the product rather than rate. Initially, after an enzyme is added to some substrate, the reaction is fast because there is plenty of substrate to form enzyme–substrate complexes. Once some of the substrate has been used, the reaction begins to slow down.

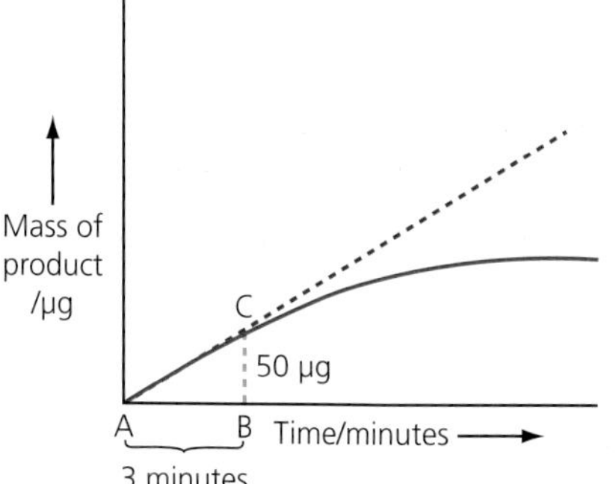

Figure 13.9 The progress of an enzyme-catalysed reaction.

You will recall that you can find the initial rate by drawing a **tangent** at the start of the curve. A tangent is a straight line that matches the gradient on the curve. The rate can be found by measuring AB and BC and then calculating BC/AB. This gives the rate at which product is formed per minute. So if 50 µg of product were formed in 3 minutes, the rate would be $\frac{50}{3} = 16.7\,\mu\text{g minute}^{-1}$.

TIP

Draw a tangent by placing a ruler on a short part of the start of the curve. Start from the origin and match the gradient. Make sure the ruler is positioned carefully. A slight change can alter the position of the tangent a lot.

Interpreting data from a sample of measured values

As you know, whenever a sample is taken from a larger group, the sample should be representative of the group as a whole. When these sorts of data are collected, a large number of observations or measurements is made and so summarising the data is often helpful. Biological data from samples can be summarised in a variety of ways, often called **descriptive statistics**.

Arithmetic means

One way of summarising sample data is by finding the **mean**. You will already have calculated mean values in practical work and will already know that the mean is found by adding up all the measurements and then dividing this total by the number of measurements.

$$\text{Mean} = \frac{\text{sum of all measurements}}{\text{number of measurements}}$$

Standard deviation

You will also know from last year that a mean becomes more useful if a standard deviation is also given with it. You will recall that standard deviation is found from the formula

$$SD = \frac{\Sigma(x - \bar{x})^2}{(n - 1)}$$

where Σ means 'the sum of', $(x - \bar{x})$ is the difference between any measurement and the mean and n is the number of measurements. To do the calculation, you need to find the difference between each measurement and the mean, square it and then add all the squared values together. Then divide this by one less than the number of measurements.

Standard deviation can be calculated manually, by using a scientific calculator or by using spreadsheet software.

Range, median and mode

Other ways of summarising the data are the **range**, the **mode** and the **median**.

The range is the difference between the highest and lowest values. Range and standard deviation are both measures of dispersion for a given set of data, but they both have advantages and disadvantages. Range is easier and quicker to calculate, but it is heavily influenced by outlying values at the extremes of the range. An outlier can increase the range significantly but the outlier is just one result in the whole set of data. The range simply uses the highest and lowest values so takes no account of the rest of the data. In this situation, standard deviation can be more useful. It takes longer to calculate, but it takes account of all of the results in a set of data and so enables a better comparison of the dispersion of two sets of data.

As you saw in *AQA A-level Biology 1 Student's Book*, the median is the middle value of the data arranged in rank order. The mode is the value that appears most often.

Frequency tables and histograms

One way of summarising data is a frequency table (see *AQA A-level Biology 1 Student's Book*, Chapter 14, Page 252). A frequency table is made by dividing the observations or measurements into a number of **classes**. Classes are either categories or ranges of measured values to which observations can be allocated. The **class frequency** or number of observations or measurements in each class can then be tallied up.

You will remember the difference between histograms (also called histographs) and bar charts from *AQA A-level Biology 1 Student's Book*. A histogram is plotted when the frequency table shows measured data. Counted data in categories would have to be plotted as a **bar chart.** In a bar chart, the bars are each separated by a space because the data are in discrete categories.

Statistical tests

During your A-level Biology course you have to choose one of three different statistical tests to analyse the data you collect in different practical situations. You will need to be able to recognise the type of data presented in a question, suggest how it could be presented on a graph and say which of the three different statistical tests best applies to the data. Using a statistical test in this way is called **inferential statistics**.

Why do scientists use statistics? Look at a roadside verge in early spring and you will almost certainly see dandelions in flower. Examine a particular length of verge more closely and you will see that most of the dandelions are growing near the road.

If you were to carry out a transect, you might find that the dandelions are less common further from the road and other species are more numerous. How can you explain this observation? Perhaps the road was treated with

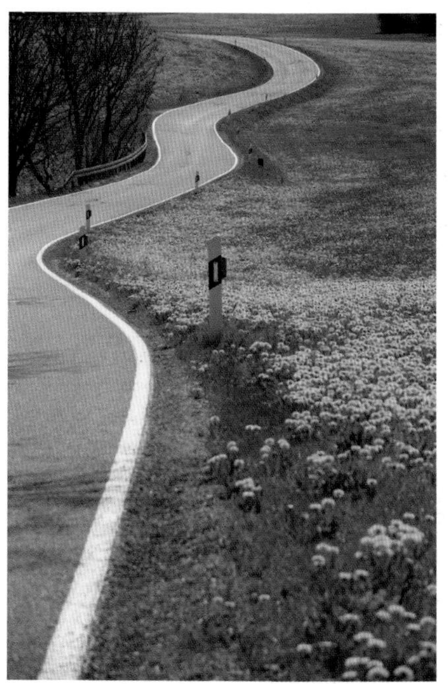

Figure 13.10 Dandelions growing near a roadside.

salt during the winter and dandelions can tolerate high salt concentrations better than other plants. There is another explanation, however, that has very little to do with biology. Maybe the differences in distribution of the various species is simply due to chance. It could be that you just picked an area where there were more dandelions growing closer to the road. A statistical test can help you judge how likely it is that there is a significant difference in the density of dandelions nearer or further away from the road.

As biologists, we accept that we can never completely rule out the effect of chance. There is still a small probability that the distributions you observed were due to chance. But it seems quite reasonable to say that the probability of different densities of dandelions nearer and further from the road arising due to chance is so low that we can safely reject chance as an explanation. We need a cut-off point and, for biological investigations, we normally accept this as a probability of 1 in 20 or 0.05 ($p = 0.05$). If there is a probability of less than 1 in 20 ($p < 0.05$) that the higher dandelion density near the road could have arisen by chance, there is another explanation. In this case we could say that it might be the salt from the road.

Different statistical tests for different purposes

A statistical test is a tool and, like all tools, you don't need to know how it works to use it. What you need to be able to do, however, is to select the right tool for the right job.

The practical work that you are likely to carry out this year could generate data that you might need to look at in different ways. It is likely that you will want to do one of the following.

- See if there is a significant difference between the means of two variables: suppose you were investigating the effect of temperature on the growth of duckweed. You could present the data as a line graph or a histogram. But you might also want to see if there was a significant difference between the mean numbers of duckweed plants when the plants were kept at two different temperatures.
- See if there is a significant correlation between two variables: if you investigated the number of stonefly larvae at different points in a stream, you could plot the data as a scattergram. But you might also want to see if there was a significant correlation between the number of stonefly larvae and the velocity of the water.
- Investigate data that fall into distinct categories. Suppose you were looking at whether maggots moved to light or dark conditions or offspring of a genetics cross matched a certain ratio. You could plot the data as a bar graph. But you might also want to know if there is a significant difference between the data you have collected and the data which your hypothesis predicts.

Each of these situations requires a different statistical test. The decision chart in Figure 13.11 should help you to choose which one to use.

Making the decision over which test relates to your data is very important. The way you plot your data on a graph and which statistical test you use are closely linked.

The null hypothesis

Each of the statistical tests that you will use is based on what is called a null hypothesis. It is very important that you understand what a null hypothesis is and why it is used. Suppose you had a table of data showing the

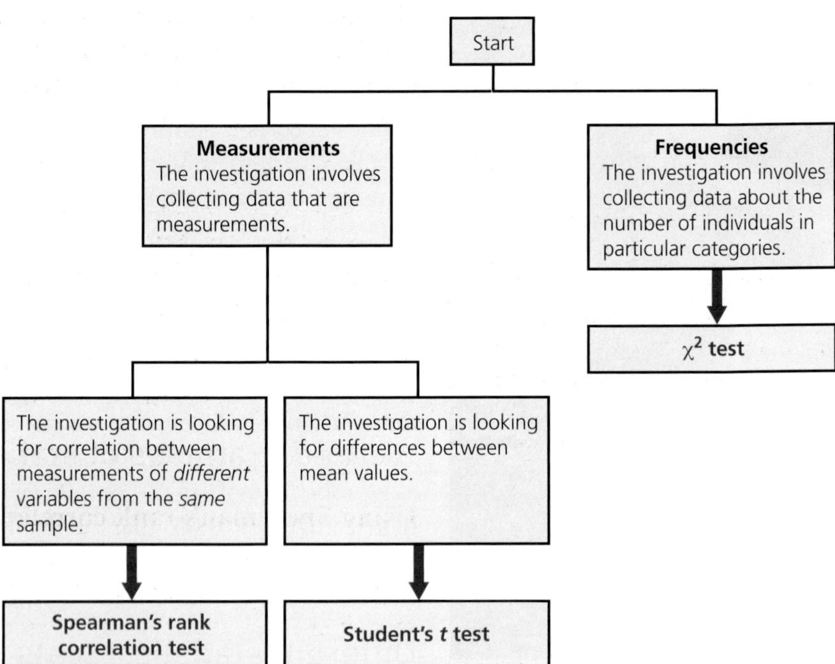

Figure 13.11 Decision chart to help you choose the appropriate statistical test.

distribution of aphids on the upper surface and the lower surface of nettle leaves. In this example, your null hypothesis could be:

There is no difference between the numbers of aphids on the upper surface and on the lower surface of nettle leaves.

You can then use the results of a statistical test to come to a conclusion about whether or not to accept the null hypothesis.

- If the test indicates that you should accept the null hypothesis, then all you can say is that there is no significant difference in the numbers of aphids on the different surfaces and that the differences you found could be explained purely by chance.
- If the test indicates that you should reject the null hypothesis, then you can say that the probability that the differences in numbers of aphids on the upper and lower leaf surfaces are due to chance is low and you can suggest a biological explanation.

> **TIP**
> You will not be asked to carry out statistical calculations in written exams. Instead you could be asked to suggest what sort of graph or what sort of statistical test is most appropriate for the data and why.

> **TIP**
> 'There is no difference in the result' is **not** an acceptable null hypothesis. You must mention both variables.

Calculating the test statistic

Once you have set up your null hypothesis, you can carry out the test you have selected. As a biologist, carrying out statistical tests is like using a recipe book to cook a meal. If you work methodically, and can use a standard scientific calculator, there is nothing to worry about. Remember that calculating test statistics will not be required in written exams but you may need to do a statistical calculation following practical work. If you look in the section at the end of this chapter, you will find worked examples of the three tests that you need to be able to use.

> **TIP**
> If you are given the result of a statistical test, or just the *p* value, you need to be able to interpret what this means.

Interpreting the results of a statistical test

Calculating the test statistic gives you a number. What does this number mean? You always need to interpret this number in terms of probability. You can then make a decision as to whether to accept or reject your null hypothesis. In order to do this, you need to look up the value that you have calculated in a table to find the probability level, or '*p* value'.

Biologists use a probability level of 0.05 (often written as $p = 0.05$) against which to interpret χ^2, *t* test or correlation coefficient values. The *p* value is like a hurdle, which is set to determine if the results might simply have been obtained by chance.

If the probability is less than 0.05 ($p < 0.05$), then the difference between the mean values, or the difference between the observed and expected results or the correlation, can be regarded as statistically significant. The null hypothesis can then be rejected. If the probability is 0.05 or greater than 0.05 ($p > 0.05$), then the null hypothesis must be accepted.

If you look at the following worked examples of each of the statistical tests, you will see how the test statistic has been interpreted in each case.

TIP

It is the difference between results that is significant or non-significant. You should never just write that 'the results are significant'. You must say what it is that is significant and refer to the variables involved.

Worked examples of statistical tests

Using Spearman's rank correlation test

You should choose this test when you wish to find out if there is a significant correlation between two sets of variables. You need between 7 and 30 pairs of measurements, such as the nitrate data in Chapter 2 (page 42).

- Start by producing a null hypothesis.

Your null hypothesis is that there is no correlation between the total mass of nitrogen added to the surrounding fields and the mean concentration of nitrate in the stream.

- Calculate the correlation coefficient.

Start by ranking the total mass of nitrogen and the mean concentration of nitrate (Table 13.2). Note that when two or more values are of equal rank, each of the values is given the mean of the ranks that would otherwise have been allocated.

Table 13.2 Rank values of total mass of nitrogen added to the surrounding fields and the mean concentration of nitrate in the stream.

Total mass of nitrogen/kg ha^{-1}	Rank	Concentration of nitrate/mg dm^{-3}	Rank	Difference in rank (*D*)	D^2
41	1.5	1.2	1	0.5	0.25
41	1.5	1.3	2	0.5	0.25
51	3	1.5	3	0	0
56	4	1.8	5	1.0	1
63	5	1.6	4	1.0	1
69	6	1.9	6	0	0
72	7	2.0	7	0	0

Calculate the difference between the rank values (*D*) and square this difference (D^2).

Find the sum of the squares of the differences: $\Sigma D^2 = 2.5$.

Now calculate the value of the test statistic, R_s, from the equation:

where *n* is the number of pairs in the sample.

$= 1 - 0.04$
$= 0.96$

- Interpret the value of R_s.

The value of R_s will always be between 0 and +1 or −1. A positive value indicates a positive correlation between the variables concerned. A negative value shows a negative correlation.

Table 13.3 shows the values of R_s. Look in the table under the correct number of pairs of measurements for our data. You had seven pairs of values, so the critical value is 0.79.

Your calculated value of R_s is larger than the critical value so you reject your null hypothesis and say that there is a significant correlation between the total mass of nitrogen added to the surrounding fields and the mean concentration of nitrate in the stream.

Although in this case you have identified a significant correlation between the variables concerned, you still have to be careful about the conclusions you draw. A correlation between these variables does not necessarily mean that the application of fertiliser causes pollution in the stream.

Table 13.3 Critical values of R_s for different numbers of paired values.

Number of pairs of measurements (n)	Critical value of R_s
5	1.00
6	0.89
7	0.79
8	0.74
9	0.68
10	0.65
12	0.59
14	0.54
16	0.51
18	0.48

TIP

- In an exam you might be asked to interpret a given value of R_s.

Using the chi-squared test

You should choose this test to determine whether the data you obtain is significantly different from what you would have expected. The null hypothesis in the woodlice example from Chapter 3 (page 54) would be that there is no significant difference between the number of woodlice going to the dark side and the number going to the light side.

O represents the observed results. If you add together the numbers in the light and dark from the 10 trials you will find that there were 61 woodlice on the dark side and 39 in the light. If the null hypothesis is true, you would expect equal numbers on each side, i.e. 50. This is represented in the equation by E. Table 13.4 shows the results of the calculations.

Table 13.4 Using the chi-squared (χ^2) test to determine whether the results are statistically significant.

	Light side	Dark side
Observed results (O)	39	61
Expected results (E)	50	50
$(O - E)^2$	121	121
$(O - E)^2/E$	2.42	2.42

TIP
You may need to work out the number of degrees of freedom in a given investigation.

Now, taking the sum of the values for the light and dark sides, you will find that $\chi^2 = 4.84$. You now need to look up the values for χ^2 in a table of critical values (Table 13.6, below). Since there are only two categories, light and dark, there is only one degree of freedom. The critical value for rejecting the null hypothesis with one degree of freedom at a probability level of 0.05 is 3.84, so you can have over 95% confidence that there is a significant difference in the number of woodlice moving to the dark. In other words, the probability of getting these results by chance is less than 5%.

EXAMPLE

Using the χ^2 test in examining genetic crosses

Data from genetic crosses are often categoric, for example data from the inheritance of the human ABO blood group. You should choose this test if you wish to check that the numbers of offspring obtained from a genetics cross are a close enough match to the expected ratio.

Geneticists investigated the ABO blood group phenotype of children born to a large number of parents. For each pair of parents, the blood group genotype of one was heterozygous I^AI^O and the other was heterozygous I^BI^O. The geneticists expected the blood groups of children in the investigation to be A, B, AB and O in a ratio of 1 : 1 : 1 : 1.

1 Explain why the geneticists involved a large number of parents in this investigation.
A large number of parents was involved so that the results would be reliable.

2 Explain why the geneticists expected a ratio of 1 : 1 : 1 : 1 in the offspring.
This is the ratio that would be predicted by a genetic diagram from a cross between parents of these genotypes.

Table 13.5 shows the results the geneticists obtained.

Table 13.5 The results of an investigation into the inheritance of human blood groups.

Blood group	Observed (O)	Expected (E)	(O – E)	(O – E)²/E
A	26			
B	31			
AB	39			
O	24			
				$\chi^2 = \sum\frac{(O-E)^2}{E}$

3 Copy the table and use the prediction of the geneticists to complete the column showing the expected number of children with each blood group (E). Explain how you arrived at your answer.
The expected number is found by adding up the total number of offspring (120) and then dividing by 4 (as the offspring are expected in a 1 : 1 : 1 : 1 ratio). The answer is 30% for each blood group.

4 Calculate the values for (O – E), (O – E)², $\frac{(O-E)^2}{E}$ and $\sum\frac{(O-E)^2}{E}$. Enter these values in your copy of Table 13.5.

Blood group	Observed (O)	Expected (E)	(O – E)	(O – E)²	(O – E)²/E
A	*26*	*30*	*4*	*16*	*0.53*
B	*31*	*30*	*1*	*1*	*0.03*
AB	*39*	*30*	*9*	*81*	*2.7*
O	*24*	*30*	*6*	*36*	*1.2*
	120				*Σ = 4.46*

Note that when (O – E) is calculated the minus sign can be ignored, because in the next stage the result is squared.

The geneticists calculated a value of $\chi^2 = 4.46$. They looked up their calculated value of χ^2 in a probability table. Table 13.6 shows part of this table.

Table 13.6 Values of χ^2 at three probability levels.

Degrees of freedom	Values of χ^2 at each probability level		
	$p = 0.10$	$p = 0.05$	$10 = 0.01$
1	2.71	3.84	6.64
2	4.61	5.99	9.21
3	6.25	7.82	11.34
4	7.78	9.49	13.28

5 How many degrees of freedom were there in the geneticists' data? Explain your answer.
Three degrees of freedom. Degrees of freedom = (number of categories) – 1 and there are four categories.

6 What does the calculated value of χ^2 tell us about the results of this investigation? Explain your answer.
In biology we use a probability of p = 0.05. The critical value of χ^2 is 7.82 for 3 degrees of freedom. The χ^2 value of 4.46 is less than the critical value of 7.82. Therefore we must accept the null hypothesis.

7 State a null hypothesis for this analysis.
The null hypothesis is that this genetic cross gives a 1 : 1 : 1 : 1 ratio of phenotypes in the offspring.

TIP
This example shows how to work out a *t* test manually but there are plenty of online programmes that will calculate the value for you.

Using the *t* test

You should choose this test to compare the means of two samples of measured data to see whether they are significantly different.

The *t* test will tell you whether there is a significant difference between the means of two samples, or not.

The *t* test formula is:

$$t = \frac{|\overline{x}_1 - \overline{x}_2|}{\sqrt{\frac{S_1^2}{n_1} + \frac{S_2^2}{n_2}}}$$

where

$\overline{x}_1$ is the mean of sample 1
$\overline{x}_2$ is the mean of sample 2
n_1 is the number of subjects in sample 1
n_2 is the number of subjects in sample 2
S_1 is the variance of sample 1, $\frac{\Sigma(x_1 - \overline{x}_1)^2}{n_1}$
S_2 is the variance of sample 2, $\frac{\Sigma(x_2 - \overline{x}_2)^2}{n_2}$.

TIP
You do not need to recall the formula for the *t* test.

Thinking about the example of limpets in Chapter 9 (page 177), first you need to calculate the variance of each sample.

TIP
Standard deviation is the square root of the variance.

Table 13.7

Sheltered shore			Exposed shore		
x_1	$x_1 - \overline{x}_1$	$(x_1 - x_1)^2$	x_2	$x_2 - \overline{x}_2$	$(x_2 - x_2)^2$
0.44	0.013	0.00017	0.32	0.019	0.000361
0.41	0.043	0.00185	0.32	0.029	0.000841
0.55	0.097	0.00941	0.40	0.053	0.002601
0.43	0.023	0.00053	0.47	0.121	0.014641
0.40	0.053	0.00281	0.32	0.029	0.000841
0.50	0.047	0.00221	0.31	0.039	0.001521
0.39	0.063	0.00397	0.37	0.021	0.000441
0.54	0.087	0.00757	0.30	0.049	0.002401
0.43	0.023	0.00053	0.42	0.071	0.005041
0.39	0.063	0.00397	0.35	0.001	0.000001
0.44	0.013	0.00017	0.30	0.049	0.002401
0.47	0.02	0.00040	0.36	0.011	0.000121
0.47	0.02	0.00040	0.36	0.011	0.000121
0.49	0.037	0.00137	0.31	0.039	0.001521
0.45	0.003	0.00001	0.32	0.029	0.000841
$\overline{x}_1 = 0.453$	$n_1 = 15$	$\Sigma = 0.03537$	$\overline{x}_2 = 0.349$	$n_2 = 15$	$\Sigma = 0.033695$

Now you calculate *t*.

$$S_1 = \frac{\Sigma(x_1 - \overline{x}_1)^2}{n_1}$$
$$= 0.002358$$

$$S_2 = \frac{\Sigma(x_2 - \overline{x}_2)^2}{n_2}$$

$= 0.002246$

$= 5.397$ (3 d.p.)

You now need to look up this calculated value of t on a table of t values (Table 13.8).

Degrees of freedom $= (n_1 + n_2) - 1 = 29$

Table 13.8 Table of t values.

Degrees of freedom	Significance level					
	20% (0.02)	10% (0.01)	5% (0.05)	2% (0.02)	1% (0.01)	0.01% (0.001)
28	1.313	1.701	2.048	2.467	2.763	3.674
29	1.311	1.699	2.043	2.462	2.756	3.659
30	1.310	1.697	2.042	2.457	2.750	3.646

For $p = 0.05$ and 29 degrees of freedom, the critical value of t is 2.043.

Your calculated value of t is 5.937, which is higher than this value.

Therefore, you must reject your null hypothesis. There is a significant difference between the mean height: base diameter ratio of the limpets on the two shores. Can you suggest a hypothesis to explain this?

The limpets on the exposed shore have a lower height : base diameter ratio than those on the sheltered shore, which means they have a greater area of 'foot' to hold them against the rock than the limpets on the sheltered shore of the same height. This means they can grip on to the rock better and are less likely to be washed off by waves. Also, a smaller height and wider base makes them more streamlined when waves hit them, again making it easier for them to stay attached to the rocks.

TIP

It is the difference between results that is significant or non-significant. You should never just write that 'the results are significant'. You must say what it is that is significant and refer to the variables involved.

14 Developing practical skills

Alongside developing your knowledge and understanding of biology, you are expected to become familiar with a range of apparatus and competent in a number of practical techniques. You will have already gained some experience of these apparatus and techniques during routine practical work and this year you will gain more. Your teacher will probably include a number of different practical investigations and activities in the course but there are a further six specified required practical activities that you must complete. Together with the six you carried out last year, written papers will assess your knowledge and understanding of the apparatus and techniques in some of these practicals. A list of the core practicals is given at the end of this chapter. But you can demonstrate your competence in these practical techniques during any of the practical work you do.

Making quantitative measurements

You will already know from the first year of your course that making and recording quantitative measurements is a key practical skill. The measurements you are most likely to make include time, volume, length, mass, temperature and pH. However, if you have already used instruments such as a colorimeter you will have come across some less familiar measurements. A colorimeter measures the optical density of a solution. You may also have come across less familiar units such as micrometres (μm) or micrograms (μg). This year you will see further new units such as millivolts (mV) or unfamiliar combinations of units such as micrograms per square metre per day ($\mu g\, m^{-2}\, day^{-1}$).

TIP

Quantitative results involve making measurements and recording numbers. Qualitative results, such as those you record if using reagents to identify biological molecules, are descriptive. They simply tell you if a particular substance is present or not, rather than how much is present.

Appropriate equipment is required to give a suitable amount of resolution when making quantitative measurements. If you are trying to find the biomass of some small animals such as freshwater shrimps, a balance reading to 0.1 g or even 0.01 g would be more suitable than one reading to just the nearest 1 g.

Figure 14.1 Digital balance being used to weigh freshwater shrimps.

You may need more specialised instruments to make some measurements. This year, you will probably come across different instruments. The pH of the water in a stream might be measured using a **pH meter** whereas a **light meter** might be used to measure the light intensity in a wood. A **respirometer** or a **photosynthometer** measures the volume of gases involved in gas exchange. You might have already used a **colorimeter**, which measures how much light can pass through a coloured solution (Figure 14.2).

If you have used a colorimeter you will know that it needs to be zeroed, rather like a balance, before making measurements. Zeroing a balance is done with nothing on the balance, but a colorimeter is zeroed using a clear solution, often water, called a **blank**. Samples are placed into a colorimeter in rectangular plastic tubes called cuvettes. Just as it is important not to spill material or solutions onto a balance pan to avoid errors, it is important to avoid spilling drops of liquid on the outside of the cuvette. This is because drops of liquid can cause errors by scattering the light. Knowing and avoiding likely causes of error with each piece of equipment helps to ensure accurate measurements.

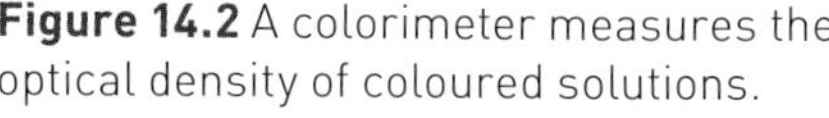

Figure 14.2 A colorimeter measures the optical density of coloured solutions.

TIP

Put the blank back into the colorimeter every so often between measurements and check that it still reads zero. If not, re-zero it. This ensures that all of your measurements remain as accurate as possible.

Most colorimeters have a method of placing coloured filters in the path of the light. This is because light of a particular wavelength is more suitable for measuring the optical density of solutions of particular colours. For example, red light is most suitable for measuring the optical density of blue Benedict's reagent after it has reacted with glucose. In general, the filter that gives the highest optical density reading for a solution of a certain colour is the best one to use.

A respirometer measures the volume of oxygen used by living material during respiration. It works by measuring how far a bead of coloured liquid in a capillary tube travels along a scale in a certain time. The bead of coloured liquid moves because as oxygen is used up the pressure inside the apparatus falls, provided a carbon dioxide absorber is present inside. Leaks at the joints of a respirometer or changes in the room temperature or air pressure can cause errors in the volume measurements.

Once respirometers are working without leaking, it is best not to disturb them. They sometimes have a three-way tap and a syringe to enable the bead of coloured liquid to be pushed back to the start. This enables repeat readings to be made without having to reassemble the apparatus every time.

A photosynthometer measures the volume of gas produced by plant material during photosynthesis and can be used to investigate the effect of variables like temperature and light intensity on the rate of photosynthesis in aquatic plants or algae (Figure 14.3).

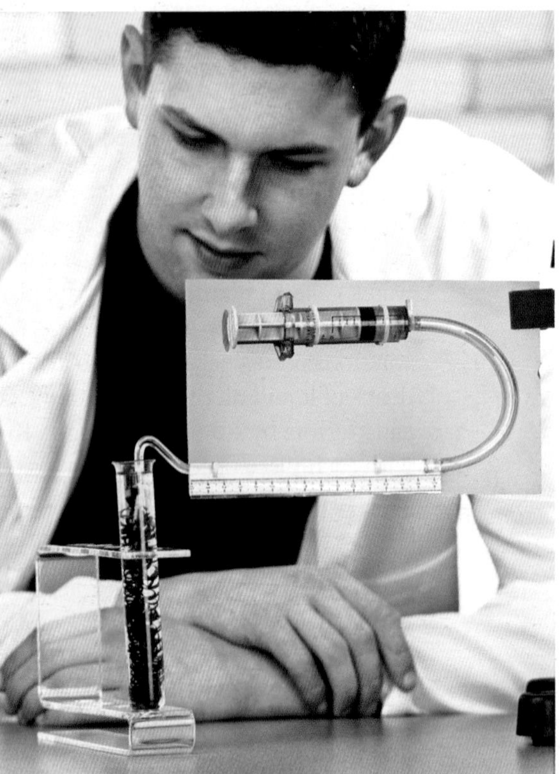

Figure 14.3 A photosynthometer being used to measure the volume of gas produced by plant material during photosynthesis.

TIP

Photosynthesis takes a while to respond to changed environmental factors. A sensible precaution is to wait a while for the plant material to adjust to the new conditions before starting to make measurements.

Collecting the gas produced by photosynthesising plant material can be done with a simple upturned tube but this does not allow volume measurements. It is better if the gas can be collected in a graduated container or collected and then pulled along a capillary using a syringe. The length of the bubble of air can then be measured. The volume of gas used or produced in respirometers and photosynthometers is calculated from the distance the bead or bubble has moved, in the same way as you will have already seen for a potometer (see Chapter 14, page 242 in *AQA A-level Biology 1 Student's Book*).

Using glassware

TIP

If you are making up a dilution series, the volume in each of your tubes or beakers should end up the same. Line them up and compare the position of the meniscus in each. They should line up exactly. If they do not, you have made a measuring error.

Measuring volumes accurately using glassware such as measuring cylinders is a key practical skill for both years of the course. This year, you will need to make a dilution series of a glucose solution to produce a calibration curve for a colorimeter. Selecting the appropriate size of measuring cylinder depending on the volume to be measured is crucial. The smaller graduations on the measuring cylinder, the better the resolution of the measurement can be. But obviously the total volume required has to be able to fit into the measuring cylinder selected. Ensuring that the full volume is drained out of the measuring cylinder is also important, as is the need to avoid splashes and spillages, which lead to error.

Using a pipette accurately to deliver a fixed volume of reagent to a solution is another practical skill. If you were making a calibration curve to find the concentration of glucose in an unknown sample with a colorimeter, you would need to add exactly the same volume of Benedict's reagent to each glucose dilution. Careful use of a graduated pipette or syringe is vital if the volume of Benedict's reagent is to be properly controlled.

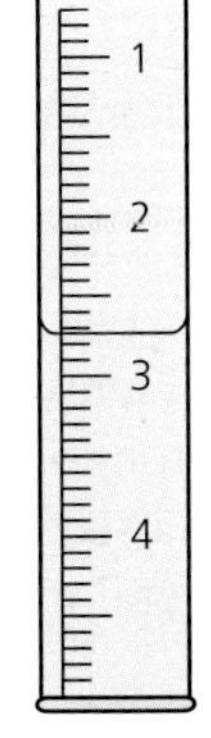

Figure 14.4 The meniscus formed by collecting carbon dioxide gas in a graduated fermentation tube.

Reading the position of a **meniscus** accurately is also important for accurate volume measurement. You may need to do this to measure the volume of carbon dioxide produced by respiring yeast cells. Fermentation tubes allow the carbon dioxide produced to be collected and measured. This can be done by placing upturned ignition tubes in a yeast culture. These have no graduations, but the volume of gas produced can still be compared qualitatively.

Alternatively, specialised pieces of graduated glassware can be used which allow the volume of gas to be measured quantitatively. The correct way to do this is to line up the bottom of the meniscus with the graduation mark, at eye level (Figure 14.4).

Using an optical microscope

TIP

Even if you are going to look at your material under high power, always start by focusing on low power. It is easier to see what you are looking at using low power and when you changes to the high power objective, it should be roughly focused already. This also helps to prevent damage to the lens.

From your work last year, you will know that the optical microscope is a vital tool for biologists because of the small size of many biological structures. Tissues, cells and organelles are all too small to be seen in any detail by eye so their images must be magnified. Optical microscopes work by directing light through a thin layer of biological material supported on a glass slide. This light is then focused through several lenses so an image can be seen through an eyepiece. You can switch from low to high power by rotating a different objective lens into position.

You will almost certainly have used an optical microscope several times by now. You will know that any material you are going to look at using a microscope has to be either transparent or really thin to allow light to pass through it. Some material might already be thin enough, such as the very thin leaves of moss which allow you to see their chloroplasts (Figure 14.5a). Other material such as pea leaf tissue might need to be sliced very thinly using a sharp blade. A drop of a culture of yeast under a cover slip will be spread out into a very thin layer so that individual yeast cells can be seen. Cutting animal tissue into a thin enough slice is much more difficult, so you will probably look at prepared slides of material such as skeletal muscle (Figure 14.5b).

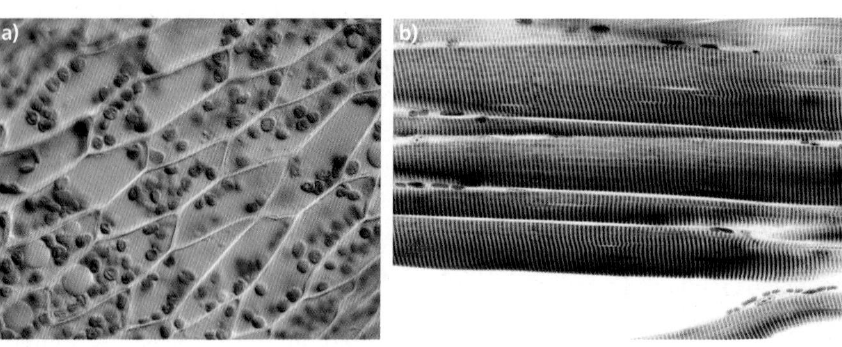

Figure 14.5 a) Chloroplasts visible in the cells of a moss leaf using a light microscope; b) skeletal muscle fibres.

Thin layers of material quickly dry out and shrivel up in the heat of the microscope lamp. This is why a drop or two of water is usually added beneath the cover slip to prevent dehydration damaging the cells.

Isolating organelles

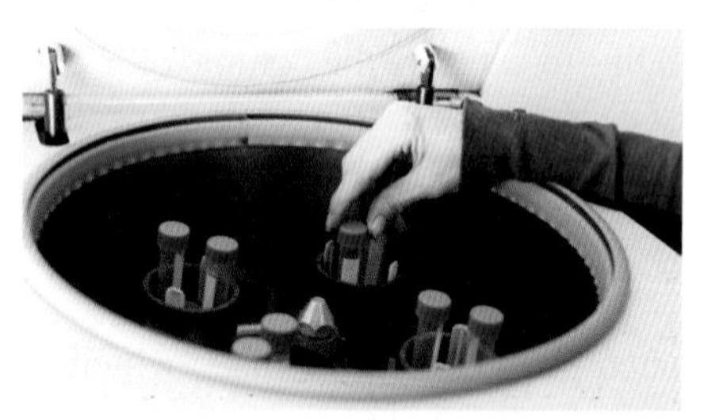

Figure 14.6 Tubes containing homogenised leaf tissue being placed in a centrifuge.

Isolating chloroplasts for microscopy, or for use in other experiments, requires the use of a centrifuge. Plant material such as chopped spinach leaves is **homogenised** in a cold solution with the same water potential in a blender. It is filtered to remove larger pieces of tissue and the filtrate is then **centrifuged** for several minutes (Figure 14.6). The nuclei are compacted down into a **pellet** at the bottom of the centrifuge tube. The supernatent is carefully poured into a second centrifuge tube and centrifuged at a higher speed for several minutes. This time, the chloroplasts form a pellet.

The **supernatant** can be poured off and the pellet can be re-suspended in an isotonic solution. The chloroplasts can then be observed using an optical microscope. They could also be used to make a chloroplast extract to investigate their dehydrogenase activity.

Making drawings

Making drawings remains an important way of recording the results of practical work this year where the outcome of an investigation is descriptive rather than quantitative. You will remember that the purpose of biological drawing is to make a clear scientific record of what you have observed rather than an artistic interpretation of the material. This means line drawings in pencil with no shading and no use of colour.

The important things are shape, proportion and scale. The structures drawn should be the same shape as those you observe, the parts of the drawing should be in proportion to one another and the drawing should be large enough to show the details clearly. If you are drawing individual cells such as skeletal muscle cells using an optical microscope, remember that you just draw two or three fibres as examples. If you are drawing tissues, such those in a section of a root with mycorrhizae, remember that you should draw a tissue map (see Chapter 15, page 260 in *AQA A-level Biology 1 Student's Book*), rather than attempting to draw all of the individual cells.

Sometimes you might make drawings from electron micrographs (see Figure 14.7).

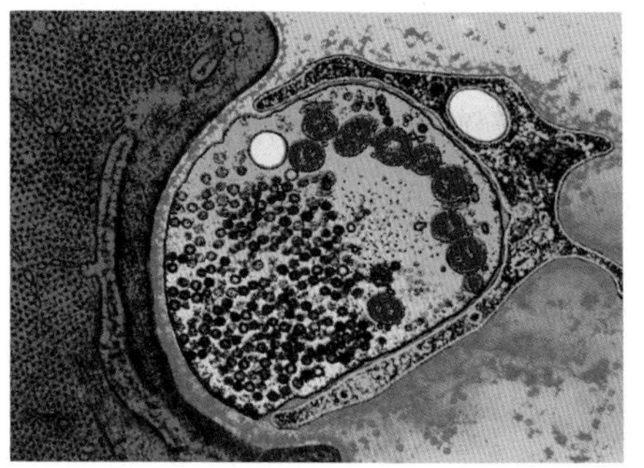

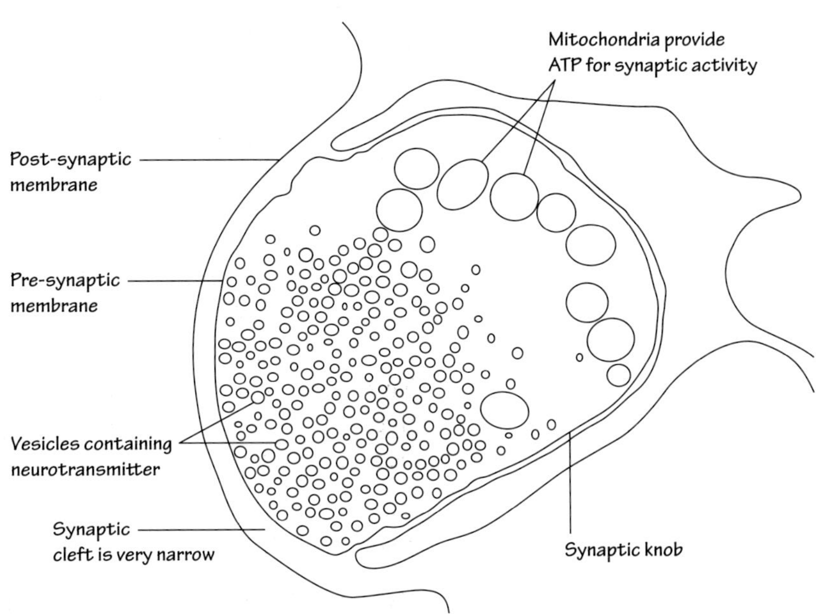

Figure 14.7 An electron micrograph of a synapse together with an annotated drawing of the image.

> **TIP**
>
> All drawings should have a title and some indication of scale or magnification. If drawing images seen using a microscope or micrograph, the drawing should show the magnification used to observe the material.

Annotating drawings makes them much more useful. **Annotations** are labels with more information than just the name of the structure. You might include the name of the structure but, equally, annotations can just be descriptive or they might include measurements. Examples of annotation are shown for features of the synapse in Figure 14.7. Remember that, for clarity, you should try to avoid label lines crossing each other or too much of the drawing. Label lines should be straight, drawn in pencil, using a ruler.

Identifying biological molecules

Being able to identify which biological molecules are present in material from organisms is a useful part of investigating how they work. You will recall that there are some simple chemical tests that you can use to identify some of the biological molecules that are present in materials or solutions. This year, you may use Biuret reagent and Benedict's reagent to analyse 'urine' samples from 'patients' with different clinical conditions. Both

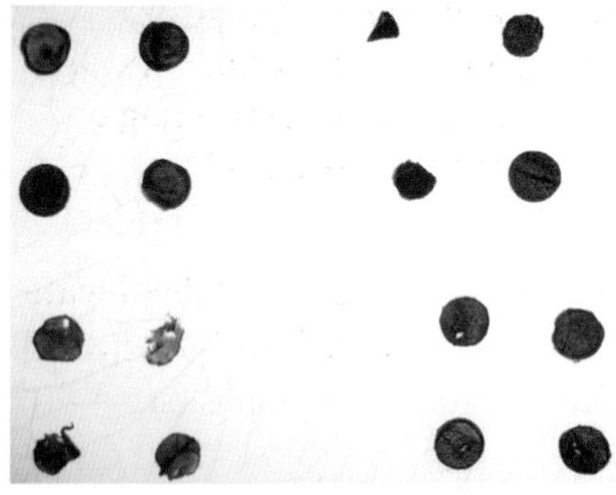

Figure 14.8 Leaf discs tested for starch after being kept in different conditions.

reagents can be used to produce a **calibration curve** (see Chapter 6, page 112) for a colorimeter to allow the concentration of protein or glucose in an unknown solution to be found. You may also use iodine solution to find out if discs cut from plant leaves and kept in different conditions have been carrying out photosynthesis.

Table 14.1 The four chemical tests for biological molecules and their positive results.

Test for	Reagent	Initial colour	Positive result
Protein	Biuret reagent	Pale blue	Violet solution
Reducing sugar	Benedict's reagent	Blue	Red precipitate
Starch	Iodine solution	Orange	Dark blue solution
Lipid	Ethanol	Colourless	Cloudy white suspension

Biuret reagent and iodine solution are simply added to the sample. A few drops are all that is needed. The colour change showing a positive result becomes apparent immediately if either protein or starch are present. Benedict's reagent must be heated for a few minutes with the sample to show a positive result. A few drops heated to 80 °C for several minutes will give a positive result if reducing sugar such as glucose or maltose is present. The test for lipid is known as the emulsion test. A volume of ethanol equal to the volume of the sample is added and the mixture shaken well. The mixture is then poured carefully into water. If lipid is present, a cloudy white suspension is formed.

TIP

You need to learn both the colour of the test reagent and the colour of a positive result.

Benedict's reagent can also be used to test indirectly for non-reducing sugars. If the sample gives a negative result when tested but a non-reducing sugar such as sucrose might be in the sample, then add a few drops of acid, warm and repeat the test. A positive result the second time around indicates that non-reducing sugar is present.

Separating biological molecules

Chromatography is a way of separating mixtures of biological molecules. If you have already done some chromatography, you will know that it is carried out on paper or on thin layers of solid media such as alumina on glass or plastic plates. A mixture of compounds, such as the photosynthetic pigments isolated from plant leaves, is very carefully pipetted onto an origin line one small drop at a time. In between each drop, the spot is dried. A small but concentrated spot is slowly built up on the origin line with repeated addition of drops and drying between each one. Some 20 or 30 small drops can be applied to the one spot.

TIP

Label your spots with pencil rather than pen and beneath the origin line. Pencil will not run when the plate or paper is added to the solvent and the labels will be out of the way below the line.

When the leaf extracts have been spotted onto the origin line, the plate or paper is held vertically in a small volume of chromatography solvent. The origin line should always be higher than the level of the solvent otherwise the spot dissolves in the solvent rather than moving up the plate or paper.

The solvent slowly moves up the paper or the solid medium on the plate. This may take several hours. As the solvent moves, the different pigments are carried up with it. But they do not all move at the same rate. Those that move faster travel further up the plate or paper so the different pigments separate out into a series of individual spots such as those in Figure 14.9.

Figure 14.9 Pigments isolated from plant leaves separated by thin-layer chromatography.

Electrophoresis is another separation technique. It separates mixtures of biological compounds on a gel by applying a voltage across the gel. Different compounds move at different rates across the gel depending on their size and charge. Electrophoresis can be used to separate mixtures of proteins or fragments of DNA.

An agarose gel is made by pouring molten agarose into a mould fitted with a plastic comb that forms holes or **wells** in the gel. The gel is allowed to set and the plastic comb is removed carefully.

TIP

Check that your solvent is shallow enough to remain below the height of the origin line when you stand your plate or paper in it. Otherwise your spots will simply be washed off into the solvent.

The gel is placed into an electrophoresis tank. An electrolyte solution is added to the tank so that the gel is submerged. Mixtures of DNA fragments in solution are pipetted carefully into the wells. A power supply is connected to electrodes at each end of the tank so that a voltage is applied across the gel. DNA fragments are negatively charged so they will move towards the positive electrode. This means that the positive wire must be connected to the tank at the end furthest from the wells.

Electrolyte A solution that will conduct electricity.

TIP

Pipetting a clear solution into wells in an almost transparent gel is very tricky. If the solution of DNA fragments has a loading dye added, it is easier to see what you are doing.

DNA fragments of different sizes will move towards the positive electrode at different rates. This may take several hours. When the separation is complete, the gel is developed using a stain such as methylene blue that will show up the positions of the bands of different sized fragments of DNA. The patterns of bands for different mixtures can then be compared.

Figure 14.10 A developed electrophoresis gel.

Using living organisms

If you use live animals such as maggots or woodlice to investigate animal movement responses you must make sure that they are not harmed during the investigation. Woodlice can be used in choice chambers and maggots can be placed in trays to investigate **taxes** and **kineses**. The effect of factors such as light or humidity on their rate or direction of movement can be discovered.

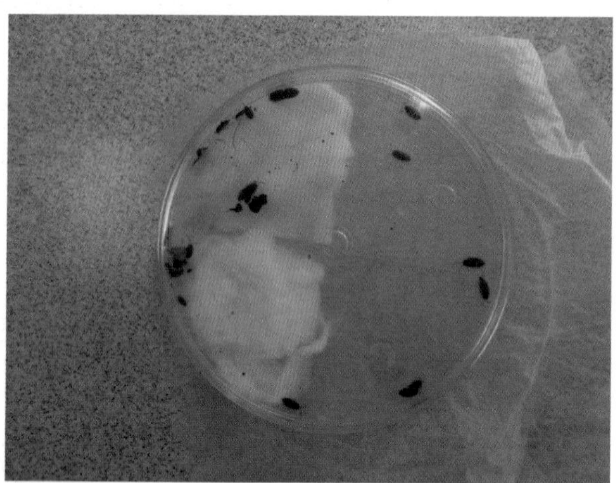

Figure 14.11 Woodlice in a choice-chamber experiment.

TIP

You should avoid using the same animal repeatedly because fatigue or learning may alter its behaviour.

If you carry out investigations on other members of your class, for example measuring pulse rate or to measure skin sensitivity or muscle fatigue, you should be very careful to avoid injury. Any exercise should be planned carefully and done in a suitable location to prevent accidents.

Using aseptic techniques to grow microorganisms

Growing bacteria successfully requires sterile conditions so that the culture is not contaminated by other microorganisms that you are not investigating, especially potential human pathogens. It is also important that none of your culture gets into the environment. Appropriate procedures to ensure **aseptic techniques** (see *AQA A-level Biology 1 Student's Book*, Chapter 15, page 263) should be followed.

This year you may grow bacteria or yeast in a broth culture to follow their population growth. In both cases, the turbidity of the suspension can be used as a way of measuring the number of cells in the culture. The turbidity is a measure of how cloudy the suspension is. As the population reproduces and grows, it becomes progressively more cloudy. A turbidity meter, or a light sensor or even a colorimeter can be used to measure how much light can get through the suspension at fixed time intervals following inoculation. This would allow a population growth curve to be plotted.

Turbidity A measure of the cloudiness of a suspension.

Figure 14.12 Yeast cultures at different stages of population growth.

Dissecting animal or plant organs

Observing biological material frequently requires that it be cut open or sectioned so the internal structures are visible. To see the distribution of chloroplasts in the different tissues of a leaf, thin cross sections have to be cut. These activities require the use of sharp instruments such as scissors and scalpels. Cutting should always be done away from you and your fingers, and on a suitable surface to avoid damage to the work bench. Sharp instruments are better than slightly blunt ones because they require less force to cut and there is therefore less risk of them slipping suddenly.

Using sampling techniques

This year there are also some outdoor practical skills you need to develop. You should know which sampling techniques to use to estimate the size of a population or to investigate the distribution of organisms. It is much easier to appreciate how the techniques work and their potential problems if you have tried them out for yourself.

Motile An organism that moves around; in contrast, non-motile, or sessile, organisms remain in the same place.

The number of individuals in a population cannot normally be counted easily. There are usually too many individuals for this to be practical. Instead, a sample of the population is used to calculate an estimate of the population size. How you do this depends on whether the organisms are non-motile, slow moving or motile. In general, animals move around whereas plants stay still. However, there are some animal species such as barnacles that are **sessile** and some that move very slowly such as snails. The population size of plants, sessile and slow-moving animals can all be estimated in the same way, but motile animals require a different approach.

The population size of plants, sessile or slow-moving animals in a given area can be estimated using randomly placed **quadrats**. It is important that the quadrats are placed at **random** to avoid bias, which would result in an over- or under-estimate of the population size. For example, this might happen if you subconsciously positioned your quadrats where there were more of the organisms or if you positioned your quadrats in places that were easier to get at.

A grid is set up across the area and random coordinates are generated using random number tables, dice or the random-number generator on a calculator. Quadrats are then placed at these coordinates.

Figure 14.13 A quadrat being used to sample a population of periwinkles.

The number of individuals in each quadrat are counted. The sample must be **representative** of the population as a whole. You can use a **running mean** to decide when enough quadrats have been used. When the mean number of individuals per quadrat stops changing the sample can be regarded as representative.

The mean number per quadrat is known as the **population density**. This is usually expressed as the number per square metre. The density is then multiplied by the number of square metres in the whole area to give an estimated **population size**.

> **TIP**
> The mark-release-recapture method makes some assumptions about the marking and the population and you should be aware of how likely they are to be true when you use this method.

If the population is mobile, the mark-release-recapture method should be used. A sample of the population is caught or trapped and marked in some way so that they can be recognised as having been caught. They are then released and allowed to mix in with the rest of the population. After a suitable time has been allowed for them to mix, another sample is caught or trapped. The technique relies on the fact that the marked individuals will be more diluted in a larger population.

The idea is to see how many marked individuals turn up in the second sample. You can then use the mark-release-recapture equation to calculate the population estimate (see Chapter 9, page 180).

If you want to find out where in a habitat a particular species is found, you will need to set up a **transect**. A transect is a line set out across the habitat along which samples are taken. It is often marked with string or a tape measure. A belt transect is a narrow strip of the habitat, perhaps 0.5 m or 1.0 m wide, alongside the line. Quadrats of the same width are placed in the strip at intervals along the line and the **abundance** of one or more species can be found at each sampling point. Using a belt transect lets you discover if a species is found more in one part of the habitat compared to another.

Some species are easily counted to find their abundance. But some species are very difficult to count, perhaps because they are very tiny and there are large numbers, or perhaps because it is not clear which individuals are which. In this sort of situation, you can measure abundance by using the techniques of **frequency** or **percentage cover**.

Figure 14.14 Quadrats placed along a transect can be used to find the distribution of a species in a habitat.

Percentage cover involves making an estimate of the percentage area of the quadrat occupied by the species. If you judge that it occupies about a quarter of the quadrat, the percentage cover is 25%. Obviously this is a little subjective so it is a good idea if the same person in a group always makes the estimate and some practice with this skill helps.

Another way to measure abundance is by finding the frequency of the species. This is the proportion of samples that contain that species. If the species is found in three out of 10 quadrats, the frequency is 30%. All you have to do to find the frequency is record if the species occurs or not at the sample point and calculate percentage of samples at which the species is found.

Frequency tells you how common a species is, but it does not give information on how much of the species is present. Percentage cover tells you how much space a species occupies in a sample, but does not give any information about how often it may be found in the whole habitat.

These two measures of abundance are related. The more space a species occupies, the more likely it is to be found in more than one sample. Using the two together can give information on how species are distributed in a habitat. A species with a high frequency but a low mean percentage cover occurs as single plants almost everywhere in the habitat. On the other hand, if you find a species with a low frequency but a high mean percentage cover, it would indicate that the species was found in groups but only in certain parts of the habitat.

TIP

Using a grid quadrat of a hundred squares reduces the subjectivity of measuring percentage cover. Each square occupied by that species counts as 1% cover.

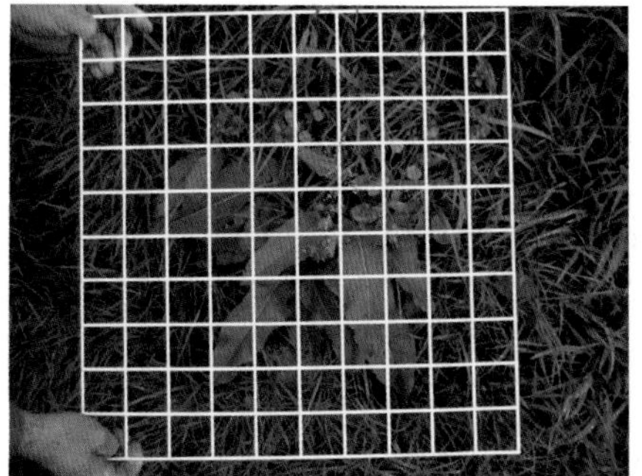

Figure 14.15 Percentage cover can be used to measure the abundance of a grass species.

TIP

Frequency and percentage cover used together can tell you different things about the distribution of a species in a habitat.

Using digital technology

Measuring human pulse rate can be done with a pulse sensor linked to a datalogger that can record the data for a specified time period and enable graphs to be produced very easily. When counting your own pulse rate, or that of a partner, it is easy to miscount. A pulse sensor avoids this problem.

The turbidity of a broth culture of microorganisms could be measured at fixed time intervals using a turbidity meter or light meter connected to a datalogger. In the same way, the changes in environmental factors such as light intensity or temperature in an outdoor location could be recorded at fixed intervals during the course of a day. The practical skills you will need to develop are those relating to positioning the sensors, setting up the time intervals for the particular datalogger and downloading the data for analysis, along with presenting and analysing the data and writing conclusions.

Computer programs that model the effects of genetic drift and natural selection are very helpful for illustrating how allele frequencies will change in large and small populations. The advantage of a simulation is that it can model the events over many generations in a short space of time. You will need to choose the conditions that you want to investigate and run the program a number of times while changing the conditions, to see what happens.

Required practical activities in Year 2 of AQA A-level Biology

TIP

Processing and presenting data are important skills. Make sure you know about these and other skills developed in the required practicals.

7 Use of chromatography to investigate the pigments isolated from leaves of different plants, e.g. leaves from shade-tolerant or shade-intolerant plants or leaves of different colours (see Chapter 1)

8 Investigation into the effect of a named factor on the rate of dehydrogenase activity in extracts of chloroplasts (see Chapter 1)

9 Investigation into the effect of a named variable on the rate of respiration of cultures of single-celled organisms (see Chapter 1)

10 Investigation into the effect of an environmental variable on the movement of an animal using either a choice chamber or a maze (see Chapter 3)

11 Production of a dilution series of a glucose solution and use of colorimetric techniques to produce a calibration curve with which to identify the concentration of glucose in an unknown 'urine' sample (see Chapter 6)

12 Investigation into the effect of a named environmental factor on the distribution of a given species (see Chapter 9)

15

Exam preparation

The advice you were given in the exam preparation chapter of *AQA A-level Biology 1 Student's Book*, Chapter 16, still applies at A-level, so do read this through again.

Now you have reached the second year of your course, you may be starting to think of the volume of work you need to revise. Don't get anxious about this: what you need to do is to prepare for revision well in advance. This is because the exam is testing your general skills and understanding from the whole course, so last-minute revision is of little benefit. You need to prepare for the exam from the first week of your Biology course. As you finish **each** topic, make succinct revision notes. Use whatever form helps you learn. Many people find annotated diagrams, spider diagrams and flow charts much more useful than pages of writing. You may find that using coloured pens and coloured paper or small pieces of card helps. Then, having made these notes, go over them regularly.

The exam

Your A-level exam will consist of three papers.

Paper 1: written exam, 2 hours

- 91 marks, worth 35% of total A-level marks
- Content from sections 1–4, including relevant practical skills. This may involve synoptic questions that make links between these different topics.
- Long- and short-answer questions, 76 marks
- Extended-response questions, 15 marks

Paper 2: written exam, 2 hours

- 91 marks, worth 35% of total A-level marks
- Content from sections 5–8, including relevant practical skills. This may involve synoptic questions that make links between all the topics you have studied over the whole of the A-level course.
- Long- and short-answer questions, 76 marks
- Comprehension question, 15 marks

Paper 3: written exam, 2 hours

- 78 marks, worth 30% of total A-level marks
- Content from sections 1–8, including relevant practical skills
- Structured questions, including practical techniques, 38 marks

- Critical analysis of given experimental data, 15 marks
- One essay from a choice of two titles, 25 marks
- Remember, sections 1–4 are the topics covered in *AQA A-level Biology 1 Student's Book*, and sections 5–8 are the topics covered in this book.

Assessment objectives

There are three different assessment objectives describing the skills that **will** be tested in the examinations.

- AO1 (35% of total marks): demonstrate knowledge and understanding of scientific ideas, processes, techniques and procedures. These are the only marks that will benefit from revision shortly before the exam.
- AO2 (35% of total marks): apply knowledge and understanding of scientific ideas, processes, techniques and procedures
  - in a theoretical context
  - in a practical context
  - when handling qualitative data
  - when handling quantitative data.
- AO3 (30% of total marks): analyse, interpret and evaluate scientific information, ideas and evidence, including in relation to issues, to
  - make judgements and reach conclusions
  - develop and refine practical design and procedures.

AO2 and AO3 skills are those you need to develop throughout the course.

The weightings of these assessment objectives are shown in Table 15.1.

Table 15.1 Weightings of A-level assessment objectives.

Assessment objective	Approximate percentage			
	Paper 1	Paper 2	Paper 3	Overall
AO1	44–48	23–27	28–32	30–35
AO2	30–34	52–56	35–39	40–45
AO3	20–24	19–23	31–35	25–30

In addition, the specification states that:

- 10% of the overall assessment of A-level Biology will contain mathematical skills equivalent to level 2 (higher-tier GCSE) or above
- at least 15% of A-level Biology will assess knowledge, skills and understanding in relation to practical work.

These are the 'rules' that examiners must follow when setting examination papers. They represent the **only** thing that you can predict about the series of examination papers you will sit.

The most important thing for you to understand is that, even if you learn your work thoroughly and can recall all the facts, this will only help you to achieve up to 35% of the marks. Therefore, it is vital that you can:

- apply your knowledge and understanding to contexts that you have not been taught
- deal with questions that contain material you have never seen before without panic, because examiners deliberately give you unfamiliar contexts to test your skills
- analyse, interpret and evaluate information you have never seen before.

Your mathematical skills will also be tested. This means you need to go through the mathematical skills chapter in this book, and ensure you understand everything in it. In addition, you need to have a thorough understanding of the practical techniques you have experienced, and be able to apply them in a new situation. This means going through all the practical techniques in this book, and, once again, ensuring you understand the rationale behind all of them. You should be able to evaluate practical designs and methodology as well. You should recognise different kinds of data, and be able to present them in a table or graph, as well as interpret the data. So you must make sure you really understand everything thoroughly. Then check your understanding by doing the questions in this book and any past paper or specimen questions that you can find. Mark your work using the exam board's mark scheme, and make sure you mark your answer very strictly.

Command words

As you were told in *AQA A-level Biology 1 Student's Book*, be very careful to learn what the key command words mean, so that you answer the question in the intended way. Be particularly careful that you know the difference between 'describe' and 'explain'.

Common pitfalls

If you are asked to do a calculation, you will often be told to show your working. This is so that, if you make a mathematical error, you can be credited for your working. Too many students do not show their working, so they fail to gain marks if their answer is wrong.

Avoid starting an answer with 'it'. For example, if the question is 'Give two differences between active transport and facilitated diffusion' and the candidate answers 'It does not use ATP' the examiner does not know what 'it' refers to. If the answer is 'Facilitated diffusion does not use ATP but active transport does' the answer is perfectly clear.

Preparing for the essay question

The essay is worth 25 marks out of a total of 78 on paper 3, so you should allow no more than 40 minutes to answer it. It is important to plan, but don't spend a lot of time on a lengthy plan. Don't spend too long selecting the appropriate topics. A good essay will cover a breadth of relevant examples. Where plant examples are relevant to the title of the essay, remember to include them. Try to choose topics that you studied at different times in your course.

Planning

Suppose you were planning the essay 'The importance of transport in living organisms'. There are different ways you could do this.

One might be to think of different aspects of biology, as shown in Table 15.2.

Table 15.2 Thinking about transport in different areas of biology.

Cells and biochemistry	Physiology
Proteins made and transported by RER to Golgi to vesicle to secretion Transport across membranes	Transport of water in plants Translocation in the phloem Transport of oxygen by haemoglobin

One method of planning might be to make a spider diagram, as shown in Figure 15.1.

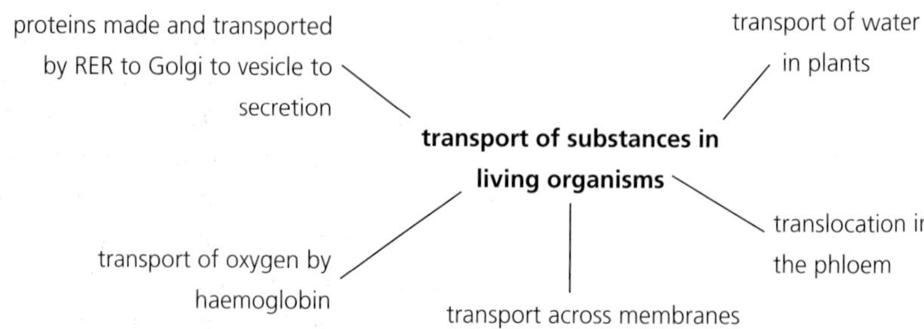

Figure 15.1 A spider diagram outlining transport in different areas of biology.

Whichever method you use, before starting to write you should:

- decide whether each of the topics in your plan is strictly relevant to the essay title and delete those that are not
- decide whether there is a logical order of topics. If there is, write your paragraphs in that order.

Writing

Remember that you need to write as much relevant detail as possible about five topics, in just 35 minutes (assuming you have spent 5 minutes planning), so there is no time to waste.

Do not waste time with an introductory paragraph, saying what you are going to cover. There are no marks for an introduction, so go straight into the first paragraph. 'One way in which transport is important in living organisms is…'. When you have done the first paragraph, it is a good idea to start the second paragraph similarly, as in: 'Another way in which transport in living organisms is important is…'. It doesn't matter if all the paragraphs start similarly, and an advantage of doing this is that it reminds you of the essay title before you write each paragraph, so it helps you to stick strictly to the title. If you include anything that isn't relevant, even if it's really good biology, you will not gain high marks. When you get to the end, just finish at the end of the last biology topic. Don't waste time by writing a final paragraph such as 'So you can see there are many ways in which transport in living organisms is important, and I have talked about some of them…'. This won't gain you any more marks, and you run the risk of being penalised for irrelevance.

TIP

In this example the essay title refers to the **importance** of transport in living organisms. So a student who simply describes transport in living organisms is not answering the essay title. Therefore, if your essay includes transport of water in the xylem, you need to make sure you are saying how this is important and not simply how it happens.

You must use appropriate technical terms in your essay. Avoid terms like 'signals' when you should say 'nerve impulses', 'levels' instead of 'concentrations' and expressions such as 'enzymes being killed' or 'energy being produced'. Write in sentences and make sure you express yourself well.

It is very important that the essay is written in **continuous prose**. This means you should not use bullet points or sub-headings.

Ideally, your essay needs to include some information that is beyond A-level, showing that you have read beyond the specification. If you have answered some of the Stretch and challenge questions in this book you may have some knowledge that goes beyond A-level. Similarly, if you read *New Scientist* magazine, or other science magazines, or regularly check the science items on the BBC News website, you may have some additional information. It is a good idea to flag this up in your essay by writing, 'In a recent *New Scientist* article I read that…'. However, make sure it is relevant. Don't be so keen to show off your additional reading that you include something that does not relate to the essay title. It is not enough to add general background material that anyone would know from watching a natural history programme on TV. Additional material must be A-level standard or higher.

How is it marked?

Examiners are given guidelines for marking the essay.

The examiner reads the essay and annotates it, indicating recall and understanding that is appropriate at A-level, errors, irrelevance, poor use of language and information that goes beyond A-level. Then they read the descriptors for each of the five 'levels' and decide which one fits the essay best. The descriptors for each level represent the middle mark at that level. If the essay doesn't completely fit any of the descriptors, the examiner uses a 'best fit' approach.

Once the examiner has chosen the level, they then decide on which mark to allocate within that level. The mark will be towards the top of that level if it is better than the descriptor, or towards the bottom of that level if it is a little worse than the descriptor.

Here are some essay titles that you might like to plan.

- The biological importance of proteins
- The importance of water in biology
- Cycles in biology
- The importance of biological polymers
- How the shape of molecules is important to their biological function
- How substances are transferred between living organisms and their environment

Answering the comprehension question

There will be a comprehension question worth 15 marks on paper 2. It will contain a passage for you to read that will relate to your A-level studies but will bring together a number of different topics. It presents these topics in a context that you are unlikely to have seen before. This question is mainly testing AO2 and AO3 skills and not AO1.

The first thing you should do is to read the passage carefully and check that you understand what it is telling you. Then read the questions quickly. This will give you an idea of the parts of the passage you need to focus on especially. Then read the passage again, carefully. Once you have understood the passage, start answering the questions. A line reference is given for each question, so make sure your answer relates to that part of the passage. You are asked to use information from the passage as well as your own knowledge to answer the questions, so make sure you do this.

Answering questions about practical procedures and evaluating data

You will be asked about one or more of the required practical activities that you carried out. When answering, give details that demonstrate you actually carried out these activities. For example, if you are asked how you carried out a Benedict's test for reducing sugar, don't say that you 'warmed' the solution. This is too vague. Give the approximate temperature you used, or say you 'boiled' the solution. If asked to devise a procedure, rather than just saying 'repeat' the investigation, suggest how many times. Say what results you would record, for example, a colour change, time taken for something to happen or a measurement you will make. Then say what you would do with the results, for example plot a graph or carry out a statistical test.

You might find a question that provides information about an investigation carried out by a student. When evaluating such an investigation, consider the student's methodology carefully.

- Did the student keep all the variables constant except the one under investigation? If not, would this affect the validity of the conclusion?
- Did the student use a large enough sample size, carry out sufficient repeats or continue the study for a sufficient length of time?
- Did the student use an appropriate control group to make the investigation valid?

In complete contrast, paper 3 carries a 15-mark question that provides information about an investigation carried out by professional scientists. When answering this question, you must assume that the professional scientists used appropriate methodology, so the above list becomes irrelevant. In this question, you might be asked to justify aspects of the methodology to show your own understanding of scientific investigations. For example:

- if the study was carried out on animals, could the conclusion be applied to humans?
- why was the control group used by the scientists appropriate?
- what was the importance of a double-blind clinical trial?
- can cause and effect be inferred from a correlation between two variables?
- how does the result of a statistical test, or the use of standard errors, help you make a valid conclusion?
- how could the scientists decide an appropriate number of repeats/samples?

Hopefully you will realise by now that success in the exam depends on skills you have acquired over the 2 years of your A-level Biology course, and not simply on revising notes for a few weeks before the exam. From the start of your course, make sure you can carry out the mathematical techniques required and, when you carry out a practical investigation in class, make sure you understand the reasons for everything you are told to do. If you think scientifically throughout the course, you will find it much easier to perform well when you take your written exams at the end of 2 years' study.

We hope you have enjoyed the A-level Biology course and found this book useful. Good luck!

Index

A

abiotic factors 168
abundance measures 268–9
accelerating centre 59–60
acetylation 199
acetylcholine (ACh) 79–80
actin filaments 92–4
actinomyosin bridge formation 93–5
action potential 72, 73–4, 75, 76
adrenaline 60–1, 107
adult stem cells 190
aerobic respiration 15–18
Agrobacterium tumefaciens 219
algae 42–3, 44–5
alleles 123, 126–7
  codominant 129, 134–7
  dominant 128–9, 130–4
  frequencies in gene pools 151–4
  locating specific alleles of genes 230–1
  multiple 137
  recessive 128–9, 130–4
allopatric speciation 160–1
all-or-nothing principle 75, 80
ammonification 36
amyloplasts 57
anaerobic respiration 15, 18–19
animal cloning 189–90
antagonistic muscle action 90
antibodies 20–1
antidiuretic hormone (ADH) 118–19
APOE gene 126–7
aquaporins 118
Arabidopsis plants 49
arithmetic mean 249
arm muscles 89–90
aromatase inhibitors 195
artificial fertilisers 38–9, 42
  timing application of 40–1
artificial selection 159
aseptic techniques 266–7
assessment objectives 272–3
athletes 88–9
ATP 5, 6–7
  hydrolysis and muscle contraction 96–7, 99
auto-immune conditions 108
automated DNA sequencing 228–9
autoradiograph 210
autosomal linkage 145–6
axons 71, 76, 77

B

bacteria 213
  growing 266
  in vivo gene cloning 211, 214–17
banded snails 151–2, 153–4, 155
bar charts 176, 250
base addition 192
base deletion 192
base substitution 192
benign tumours 202
beta cells (β cells) 104, 105, 106
β-globulin 157–8
biceps 89–90
biological molecules
  separating 264–5
  tests for 263–4
biomass 2
biotechnology bubble 206–7
biotic factors 168–9
blood glucose concentration 104–11
  diabetes 108–10
  dropped 106–7
  raised 104–6
blood water potential control 112–19
blunt ends 208
brain 112, 113
  heart rate control 59–61
  nervous coordination 69–85
breast cancer 195–6, 232–3

C

calorimeter 14
Calvin cycle 7, 8–9
cancer 200, 202–3 221, 232–3
  breast cancer 195–6, 232–3
  gene mutations and 201–3
carbon dioxide concentration 9–10, 11
cardioregulatory centre 59–61
carrying capacity 169
chain-termination technique 227
chalcone synthase 196–8
chalk grassland 184
chickens 87
  growth and food consumption 24–5
  inheritance of comb shape 143–4
chi-squared (χ^2) test 252, 254–5
chloroplasts 4, 262
  rate of dehydrogenase activity in extracts from 7
choice chambers 53, 54
cholinergic synapses 79–80
chromatography 5, 264–5
chromosome pair 23 138
climax community 183
clones 189–90, 201
  animal 189–90
  gene cloning 211–17
  plant 189
codominant alleles 129
  monohybrid inheritance involving 134–7
colorimeters 110–11, 259–60
community 168
competition 169–71
complementary base pairing 191
complementary DNA (cDNA) 207
composting 29–31
cone cells 64–6
consumers 22
  conversion efficiencies 23–4
  net production 24–5
contraction of muscles 93–8
copper tolerance 161–2
correlation 247
counting 174–80
cross-over values 124
curare 83
cyclic adenosine monophosphate (cAMP) 107–8
cystic fibrosis 220–1

D

danger, response to 60–1
data interpretation 249–57, 276
decomposers 22, 31, 34
decomposition 31–3
defensive chemicals 49, 55
dehydrogenase activity 7
denitrification 36
depolarisation 73, 75, 76
descriptive statistics 249–50
diabetes 108–10
dideoxyribonucleotides 227, 228, 229–30
differential reproductive success 155
dihybrid inheritance 141–6
  with autosomal linkage 145–6
  with no linkage 141–2
directional selection 156
disruptive selection 157
dissection 267
distal convoluted tubule 114, 118–19
DNA amplification 211–17
DNA fingerprinting 234–7
DNA hybridisation 230–1, 234–6
DNA methylation 199
DNA probes 231, 234–6
DNA profiling 236–7
DNA sequencing 227–30
DNA technology 206–24
  analysing restriction fragments 209–10
  ethics 221–2
  gene cloning 211–17
  gene therapy 220–1
  recombinant DNA technology 207–9
  usefulness of transgenic organisms 218–20
dominant alleles 128–9
  monohybrid inheritance involving 130–4
dopamine 83
drawings 262–3
Drosophila (fruit flies) 123, 133, 160
drugs 83

E

e4 allele 126–7
ecosystems 166–87
 competition 169–71
 counting and estimating populations 174–80, 267–9
 organisms and environment 168–9
 predators and prey 172–3
 succession 181–5
effectors 50
efficiency of energy transfer 2–3, 12–14, 23–4
electrolytes 265
electron transfer chains 6–7, 17–18
electrophoresis 209–10, 265
ELISA tests 20–1
embryonic stem cells 190
energy transfer 1–28
 efficiency 2–3, 12–14, 23–4
 food 3, 19–25
 photosynthesis 1–2, 3–14
 respiration 2, 3, 15–19
environment
 effect on distribution of a species 177–8
 fertilisers and 41–5
 GM plants and 219–20
 organisms and their environment 168–9
 phenotype, genotype and 125–6
 response to 51–8
epigenetic imprinting 199–200
epigenetics 198–200
epistasis 143–6
epithelial cell lining 116
equations 246–7
ER alpha (ER α) oestrogen receptors 194–5
Escherichia coli (*E. coli*) 216
estimating 174–80
ethics 221–2
Euglena 52
eukaryotes 198–200
eutrophication 42–5
eye 64–6
excitatory synapses 81–2

F

fast muscle fibres 86–7, 88, 98–9
Fast plants 130–3
feeding 3, 19–25
fertilisers 38–45
fieldwork data 43–4
filtrate 110, 111
fixed dunes 181, 183
flower colour 134–6
food
 energy transfer 3, 19–25
 genetically modified 219–20, 222
food chains 19, 22
food webs 19, 20, 22
fore dunes 181–2
founder effect 163
fractions 243–4
frame-shift mutation 192
frequency 176, 269
frequency tables 250
fruit flies (*Drosophila*) 123, 133, 160
functional magnetic resonance imaging (fMRI) 70
fungi 34, 37–8

G

gated ion channels 73
gel electrophoresis 209–10, 265
gene cloning 211–17
gene duplication 192
gene expression, control of 188–205
 epigenetic 198–200
 stem cells 190–1
 transcription 193–6
 translation 193, 196–8
gene inversion 192
'gene machine' 209
gene mutations 160, 161, 191–3
 and cancer 201–3
gene pools 151–4
generator potential 63
genes 123, 126–7
 locating specific alleles of 230–1
gene therapy 220–1, 223
genetically modified (GM) food 219–20, 222
genetic bottlenecks 163
genetic counselling 232–3
genetic diagram construction 130–3
genetic drift 162–3
gene translocation 192
genome projects 124, 225–7
genotypes 125–6, 140–1
 homozygous and heterozygous 128
geographic isolation 160–1
germ cell therapy 221, 222
giant axons 77
glomerular filtrate 115
glomerulus 114–15
glucagon 106, 107
gluconeogenesis 106
glucose concentration 110–11
 in blood *see* blood glucose concentration
glucose tolerance test 110
glycogenesis 105
glycogenolysis 106
glycolysis 15
glyphosate 219
grana 4
graphs 247–9
gravitropism 55, 57
grey dunes 181, 183
gross primary production (GPP) 12–13
growth responses 55–7

H

habitat 168
haemophilia 138–9
Hardy–Weinberg equation 152–4
Hardy–Weinberg principle 152, 155
hares, snowshoe 173
heart rate control 58–61
heather 13–14
herbivores 22, 24
herceptin 195–6
heterozygous genotypes 128
Himalayan rabbit 125
histograms 250
histones 199
homeostasis 103–4
homozygous genotypes 128
horizontal gene transfer 220, 222
hormones 102–21
human eye 64–6
Human Genome Project (HGP) 124, 225–6
human proteins 218
hyperpolarisation 73, 75
hyphae, fungal 34, 37–8

I

identical twins 198
impulses 71–7
indoleacetic acid (IAA) 55–7
induced pluripotent stem cells (iPS cells) 191
inferential statistics 250–7
inheritance 122–48
 dihybrid 141–6
 dominant, recessive and codominant alleles 128–9
 epistasis 143–6
 genes and alleles 126–7
 homozygous and heterozygous genotypes 128
 interpreting pedigrees 139–41
 monohybrid 130–9
 phenotype and genotype 125–6
 using the chi-squared test 255
inhibition 81–2
inhibitory centre 59–60
inhibitory synapses 82
insulin 104–6, 109
internal control 102–21
 blood glucose concentration 104–11
 diabetes 108–10
 homeostasis 103–4
 osmoregulation 112–19
 second messenger model 107–8
internal stimuli, responses to 58–61
interspecific competition 170, 171
intraspecific competition 170–1
in vitro gene cloning 211–13, 217
in vivo gene cloning 211, 214–17
ion uptake 37–8
islets of Langerhans 104, 105

K

kidneys 112–19
kinesis 51–4
kite diagrams 177–8
Krebs cycle 16

L

labelled DNA probes 231, 234–6
leaching 42
leaves
  chromatography and pigments isolated from 5
  dead and decomposition 31–3
ligation 215
light-dependent reaction 3, 4–7
light-independent reaction 3, 8–9
limiting factors 9–11
linear relationships 248
line graphs 247–9
linkage groups 146
link reaction 15–16
liposomes 221
logarithms 245
loop of Henle 114, 117
Lotka–Volterra model 172–3
Lund tube 45
lynx 173

M

magnification 244–5
malaria 158, 226–7
manual DNA sequencing 229–30
mark-release-recapture 179–80, 268
marram grass 182–3
mean 249
measurements
  interpreting data from 249–50, 252
  making 258–61
meat consumption 202–3
median 250
medical screening 232–3
mesophils 30
microscope, optical 261–2
mitochondria 17–18
mode 250
monohybrid inheritance 130–9
  involving codominant alleles 134–7
  involving only dominant and recessive alleles 130–4
  involving a sex-linked character 138–9
motile organisms 267
motor group 98
motor neurones 50–1, 70, 71
movement
  effect of an environmental variable on 54
  muscles and 86–101
multiple alleles 137
muscle fibres 91–9
  fast and slow 86–7, 88, 98–9
muscles 86–101
  contraction 93–8
  mechanics 89–90
  myofibrils 91, 92–3
  skeletal 88–99
  structure 91
mutagenic agents 193
mycorrhizae 37–8
myelinated neurones 70, 76
myelin sheath 71, 76–7
myofibrils 91, 92–3
myogenic muscle cells 58
myoglobin 87
myosin filaments 92–4
myxomatosis 149–51, 184

N

NAD 15, 18
natural fertilisers 38–9, 42
natural selection 155–8
negative feedback 103, 106
nephrons 112, 113–19
nervous coordination 69–85
  impulses 71–7
  neurones *see* neurones
  synapses 77–83
net primary production (NPP) 13
  fertilisers and 38–45
net production of consumers 24–5
neuromuscular junctions 71, 95–6
neurones 60, 70–1, 78
  movement of impulses along 75–6
  myelinated 70, 76
  reflex arc 50–1
niches 169
nitrate ions 42–4
nitrification 36
nitrogen-containing fertiliser 39–41
nitrogen cycle 35–7
nitrogen fixation 36–7
nodes of Ranvier 70, 71, 76
noradrenaline 60
null hypothesis 251–2
nutrient cycles 29–47
  composting 29–31
  decomposition 31–3
  fertilisers and NPP 38–45
  mycorrhizae and ion uptake 37–8
  nitrogen 35–7
  phosphorus 35

O

oestrogen 194–5
oestrogen-dependent breast tumours 195
optical microscope 261–2
orders of magnitude 244–5
organelles, isolating 262
organisms
  and their environment 168–9
  using living organisms 266
osmoreceptors 118–19
osmoregulation 112–19
ovarian cancer 232–3
oxidation 15
oxidative phosphorylation 18

P

Pacinian corpuscles 62–3
parasympathetic neurones 60
Parkinson's disease 83
pea plants 141–2
pedigrees 139–41
percentage cover 176, 269
percentages 244
petunias 196–8
phenotypes 125–6, 140
phenotypic ratios 243
phosphate concentration 44
phosphate ions 42–5
phosphocreatine 96–7
phosphorus cycle 35
phosphorylation 15, 16, 18
photoautotrophic organisms 2
photoionisation 4, 6
photolysis 6
photosynthesis 1–2, 3–14
  efficiency 12–14
  limiting factors and 9–11
photosynthometers 260–1
phototaxis 52, 53
phototropism 55, 56–7
pioneer species 181
planarians 51–2
plants
  copper tolerance 161–2
  photosynthesis *see* photosynthesis
  responses 48, 49, 55–8
  succession when plant cover has been destroyed 185
  transgenic and GM food 219–20, 222
plasmids 214–15, 216
Plasmodium falciparum 158, 226
pluripotent stem cells 190
  induced 191
polydactyly 163
polymerase chain reaction (PCR) 236–7
  gene cloning 211–13, 217
population density 176, 268
population growth 266–7
populations 151
  allele frequencies 151–4
  counting and estimating 174–80, 267–9
  in ecosystems 166–87
positive feedback 104
poultry farming 24–5
Prader–Willi syndrome 200
predators 172–3
prey 172–3
primary consumers 22, 23
primers 211, 212, 213
producers 22, 23
promoter regions 193–4
proteins
  ELISA tests 20–1
  human from transgenic organisms 218
proteome 229
proto-oncogenes 201
proximal convoluted tubule 114, 115, 116
Punnett squares 132

Q

quadrats 175–8, 267–8, 268–9
quantitative measurements 258–61

R

rabbits 125, 184
  biological control 149–51
radioactive marker 210
random sampling 175
range 250
rate of change measurements 249
ratios 243
reabsorption, selective 116–19
receptors 50, 62–6
recessive alleles 128–9
  monohybrid inheritance involving 130–4
recognition sequences 207–9
recombinant DNA 215
  producing DNA fragments 207–9
recombinant offspring 145–6
reduction 15
reflex arc 50–1
refractory period 75
relay neurones 50–1, 71
renal capsule 114, 115
repolarisation 74, 75, 76
reproductive isolation 160–1
Required practicals 270
respiration 2, 3, 15–19
respiratory substrates 2
respirometers 260–1
response 48–68
  to internal stimuli 58–61
  kinesis and taxis 51–4
  plant responses 48, 49, 55–8
  receptors 50, 62–6
  survival and 49–51
resting potential 72, 75, 76
restriction endonuclease 207–9
restriction fragments 207–9
  analysing 209–10
retina 64–6
reverse transcripterase 207
RNA-dependent RNA polymerases (RDRs) 197
RNA interference (RNAi) 196–8
rod cells 64–6
rubisco 1–2, 8

S

saltatory conduction 76
sample size 174
sampling 174–80, 267–9
sand dunes 181–4
saprobionts 31, 34
sarcomeres 92–3
sarcoplasmic reticulum 91, 96, 99
scale bar 245
scattergrams 247
schizophrenia 83
Schwann cells 71
screening, medical 232–3
secondary consumers 22, 23
second messenger model 107–8
selection, natural 155–8
selective reabsorption 116–19
sensitivity to light 64
sensory neurones 50–1, 71
sequencing
  DNA 227–30
  of genomes 226–7
sessile organisms 267
sex-linked characters 138–9
sickle-cell anaemia 157–8, 232
significant figures 241
simple reflexes 50–1
single-celled organisms 19
sinoatrial node (SAN) 58, 60
skeletal muscles 88–99
skin 62–3
sliding filament hypothesis 93
slow muscle fibres 87, 88, 98–9
small interfering RNA (siRNA) 197–8
snails, banded 151–2, 153–4, 155
somatic cell nuclear transfer (SCNT) 189–90, 221–2
somatic cell therapy 220–1
sowing density 170–1
spatial summation 81
Spearman's rank correlation test 44, 252, 253–4
speciation 159–62
specific growth factors 55–7
spider diagrams 274
stabilising selection 156
standard deviation 249–50
standard form 242–3
statistical tests 250–7
stem cells 190–1
sticky ends 208
stimuli 50
  contraction of skeletal muscle fibres 95–8
  responses to internal stimuli 58–61
substrate-level phosphorylation 16
succession 181–5
  managing 184
  in sand dunes 181–4
  when plant cover has been destroyed 185
summation 66, 80–1
surface area to volume ratios 24, 243
survival 49–51
sustainability 167
symbols 246–7
sympathetic neurones 60
sympatric speciation 161–2
synapses 77–83
synaptic cleft 77–8

T

tamoxifen 195
taxis 51–4
temporal summation 80, 81
tertiary consumers 22, 23
thermal cycler 211–12
thermophils 30
three-neurone reflex arc 50–1
thylakoid membrane 7
thylakoids 4
totipotent cells 190
touch 62–3
transcription, control of 193–6
transcription factors 193–5
transects 175–8, 268
transformed bacteria 215–16
transgenic organisms 215
  mice 201
  moral and ethical issues 221–2
  usefulness 218–20
translation, control of 193, 196–8
triceps 89–90
trophic levels 22, 23
tropisms 55
t test 252, 256–7
tumours 191, 201–3
tumour-suppressor genes 201
turbidity 266

U

ultrafiltration 115
unidirectionality 80
unipotent cells 190
units 240
urinary system 112, 113
urine 112, 116, 118–19

V

variable number tandem repeats (VNTRs) 234
  PCR of 236–7
vectors 214–15
vestigial wings 133, 160
vision 64–6
visual acuity 65–6
visualisation of DNA bands 210

W

water potential, blood 112–19
winter wheat 40–1
woodland 181, 183
woodlice 32–3, 53–4
World Wildlife Fund (WWF) Living Planet Index 166–7

X

X chromosome 138–9

Y

Y chromosome 138, 139
yellow dunes 181, 182–3
yield 9, 10, 11
  fertilisers and 39–40

Free online resources

Answers for the following features found in this book are available online:

- Test yourself questions
- Activities

You'll also find an Extended glossary for each chapter to help you learn the key terms and formulae you'll need in your exam.

Scan the QR codes below for each chapter.

Alternatively, you can browse through all resources at:

www.hoddereducation.co.uk/AQABiology2

How to use the QR codes

To use the QR codes you will need a QR code reader for your smartphone/tablet. There are many free readers available, depending on the smartphone/tablet you are using. We have supplied some suggestions below, but this is not an exhaustive list and you should only download software compatible with your device and operating system. We do not endorse any of the third-party products listed below and downloading them is at your own risk.

- for iPhone/iPad, search the App store for Qrafter
- for Android, search the Play store for QR Droid
- for Blackberry, search Blackberry World for QR Scanner Pro
- for Windows/Symbian, search the Store for Upcode

Once you have downloaded a QR code reader, simply open the reader app and use it to take a photo of the code. You will then see a menu of the free resources available for that topic.

1 Energy transfer

3 Response

2 Nutrient cycles

4 Nervous coordination

5 Muscles and movement

9 Populations in ecosystems

6 Internal control

10 The control of gene expression

7 Genes, alleles and inheritance

11 Gene cloning and gene transfer

8 Gene pools, selection and speciation

12 Using gene technology